Lecture Notes in Civil Engineering

Volume 117

Lecture Notes in Civil Engineering (LNCE) publishes the latest developments in Civil Engineering - quickly, informally and in top quality. Though original research reported in proceedings and post-proceedings represents the core of LNCE, edited volumes of exceptionally high quality and interest may also be considered for publication. Volumes published in LNCE embrace all aspects and subfields of, as well as new challenges in, Civil Engineering. Topics in the series include:

- Construction and Structural Mechanics
- Building Materials
- Concrete, Steel and Timber Structures
- Geotechnical Engineering
- Earthquake Engineering
- Coastal Engineering
- Ocean and Offshore Engineering; Ships and Floating Structures
- Hydraulics, Hydrology and Water Resources Engineering
- Environmental Engineering and Sustainability
- Structural Health and Monitoring
- Surveying and Geographical Information Systems
- Indoor Environments
- Transportation and Traffic
- Risk Analysis
- Safety and Security

To submit a proposal or request further information, please contact the appropriate Springer Editor:

– Pierpaolo Riva at pierpaolo.riva@springer.com (Europe and Americas);
– Swati Meherishi at swati.meherishi@springer.com (Asia - except China, and Australia, New Zealand);
– Wayne Hu at wayne.hu@springer.com (China).

All books in the series now indexed by Scopus and EI Compendex database!

More information about this series at http://www.springer.com/series/15087

T. G. Sitharam · Ravi Jakka · L. Govindaraju
Editors

Local Site Effects and Ground Failures

Select Proceedings of 7th ICRAGEE 2020

Springer

Editors
T. G. Sitharam
Indian Institute of Technology Guwahati
Guwahati, Assam, India

Ravi Jakka
Department of Earthquake Engineering
Indian Institute of Technology Roorkee
Roorkee, Uttarakhand, India

L. Govindaraju
Department of Civil Engineering
Bangalore University
Bengaluru, Karnataka, India

ISSN 2366-2557 ISSN 2366-2565 (electronic)
Lecture Notes in Civil Engineering
ISBN 978-981-15-9983-5 ISBN 978-981-15-9984-2 (eBook)
https://doi.org/10.1007/978-981-15-9984-2

This Springer imprint is published by the registered company Springer Nature Singapore Pte Ltd.
The registered company address is: 152 Beach Road, #21-01/04 Gateway East, Singapore 189721, Singapore

Preface

The local soil condition can amplify the seismic waves passing through the soil and this effect is termed as local site effects. This book volume contains the latest research papers on ground response analyses and local site effects, seismic slope stability and landslides, and GIS and remote sensing applications for geo-hazards, selected from the proceedings of the 7th International Conference on Recent Advances in Geotechnical Earthquake Engineering and Soil Dynamics, 2021.

We thank all the staff of Springer for their full support and cooperation at all the stages of the publication of this book. We do hope that this book will be beneficial to students, researchers, and professionals working in the field of earthquake hazards. The comments and suggestions from the readers and users of this book are most welcome.

Guwahati, India — T. G. Sitharam
Roorkee, India — Ravi Jakka
Bengaluru, India — L. Govindaraju

Acknowledgements

We (editors) want to thank all the authors, who have contributed to the book. We could bring this book out due to all the authors' timely contribution and cooperation. We thank and acknowledge the service of the following reviewers for their valuable time and efforts.

Ajay Chourasia, CSIR-CBRI
Amarnath Hegde, Indian Institute of Technology Patna
Amit Verma, IIT (BHU)
Anil Cherian, Strainstall
Anitha Kumari S. D., Ramaiah University of Applied Sciences
Arvind Kumar Jha, Indian Institute of Technology (IIT) Patna
Asha Nair, CMR Institute of Technology, Bengaluru
Babloo Chaudhary, NITK Surathkal
Bal Rastogi, Indian Society of Earthquake Science
Chittaranjan Birabar Nayak, Vidya Pratishthan's Kamalnayan Bajaj Institute of Engineering & Technology
Dauji Saha, Bhabha Atomic Research Centre; Homi Bhabha National Institute
Deepankar Choudhury, Indian Institute of Technology Bombay
Dhanaji Chavan, IISc
Gopal Santana Phani Madabhushi, University of Cambridge
Jagdish Sahoo, IIT Kanpur
Kalyan Kumar G., NIT Warangal
Karthik Reddy Konala S. K., IIT Hyderabad
Ketan Bajaj, Risk Management Solutions
Manas Kumar Bhoi, PDPU
Md. Mizanur Rahman, University of South Australia
Padmanabhan G., Indira Gandhi Center for Atomic Research
Pradeep Kumar Singh Chauhan, CSIR-Central Building Research Institute, Roorkee
Premalatha Krishnamurthy, Anna University
Prishati Raychowdhury, IIT Kanpur
Purnanand Savoikar, Goa Engineering College
Rajib Saha, NIT Agartala

Rajib Sarkar, IIT(ISM) Dhanbad
Ramkrishnan R., Amrita Vishwa Vidyapeetham
Rangaswamy K., NIT Calicut
Ravi K., IIT Guwahati
Renjitha Varghese, National Institute of Technology, Calicut, Kerala, India
Sanjay Verma, Indian Geotechnical Society, Jabalpur Chapter
Sarat Kumar Das, Indian Institute of Technology (ISM) Dhanbad
Shreyasvi C., National Institute of Technology Karnataka
Snehal Kaushik, Girijananda Chowdhury Institute of Management and Technology, Guwahati
Sreevalsa Kolathayar, National Institute of Technology Karnataka, Surathkal
Supriya Mohanty, Indian Institute of Technology (BHU), Varanasi
Surya Muthukumar, Amrita School of Engineering, Amrita Vishwa Vidhyapeetham
Vinay Srivastava, Retd. IIT ISM Dhanbad
Vipin K. S., Swiss Re
Visuvasam Joseph Antony, Vellore Institute of Technology

Contents

About the Editors

Prof. T. G. Sitharam is currently the Director of Indian Institute of Technology Guwahati, India. He is also a KSIIDC Chair Professor in the area of Energy and Mechanical Sciences and Senior Professor at the Department of Civil Engineering, Indian Institute of Science, Bengaluru (IISc). He was the founder Chairman of the Center for Infrastructure, Sustainable Transport and Urban Planning (CiSTUP) at IISc, and is presently the Chairman of the AICTE South Western Zonal Committee, Regional office at Bengaluru and Vice President of the Indian Society for Earthquake Technology (ISET). Prof Sitharam is the founder President of the International Association for Coastal Reservoir Research (IACRR). He has been a Visiting Professor at Yamaguchi University, Japan; University of Waterloo, Canada; University of Dolhousie, Halifax, Canada; and ISM Dhanbad, Jharkhand, and was a Research Scientist at the Center for Earth Sciences and Engineering, University of Texas at Austin, Texas, USA until 1994.

Prof. Ravi Jakka is working as Associate Professor in the Department of Earthquake Engineering, Indian Institute of Technology, Roorkee. He is also currently serving as Secretary, Indian Society of Earthquake Technology(ISET). He has graduated in Civil Engineering from Andhra University Engineering College in the year 2001. He has obtained masters and doctorate degrees from IIT Delhi in the years 2003 and 2007 respectively. His areas of interest are Dynamic Site Characterization, Soil Liquefaction, Seismic Slope Stability of Dams, Landslides, Foundations & Seismic Hazard Assessment. He has published over 100 articles in reputed international journals and conferences. He has supervised over 35 Masters Dissertations and 6 Ph.D. thesis, while he is currently guiding 10 PhD Thesis. He has received prestigious DAAD and National Doctoral fellowships. He has obtained University Gold Medal from Andhra University. He also received 'Young Geotechnical Engineer Best Paper Award' from Indian Geotechnical Society. He was instrumental in the development of Earthquake Early Warning System for northern India, a prestigious national project. He is also the Organizing Secretary to 7th International Conference on Recent Advances in Geotechnical Earthquake Engineering.

Prof. L. Govindaraju is currently Professor in the Department of Civil Engineering, University Visvesvaraya College of Engineering (UVCE), Bangalore University, Bangalore. He obtained his Bachelor's in Civil Engineering from Mysore University, India (1986), Master's in Geotechnical Engineering from National Institute of Technology Karnataka (NITK), India (1994) and Ph.D. in Civil Engineering from Indian Institute of Science (IISc), India (2005). He was awarded fellowship from National Programme on Earthquake Engineering Education (NPEEE) instituted by the Ministry of Human Resources & Development, Government of India to pursue post doctoral research in the area of earthquake engineering at the Department of Engineering Science, University of Oxford (UK) from March 2007 to August 2007. He is the member of many professional bodies including IGS, ISTE and ISET. He has published more than 150 technical papers and two books.

Ground Response Analysis with Deep Bedrock Depth in Indo-Gangetic Plains

Priyanka Sharma, M. L. Sharma, and V. A. Sawant

Abstract Local site effects play an important role in causing damage to structures during earthquakes. Thus, one of the aim of seismologists and geotechnical engineers is to characterize the soil for the region prior to seismic hazard assessment. In the present work, an endeavor is made to study the depth of bedrock in Indo-Gangetic Plains from Seismotectonic Atlas of India [6]. A huge variation of bedrock depth ranging from 0 to 4000 m indicates the presence of thick soil cover in the study region. Roorkee city, situated in the foothills of Himalayas has a bedrock depth of around 3000 m and due to the presence of this huge soil cover, the occurrence of any great seismic event will pose a threat to both life and property as properties of propagating waves change as they travel toward the surface. The site characterization is carried out by MASW and Microtremor methods and shear wave velocity profiles are estimated with the thickness of soil. The region has lower values of shear wave velocity and falls in Site Class D. Ground Response Analysis (GRA) of the site is performed using STRATA [9] to Uttarkashi Earthquake (1991) and Chamoli Earthquake (1999) and is compared with the suggested response spectrum in (IS, in (Part 1): 2002 Indian Standard, Criteria for earthquake resistance design of structures, Fifth Revision, Part-I, Bureau of Indian Standard, New Delhi, 1893):2002 (Part 1). The comparison of the response spectrum from two earthquakes inhibits higher variation as provided in the IS Code.

Keywords MASW · Microtremor · Joint fit inversion · Response spectra

P. Sharma (✉) · M. L. Sharma
Department of Earthquake Engineering, IIT Roorkee, Roorkee 247667, India
e-mail: pssharma.iitr@gmail.com

V. A. Sawant
Department of Civil Engineering, IIT Roorkee, Roorkee 247667, India

T. G. Sitharam et al. (eds.), *Local Site Effects and Ground Failures*, Lecture Notes in Civil Engineering 117, https://doi.org/10.1007/978-981-15-9984-2_1

1 Introduction

About 59% of India's land mass is at a risk of moderate to rigorous seismic hazard covering many important cities, as given in seismic zoning of the country. The surface ground motion in comparison to bedrock motion is highly influenced by the local site conditions. The presence of loose material on the bedrock amplifies the incoming waves as they travel toward the surface causing damage to both life and property. The recently experienced large earthquakes caused heavy damage at epicentral distances ranging from 250 to 500 km or so due to local site conditions. For example, in 1985 Michoacan earthquake, major damage was observed at 200 miles away from the epicenter in Mexico City. Similarly, in Bhuj earthquake, 2001, heavy damage was seen in multistory building at an epicentral distance of 350 km in Ahmedabad City.

Himalayan region is one of the most seismically active regions due to the accumulation of heavy strains. Many researchers [5, 2] have predicted the presence of a seismic gap in this region. Moreover, the probability of occurrence of a great seismic event is 0.59 in the next 100 years [8]. The Indo-Gangetic Basin (IGB) lies parallel to the Himalayas and is highly populated. The basin is formed by the loose soil deposits and any future earthquake in the Himalayas will lead to massive destruction and will pose a threat to both life and property. Many studies in India to estimate the effect of local site conditions have been done by many authors [1, 4]. The site response study for Delhi region was done by [10]. Kirar et al. developed a correlation between V_S and N for all types of soils, clay soils and sandy soil for Roorkee region. The insitu tests and laboratory tests were used to obtain the dynamic properties of soil in Italy by Cavallaro et al. (2017). Ray et al. (2019) used the GR relationship and the Gumbel's annual extreme value method to see the effect on *b*-value of Sylhet region. Thuamthansanga et al. (2019) reported the anomalous behavior of Radon isotope pairs at Mat fault, Mizoram (India) in different depths of soil. Hence, it becomes necessary to consider these local site conditions in order to predict the seismic hazard before the occurrence of an earthquake and proper mitigations measures can be taken for the infrastructures.

In the present paper, firstly, the variation of bedrock depth in Indo-Gangetic Plains is estimated using GSI [6]. Secondly, shear wave velocity is estimated up to depth 500 m using the joint fit inversion technique from dispersion curves obtained from multichannel analysis of surface wave (MASW) and horizontal to vertical spectral ratio curves from microtremor measurements using [15] method for Roorkee city having deep bedrock depth of around 3000 m. The analysis is done at different sites of the city and the city falls in Site Class D as per NEHRP Site Classification based on the shear wave velocity at a 30 m depth of soil. After obtaining the velocity profiles, ground response analysis (GRA) is performed using STRATA [9] by equivalent linear method. The response spectra from the two earthquakes, namely, Uttarkashi Earthquake (1991) and Chamoli Earthquake (1999) is compared with that given in [10]:2002 (Part 1).

2 Estimation of Bedrock Depth in Indo-Gangetic Plains

The depth of bedrock is estimated from Seismotectonic Atlas of India [6] in which contours represent the bedrock depth of a place. In AutoCAD, the study area (longitude 77.4–78.4° E and latitude 28.5–30.0°N) covering an area of 18,150sq km (length and breadth are 165 km and 110 km, respectively) was marked. Further, this rectangular area was divided in such a way that six profiles, namely, AG, BF, CE, AC, HD, GE were obtained. Figure 1 represents the variation of bedrock depth of the six profiles obtained. It can be observed, that a huge variation of bedrock depth ranging from 0 to 4000 m indicates the presence of thick soil cover in the study region. The length of the AG profile is 165 km. The location of this profile is from 30°00′20″N, 77°81′E to 28°50′45′N, 77°40′E. A wide variation in depth of bedrock from 100 to 3000 m is seen in this profile. From 100 m depth at Shahdara, the depth varies to around 3000 m at Saharanpur. In Khatauli, depth varies from 1000 to 2000 m. The length of BF profile is 165 km. The location of this profile is from 30°09′11"N, 78°20′12″E to 28°62′87″N, 77°74′92″E. A wide variation in depth of bedrock is observed ranging from 0 to 3000 m. There is a sharp increase in the bedrock depth

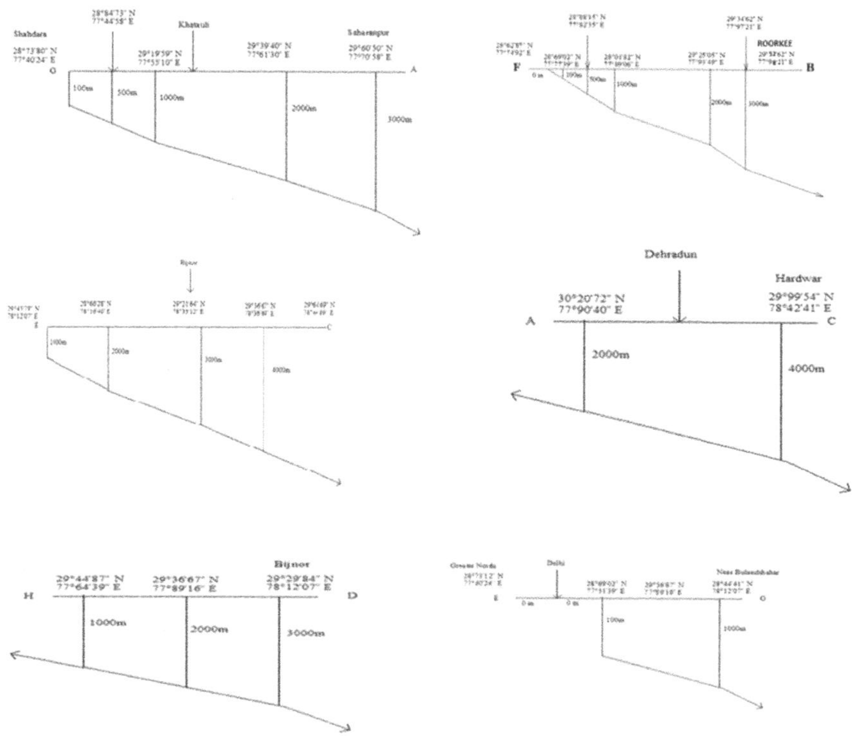

Fig. 1 Variation of bedrock depth for six profiles AG, BF, CE, AC, HD and GE

from 0 to 1000 m and then a slow rate is seen from 1000 to 2000 m and again a steep increase to 3000 m bedrock depth near Roorkee.

The length of CE profile is 165 km. The location of this profile is from 29°64′69″N, 78°44′89″E to 28°45′78″N, 78°12′07″E. A wide variation is observed ranging from 1000 to 4000 m in this profile. A sharp increase in bedrock depth is observed from 1000 to 2000 m, but a slow trend is observed from 2000 to 3000 m and again a sharply bedrock depth increases to 4000 m. The length of AC profile is 110 km. The location of this profile is from 30°20′72″N, 77°90′40″E to 29°99′54″N, 78°42′41″E. It can be seen from the Figure, that bedrock depth of Dehradun city lies between 2000 and 4000 m and at Hardwar, bedrock depth is around 4000 m. The length of HD profile is 110 km. The location of this profile is from 29°44′87"N, 77°64′39″E to 29°29′84″N, 78°12′07″E. An increase in bedrock depth is observed from 1000 to 3000 m toward Bijnor. The length of HD profile is 110 km. The location of this profile is from 28°44′41"N, 78°12′07″E to 29°73′12″N, 77°40′24″E. In this profile, the contour of 0 m intersects the profiles at two places where there is an exposure of bedrock. The variation of bedrock in this profile is from 0 to 1000 m. The depth of bedrock from 0 m near Delhi varies to 100 m in about 10 km of distance and then, increases gradually to 1000 m near Bulandshahar. The absence of soil cover in this region will not cause any amplification of seismic waves.

3 Study Region: Roorkee (Uttarakhand)

In the present study, the region selected for site characterization is Roorkee attributing to its high seismicity and having local site characteristics of soil. Roorkee is situated at latitude 29° 51′ and longitude 77° 54′ is a city in the state of Uttarakhand. Himalayan Frontal Thrust lies very close to Roorkee and to the north of it lies the Main Boundary Thrust and Main Central Thrust. Moreover, it is located around 30 km from Shiwaliks range which can cause the city to experience strong ground motion, if, a moderate earthquake occurs on Main Boundary Thrust or on Main Central Thrust. Karakoram Fault is partly exposed in the study area and extends for almost 1000 km from Central Pamir to Kumaon Himalaya [6]. The Main Central Thrust (MCT) is among the important tectonic features of the Himalayas which terminates against the Kishtwar fault in Jammu and Kashmir. MCT acts as a geological boundary between the Lesser and Higher Himalaya. The geology of the area is represented by alluvial soil and forms a part of Gangetic alluvial plains. Important tectonic features which are Almora, Berinag, Ramgarh, Munsiary Thrust, etc., also affect the seismicity of this region. As per the seismic zoning map of India as incorporated in Indian Standard Criteria for Earthquake Resistant Design of Structures: General Provisions and Buildings, Roorkee lies in seismic Zone IV (IS: 1893-Part I; 2002). Indian Meteorological Department has reported that numerous earthquakes of medium to large size have occurred in the study region as per the data obtained from historical and instrumental records.

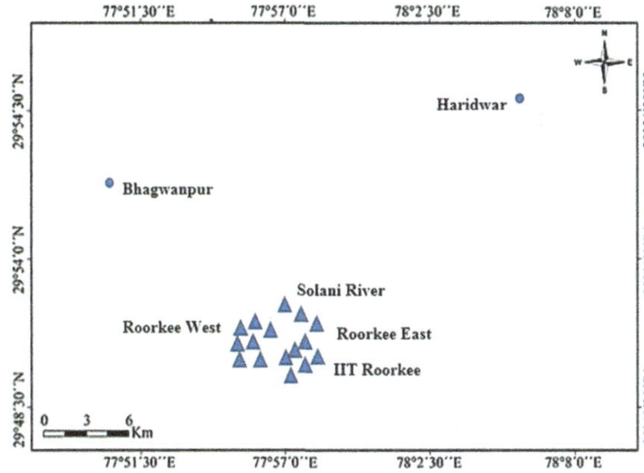

Fig. 2 Roorkee city division into three main areas (i) Roorkee East (ii) Solani River and (iii) Roorkee West

Shear wave velocity is an important parameter for understanding the dynamic behavior of soil. Moreover, it can be used for the determination of shear modulus (G) of soil as well as for the site characterization applications of geotechnical earthquake engineering. In this paper, 15 sites of Roorkee city, having a deep bedrock depth of 3000 m is explored using geophysical methods. These sites were located at a distance of about 2–3 km from each other. Figure 2 shows the division of Roorkee city namely, as, Roorkee East, Solani River and Roorkee West. In Roorkee East, six sites are taken (Sites 1–6), in Solani River, three sites are selected (Sites 7–9) and in Roorkee West, six sites are chosen (Sites 10–15). In the field set up, two tests were performed. Firstly, active MASW consisting of Soil Spy Rosina as a data acquisition system. Secondly, HVSR of Microtremor was performed in which Tromino was used as a data acquisition system.

4 Methodology

4.1 Multichannel Analysis of Surface Waves (MASW)

The setup used for investigating the subsurface materials is 9-channel MASW setup. For many applications, such as, microzonation and dynamic response analysis of sites, Multichannel Analysis of Surface waves (MASW) is increasingly being used in earthquake geotechnical engineering. It is generally used for the estimation of shear wave velocity and dynamic properties. Moreover, subsurface material boundaries identification and spatial variations of shear wave velocity can also be measured. The basic steps are as follows:

- In this technique, the recorded Rayleigh waves are transformed into the frequency domain from the time domain using the Fast Fourier Transform (FFT) techniques.

- The corresponding phase difference $\Delta\Phi(f)$ between each receiver pair of channels is computed for every component of frequency measured.
- The travel time $\Delta(t)$ between receivers is computed using the following relation

$$\Delta t(f) = \Delta\Phi(f)/2\pi f \tag{1}$$

- The distance between each pair of receivers $\Delta d = d_2 – d_1$ is known which can be used in estimation of the Rayleigh wave phase velocity [*VR*(*f*)] as follows:

$$V_R(f) = \Delta d/\Delta t(f) \tag{2}$$

- After the estimation of Rayleigh velocity, the wavelengths (λ_R) are determined as

$$\lambda_R(f) = V_R(f)/f \tag{3}$$

- Finally, a dispersion curve can be plotted between V_R-λ or V_R-depth.

4.2 Micro Tremor Method

Nakamura [15] proposed the use of microtremors for microzonation and in the study of side effects, and, is therefore also known as Nakamura technique. In the Horizontal-to-Vertical Spectral ratio (HVSR) technique, a sharp peak of the H/V spectrum at a particular site corresponds to the fundamental frequency of the sediments. In the present scenario, it is widely accepted that this frequency of the HVSR peak reflects the fundamental frequency of the underlying sediments and the amplitude is mainly dependent on its impedance contrast with the bedrock. The major advantages of HVSR method included its low-cost and simple measurement technique and providing the direct estimates of the resonance frequency of sediments. The shear-wave velocity structure cannot be obtained directly from the Microtremor HVSR method, but can be derived by synthetic modeling of spectral ratio curve. It can be inferred from the various researches carried out in the field of H/V spectrum that by applying the inversion technique of H/V spectrum, shear wave velocity structure can be obtained by estimating the thickness of every soil layer. H/V can be efficiently used to estimate the shear wave velocities upto bedrock if there is some methodology adopted to constraint the shear wave velocity of shallow layers [3]. Parolai et al. [17] used the surface wave technique for constraining the estimation of shear wave velocity in shallow layers. Hence, in the present study, the same methodology has been adopted, where for constraining the shear wave velocity profile in shallow layers MASW technique is used and for deeper depths HVSR technique is used for the estimation of shear wave velocity.

4.3 Experimental Setup

In the field setup, in order to characterize the sites of the present study, two tests were performed. Firstly, active MASW consisting of Soil Spy Rosina as a data acquisition system. Secondly, HVSR of Microtremor was performed in which Tromino was used as a data acquisition system. The basic components and the methodology are briefly explained as follows:

- Soil Spy Rosina [12] is the hardware and software platform of multichannel digital system for carrying out active and passive seismic surveys.
- The setup used in the present study consisted of nine sensors with 2 m inter-geophone spacing. For setting out a trigger, first geophone from the source was used.
- Tromino user's manual [13] has three channels connected to three orthogonal electrodynamic velocity meters with selectable gain for seismic tremor acquisition. For HVSR analysis, this instrument has been used to record ambient vibrations for 20 min at each location.

The setup used for investigating the subsurface materials for upper layers was 9channel MASW setup. It is generally used for the estimation of shear wave velocity and dynamic properties for shallow depths (upto 30 m). In order to characterize deeper soil, the use of microtremors with passive source was explored. In this method, a sharp peak of the H/V spectrum at a particular site corresponds to the fundamental frequency of the sediments is obtained. The major advantages of HVSR method included its low-cost and simple measurement technique and providing the direct estimates of the resonance frequency of sediments. By applying the inversion technique of H/V spectrum, shear wave velocity structure can be obtained by estimating the thickness of every soil layer. For more deeper depths, the available bore hole data has been used. The information was obtained from the construction division (IIT Roorkee) for bore holes that were drilled for installation of tube wells in IITR campus. In Roorkee East, the bore hole data is available upto depth of 150 m. It represents that, poorly graded sands, sands with gravels with fines or without fines are overlaid on a layer of gravelly clays, inorganic clays, silty clays of low plasticity of 10 to 15 m thickness. In Roorkee West and Solani River, the available standard penetration tests (SPT) data shows inorganic silts consisting of very fine sand rock flour, clayey silts with zero to low plasticity at the surface laid on silty sands. The SPT N values have been used for estimating the shear wave velocity profiles by using the available empirical relations. The empirical relationships considered in the present study for different types of soils are by Shibata [18], Ohba and Toriumi [16] and Lee [11].

5 Results and Discussions

Figure 3a, b and c shows the graphical representation of shear wave velocity profile of all the sites where vertical axis represents the estimated depth (m) and the horizontal axis represents the shear wave velocity V_S (m/s). Figure 3a shows the estimated shear wave velocity profiles for Roorkee East. As expected and seen from the Figure, the shear wave velocity increases as depth increases. The shear wave velocity (V_{s30}) of all the sites in this region is about 250–300 m/s Fig. 3b shows the estimated shear wave velocity profiles for sites near Solani River. The shear wave velocity (V_{s30}) of all the sites in this region is about 200–250 m/s. Figure 3c shows the estimated shear wave velocity profiles for Roorkee West. The shear wave velocity (V_{s30}) of all the sites in this region is about 240–300 m.

6 Ground Response Analysis

In order to estimate the response spectrum at a particular site, for any input time history at the base, it is essential to conduct the ground response analysis. The properties of the soil at a given site and the acceleration time history are required for the analysis. The shear wave velocity profiles which are obtained from the joint fit inversion technique of the Grilla Software have been used to carry out the ground response analysis. STRATA Software is used for the ground response analysis of each velocity profile. In this software, one can generate multiple profiles with variations in a single profile by incorporating the uncertainties. The input parameters are the thickness of the layer, unit weight, shear wave velocity of the layer and the bedrock depth.

All the site studies lie in the seismic zone IV having zone factor of 0.24 (Bureau of Indian Standards [7]: 2002 Part-1). For the generation of response spectra, two earthquakes (Uttarkashi Earthquake 1991 and Chamoli Earthquake 1999) have been chosen and were applied at bedrock of the Roorkee East division. A brief description about the earthquakes is given in Table 1. The acceleration time histories of these earthquakes have been applied at the bedrock as an input motion in the STRATA software for the ground response analysis. For the estimation of dynamic soil properties such as G/G_{max} and the damping ratio values, the inbuilt curves in the software have been used. The output obtained is thus compared with the response spectrum given in [7]:2002 (Part-1).

Figure 4 shows the comparison of the response spectrum obtained as output from the applied input ground motion at the bedrock of the study region with that provided in the [7]:2002 (Part 1). It can be clearly seen that the response spectrum obtained after applying the strong ground motion inhibits higher variation as seen in that provided in [7]:2003 (Part 1). Large amplification is observed for the Chamoli earthquake at period around 0.4 s in reference to that provided in the [7]:2002 (Part 1). In the case

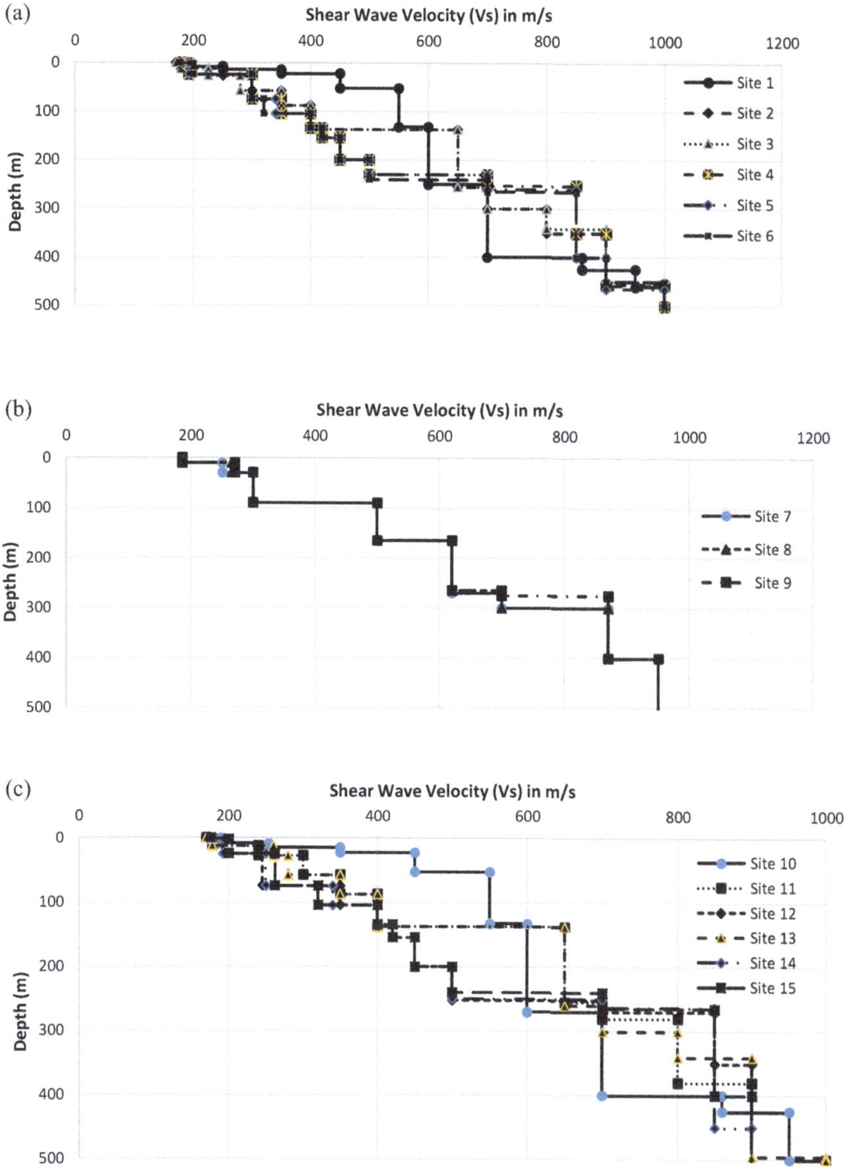

Fig. 3 **a**, **b** and **c** Graphical representation of shear wave velocity profiles where vertical axis represents the estimated depth (m) and the horizontal axis represents the shear wave velocity V_S (m/s) of all the six sites

Table 1 Description about Input Motions

Name	Date & time	Latitude (N)	Longitude (E)	Magnitude	PGA (g)
Uttarkashi Earthquake	20.10.1991 02:53	30.9	78.1	6.8	0.309
Chamoli Earthquake	29.03.1999 00:35	30.5	79.4	6.5	0.359

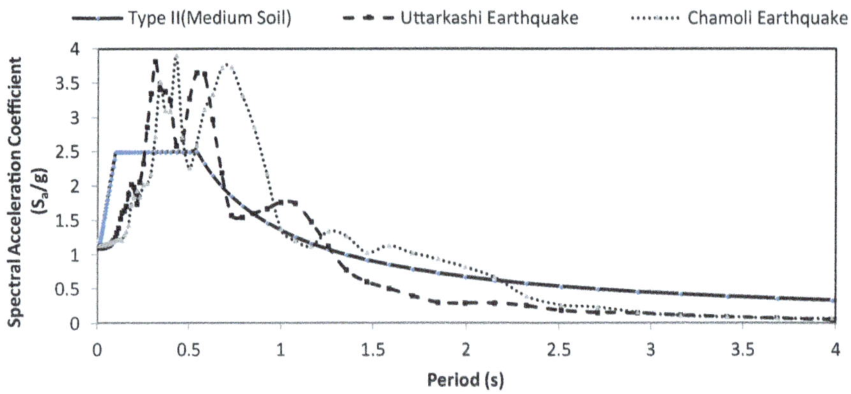

Fig. 4 Response Spectra comparison for given earthquakes by ground response analysis in STRATA Software with response spectrum given in [7]:2002 (Part 1)

of Uttarkashi earthquake, the amplification is observed around 0.3 s in reference to that provided in the [7]:2002 (Part 1).

7 Conclusion

Seismic site characterization of Roorkee (Uttarakhand) was carried out by using the geophysical methods and the available bore log data for deeper soil characteristics. The results indicate the increase in shear wave velocity with depth, which was expected. The range of V_{S30} observed lies between 224 and 357 m/s. The deepest layer observed was at a depth of 801 m from the top with half space shear wave velocity of 1000 m/s. From the analysis of the data, it was found that all the six sites fall under class D as given by NEHRP site class classification. However, according to Mittal et al. [13] all of the studied sites fall under class C which corresponds to sites having the shear wave velocity less than 375 m/s. The comparison of the response spectrum obtained from STRATA for Uttarkashi Earthquake (1991) and Chamoli Earthquake (1999) applied at the bedrock inhibits higher variation as seen in that provided in [7]:2003 (Part 1).

Large amplification is observed for the Chamoli earthquake at a period around 0.4 s in reference to that provided in the [7]:2002 (Part 1). In the case of Uttarkashi earthquake, the amplification is observed around 0.3 s in reference to that provided in the [7]:2002 (Part 1). Thus, it requires the engineer's or scientist's attention for considering the IS Code modification or site specific studies.

The above conclusions are based on the limited test data and to arrive at quantitative outcome, more number of tests needs to be conducted in the city. Nonetheless, the research presented has very much practical significance and will provide data for carrying out microzonation or hazard studies in future.

References

1. Anbazhagan P, Sitharam TG (2008) Spatial variability of the weathered and engineering bedrock using multichannel analysis of surface wave survey. Pure Appl Geophys 166(3):409–428
2. Bilham R, Blume F, Bendick R, Gaur VK (1998) Geodetic constraints on the translation and deformation of India: implications for future great Himalayan earthquakes. Curr Sci 74(3):213–229
3. Castellaro S, Mulargia F (2009) V_{S30} estimates using constrained H/V measurements. Bull Seismol Soc Am 99:761–773
4. Desai SS, Choudhury D (2015) Site-specific seismic ground response study for nuclear power plants and ports in Mumbai. Nat Hazards Rev 16(4)
5. Gahalaut VK, Chander R (1997) On interseismic elevation changes and strain accumulation for great thrust earthquakes in Nepal Himalaya. Geophys Res Lett 24:1011–1014
6. GSI (2000) Seismological atlas of India and its environs. Geological survey of India
7. IS 1893 (Part 1) (2002) Indian standard, criteria for earthquake resistance design of structures, Fifth Revision, Part-I. Bureau of Indian Standard, New Delhi
8. Khattri KN (1999) An evaluation of earthquakes hazard and risk in northern India. Himalayan Geol 20:1–46
9. Kottke AR, Wang X, Rathje EM (2013) Technical manual of STRATA. Department of Civil, Architectural and Environmental Engineering, University of Texas, Geotechnical Engineering Center
10. Kumar A, Baro, O, Harinarayan NH (2016) Obtaining the surface PGA from site response analyses based on globally recorded ground motions and matching with the codal values. Nat Hazards 81(1):543–572
11. Lee SH (1990) Regression models of shear wave velocities. J Chin Inst Eng 13:519–532
12. Micromed S. p. A. (2008a) Soil spy rosina user's manual Ver. 2.0, Micromed S.p.A., Treviso, Italy
13. Micromed S. p. A. (2012) Tromino user's manual, Micromed S.p.A., Treviso, Italy
14. Mittal H, Kumar A, Ramhmachhuani R (2012) Indian national strong motion instrumentation network and site characterization of its stations. Int J Geosci 3(06):1151
15. Nakamura Y (1989) A method for dynamic characteristics estimation of subsurface using microtremors on the ground surface. Quart Rep Railway Tech Res Inst 30(1):25–33
16. Ohba S, Toriumi I (1970) Dynamic response characteristics of Osaka plain. In: Proceedings of the annual meeting AIJ (in Japanese)

17. Parolai S, Picozzi M, Richwalski SM, Milkereit C (2005) Joint inversion of phase velocity dispersion and H/V ratio curves from seismic noise recordings using a genetic algorithm, considering higher modes. Geophys Res Lett 32(1). 1 Jan 2005
18. Shibata T (1970) Analysis of liquefaction of saturated sand during cyclic loading. Disaster Prev Res Inst Bull 13:563–570

Soil Amplification Study for Kalyani Region, Kolkata

Himanshu Shukla, Pradeep Muley, and Sajjan Kumar

Abstract The main objectives of the present paper are: (i) to identify the areas in Kalyani region, Kolkata, i.e. AIIMS Kolkata campus in which soil formations are prone to amplifying ground motions and (ii) to analytically evaluate amplification ratio using computer program for earthquake site response analysis. Peak ground acceleration at different depth is calculated for all the selected six sites with the help of soil profile. For this purpose, one-dimensional linear ground response analysis is done by the help of DEEPSOIL V-7.0 computer program. It can be observed that from the results peak ground acceleration of different site is varying from 0.159 to 0.190 g and the soil amplification value is ranging from 0.996 to 1.190 for peak ground acceleration of 0.16 g.

Keywords Amplification factors · Peak ground accelerations · Ground motion amplification

1 Introduction

Site response analysis is considered an important and primary phase of any seismic soil structure study and is also required for the microzonation studies of a region. Ground motion characteristics are greatly influenced by geological formation of soil. For the design of any structures ground response analysis and seismic analysis are compulsory to know the various factor of safety. Effects of earthquake can be estimated by predicting the site response and liquefaction potential as the local ground are affected by seismic waves when earthquake occurs [1].

H. Shukla · P. Muley (✉) · S. Kumar
Department of Civil Engineering, Madan Mohan Malaviya University of Technology Gorakhpur, Gorakhpur 273010, U.P., India
e-mail: pmce@mmmut.ac.in

T. G. Sitharam et al. (eds.), *Local Site Effects and Ground Failures*, Lecture Notes in Civil Engineering 117, https://doi.org/10.1007/978-981-15-9984-2_2

Soil formation of any site highly affects the peak ground acceleration (PGA) value so their amplification factor and de-amplification factor must be determined. Soil amplification of ground motion is generally calculated by the specifying site dependent response spectra or by specifying amplification co-efficient for various soil formations.

According to various researchers [2–4] soil classification should give more proper information of soil formation through the use of several site response parameters, e.g., soil grain size, standard penetration test blow-count (N-SPT), shear resistance modulus, material damping, seismic impedance contrast between the bedrock and the overlying soil deposits.

The most used various equivalent linear computer programs are DEEPSOIL developed by [5], EERA [6], SHAKE [7], SHAKE-91 [8], SHAKE-2000 [9], ProShake [10] and CyberQuake [11]. All these programs use for computational programming. Previously number of studies have been completed by several researchers on soil amplification for the various region of India like for Roorkee region it was carried out by [12]. Similar study on soil amplification for Bangalore region and Ahmedabad region was conducted by [13, 14] respectively.

To estimate the realistic earthquake ground motion and seismic hazard of region, evaluation of site amplification functions is mandatory. Technique used for estimation of site response requires a reference site. For ideal condition, a complete ground response analysis will decide the rapture phenomenon at the earthquake source, beneath a particular site, seismic wave traveling through the earth toward the top of bedrock and the influence soil formation above the bedrock on the ground surface motion measured using seismic wave [1]. But in actual case the phenomenon of fault rupture is very complicated, and the uncertainty is found in energy transmission between source and site. This may be the reason that this method is not used in engineering field. Various empirical relations are developed based on realistic earthquake which was recorded. These developed relations are used to predict the motion of bedrock at the site for seismic hazard analysis. So, in-ground response analysis problem motion of the bedrock is determined instead of response of soil formation. Seismic waves can penetrate 10 km through rock and are able to go through soil up to depth of 100 m.

The main objective of this study is used to evaluate the soil amplification factor for Kalyani region, Kolkata, India. This factor is used to identify the soil formations which are prone to amplification. This aim is achieved by performing analytical evaluation of amplification ratio using DEEPSOIL V-7.0 computer program for earthquake site response analysis.

To determine the response of motion of the bedrock closely below the soil site specific ground response investigation is carried out. This also shows the effect on amplification of seismic waves of local soil condition [1]. These results may help in further study and microzonation of the study area.

Table 1 Location of sites in AIIMS Campus in Kalyani region

S. No	Name of sites	Bore log depth (m)	Latitude	Longitude	Water table (m)	No of SPT sample
1	BH-01	18.00	N22°5812.62	E88°3138.44	2.8	11
2	BH-02	18.00	N22°5812.62	E88°3138.44	3.1	11
3	BH-03	18.00	N22°5812.62	E88°3138.44	1.6	11
4	BH-04	18.00	N22°5812.62	E88°3138.44	3.3	13
5	BH-05	18.00	N22°5812.62	E88°3138.44	1.4	11
6	BH-06	18.00	N22°5812.62	E88°3138.44	1.8	12

2 Study Area

In this paper soil amplification ratio is evaluated for Kalyani region, Kolkata India. AIIMS Kalyani, Kolkata Campus was selected for soil amplification study purpose. The present study will predominantly be concerned with identifying the areas in Kalyani region, Kolkata in which soil formations are susceptible to amplifying ground motions. AIIMS Kalyani, Kolkata is a medical institute which is located at Kalyani city of Nadia district located around 50 km from the capital city of West Bengal, India. Kalyani region lies in seismic zone-III [15] which is susceptible to earthquake. Six locations in AIIMS Kalyani campus were selected for this study. Locations of sites, i.e., latitudes and longitudes, borehole depth, location of water table and number of samples collected through SPT are mentioned in Table 1.

3 Geological Investigation

In this paper the amplification factors of AIIMS Kalyani campus of West Bengal are evaluated. In-situ tests are performed to collect the geotechnical data and soil samples of the study area which are required for the evaluation of soil amplification. Geotechnical investigation was performed by Centre for Advanced Engineering Kolkata [16]. For geotechnical investigation purpose Standard Penetration Test (SPT) was performed as per guidelines of Indian standard [17]. Liquefaction potential for the same locations has been evaluated by [18] and details of soil profile and SPT N value with the depth are mentioned in [18].

Based on geotechnical investigation, soil profile may be divided into three layers. First layer up to 4.5 m, second layer started around 5.0–13 m and further third layer is exploring up to depth of 20 m. First layer consists of clayey silt which is soft to medium brownish grey and presence of sand and kankar in small amount was observed. Values of SPT N values lie between 4 and 19 in this layer. Similarly soil of second layer carries loose to medium grey silty fine sand and due to presence of sand, ranges of SPT N values varying from 11 to 41. Soil deposits of the third layer

Table 2 Summary of key point of all the six sites in AIIMS Kolkata campus

S. No	Name of site	Depth (m)	N Value	Vs (m/Sec)	FC (%)	γ_d (kN/m^3)	Specific Gravity (G)
1	BH-01	18.00	5–44	145–211	10–88	1.45	2.64–2.67
2	BH-02	18.00	7–47	157–204	76–88	1.42	2.63–2.65
3	BH-03	18.00	4–29	138–190	2–86	1.45	2.64–2.68
4	BH-04	18.00	4–55	136–218	79–89	1.45	2.64–2.65
5	BH-05	18.00	10–39	170–205	3–90	1.44	2.63–2.66
6	BH-06	18.00	4–45	138–203	7–87	1.42	2.64–2.66

consists of dense to very dense grey fines and a trace of silt was found in medium sand. Exploration of last layer was done up to depth of 20 m. SPT N value is high for this layer and is varying from 30 to 55.

Geotechnical investigation of all the six sites is surmised in Table 2. N value is varying from 4 to 55 and the N values are higher for BH-04. As the depth increases value of shear wave velocity also increases due to increase in N value [18]. The values of shear wave velocity (Vs) are higher for BH-04 i.e.136–218 m/s, it can be attributed due to the fact of high SPT N values. For the calculation of the corrected shear wave velocity equations given by [19] used in this analysis (Fig. 1).

Properties of soil, i.e. unit weight, specific gravity, presence of fines content for Kalyani region, Kolkata are mentioned in Table 2. Variation in fines content for BH-04 is 79–89%.

The Shear wave velocity profile for AIIMS Campus is shown in Fig. 2. It can be observed from Fig. 2 that the maximum and minimum values of Vs are obtained at higher (15.0 m) and at shallow depth at BH-04 respectively. Also, it can be observed from Fig. 2 that the value of Vs increases with the depth for all the six sites. This variation may be due to the increasing values of SPT N with the depth [18].

4 Ground Response Analysis

Ground response analysis is used to forecast site natural period, to access ground motion amplification, in the development and design of response spectra. It can also be used in evaluation of liquefaction potential and determination of earthquake induced forces which can lead to instability of earth and earth slope [1]. For one-dimensional ground response analysis three methods are used, i.e. linear method, equivalent linear method and nonlinear method.

Linear approach to one dimensional ground response analysis of layered soil has been coded into widely used computer program named as DEEPSOIL [5]. This approach is being used in the present study for the development and analysis of the peak ground acceleration for six sites in Kalyani region, Kolkata. Linear frequency

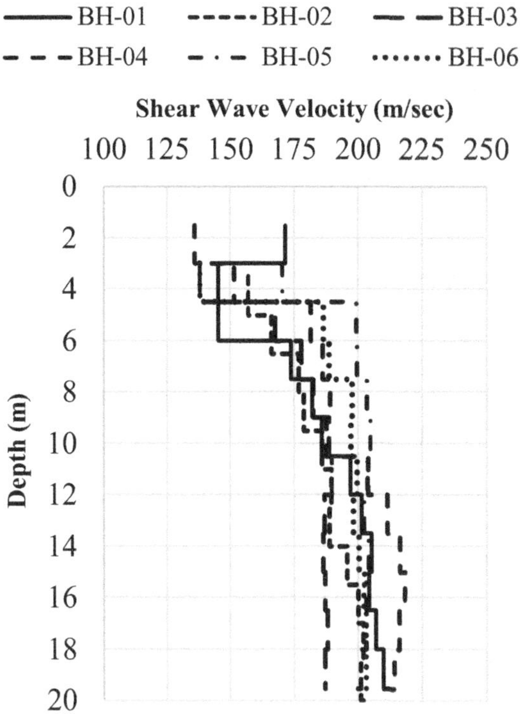

Fig. 1 Shear Wave Velocity with the depth for all the six sites in AIIMS Kalyani, Kolkata

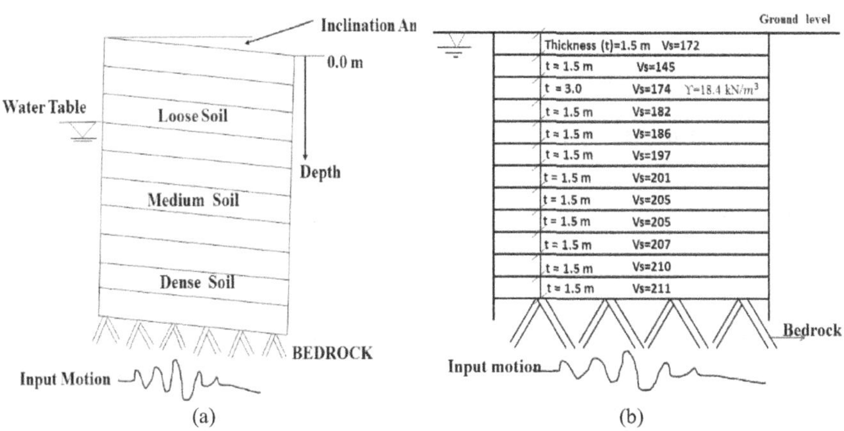

Fig. 2 Layered soil profile **a** Soil profile model **b** Layered soil profile of BH-01

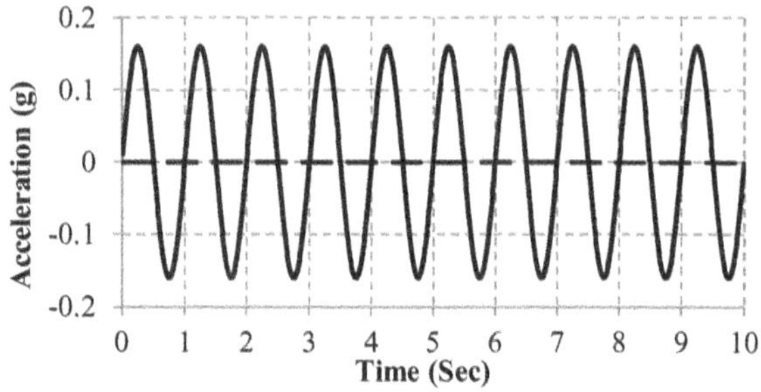

Fig. 3 Acceleration time history (sinusoidal wave for 10 s and 0.16 g)

domain analysis for damped elastic rock is being used in this study, stiffness of layer in shear wave velocity (Vs) or shear modulus (G) is also used.

DEEPSOIL computer program includes three major components—soil profile, input motion and result [5]. Soil profile model is shown in Fig. 2a. Layered soil profile of BH-01 is shown in Fig. 2b which describes the different layers of the soil according to their depth, density and shear wave velocity. It can be observed from Fig. 2b that the thickness of each layer is uniform (1.5 m) except third layer which is of 3.0 m for BH-01. Shear wave velocity plays an important role in the analysis.

Assessment of potential ground motion of each site is essential after the characterization of site. Suitable rock motion is selected to symbolize design rock motion for the study area. Kalyani region, Kolkata lies in seismic zone-III. For this seismic zone, PGA = 0.16 g [15] is considered in the present study. In the present analysis input motion is applied in the bedrock for 0.16 g acceleration for 10 s. The acceleration time history used in this analysis is shown in Fig. 3. Geotechnical investigation is done up to the depth of 20 m and further bedrock is considered in this analysis.

5 Results and Analysis

Study of ground response analysis is important which may help in controlling the damage situation during an earthquake or cyclic loading. The destruction of earthquake at any location does not only depend on its magnitude and distance from the epicenter but also on the sub-soil properties of the location [1].

In this study one-dimensional ground response analysis is done by DEEPSOIL computer program. This program can perform one dimensional nonlinear time domain analysis with and without pore water pressure generation [5]. This software was developed by [5]. Ground response analysis with DEEPSOIL includes three steps—profile tab, motion tab and analysis tab [5].

In the present study soil amplification for AIIMS Kalyani Campus is evaluated for PGA = 0.16 g. For this purpose, soil amplification factor and peak ground acceleration with the depth for all the six sites are calculated using DEEPSOIL computer program. It can be observed from Table 3a and b that the peak ground acceleration value and the soil amplification factor are maximum for all the sites at the shallow depth and are decreasing with depth.

The maximum value of soil amplification factor is 1.190 observed at top surface in case of BH-03 (Table 3a). Ranges of Peak ground acceleration for all the six sites vary from 0.190 to 0.159 g. From Table 3a it is observed that range of soil amplification factor is 1.086–1.002, 1.145–1006 and 1.190–1.012 for BH-01, BH-02 and BH-03 respectively. Similarly, from Table 3b it can be concluded that the amplification factor varies in the range of 1.053 to 0.996, 1.053–1.002 and 1.080–1.000 for BH-04, BH-05 and BH-06 respectively.

6 Summary and Conclusion

Study of ground response analysis is necessary which may help in controlling the damage situation during an earthquake. Amplification factor shows the measure of increase in certain frequency of ground motion of soft soil deposit that causes increment of earthquake damage. The one-dimensional linear ground response analysis of different location of the AIIMS Kalyani campus, Kolkata is evaluated using DEEPSOIL V-7.0 computer program. Peak ground acceleration with depth is evaluated for all the six sites for Harmonic Motion of PGA = 0.16 g for 10 s duration. It can be observed that from the results, peak ground acceleration for all the six sites is varying from 0.159 to 0.190 g and the soil amplification value is ranging from 0.996 to 1.190 respectively. This analysis may be further refined analysis using different earthquake motions and also for Non-Linear site response analysis.

Table 3 **a** PGA and Soil amplification Factor for three sites (BH-01 to BH-03). **b** PGA and Soil amplification Factor for other three sites (BH-04 to BH-06)

Site Identification	Depth (m)	Acceleration (g)	Soil amplification factor
BH-01	0.0	0.174	1.086
	1.5	0.173	1.083
	3.0	0.171	1.067
	6.0	0.168	1.05
	7.5	0.167	1.043
	9.0	0.166	1.037
	10.5	0.165	1.031
	12.0	0.164	1.025
	13.5	0.163	1.020
	15.0	0.162	1.014
	16.5	0.161	1.008
	18.0	0.160	1.002

(continued)

Table 3 (continued)

Site Identification	Depth (m)	Acceleration (g)	Soil amplification factor
BH-02	0.0	0.183	1.145
	3.5	0.179	1.116
	5.0	0.176	1.100
	6.5	0.174	1.086
	8.0	0.172	1.073
	9.5	0.170	1.063
	11.0	0.168	1.052
	12.5	0.167	1.041
	14.0	0.165	1.032
	15.5	0.164	1.023
	17.0	0.162	1.014
	18.0	0.161	1.006
BH-03	0.0	0.190	1.19
	3.0	0.185	1.155
	4.5	0.182	1.138
	6.0	0.179	1.122
	7.5	0.177	1.106
	9.0	0.175	1.093
	10.5	0.173	1.080
	12.0	0.171	1.066
	13.5	0.168	1.052
	15.0	0.166	1.039
	16.5	0.164	1.025
	18.0	0.162	1.012
BH-04	0.0	0.169	1.053
	1.5	0.168	1.051
	3.0	0.167	1.044
	4.5	0.166	1.040
	6.0	0.165	1.034
	7.5	0.164	1.027
	9.0	0.163	1.020
	10.5	0.162	1.015
	12.0	0.162	1.010
	13.5	0.161	1.007

(continued)

Table 3 (continued)

Site Identification	Depth (m)	Acceleration (g)	Soil amplification factor
	15.0	0.161	1.004
	16.5	0.160	1.000
	18.0	0.159	0.996
BH-05	0.0	0.169	1.053
	3.0	0.168	1.048
	4.5	0.167	1.045
	6.0	0.167	1.041
	7.5	0.166	1.036
	9.0	0.165	1.031
	10.5	0.164	1.026
	12.0	0.163	1.022
	13.5	0.163	1.017
	15.0	0.162	1.012
	16.5	0.161	1.008
	18.0	0.160	1.002
BH-06	0.0	0.173	1.080
	3.0	0.168	1.052
	4.5	0.167	1.046
	6.0	0.167	1.042
	7.5	0.166	1.038
	9.0	0.165	1.033
	10.5	0.165	1.028
	12.0	0.164	1.023
	13.5	0.163	1.017
	15.0	0.162	1.012
	16.5	0.161	1.006
	18.0	0.160	1.000

References

1. Kramer S (1996) Geotechnical earthquake engineering. Pearson Education Pvt. Ltd., Singapore
2. Medvedev J (1962) Engineering seismology. Academia Nauk Press, Moscow
3. Zhao J, Irikura K, Zhang J, Fukushhima Y, Somerville P, Asono A, Takashi T, Ogawa H (2006) An empirical site classification method for strong motion stations in Japan using H/V response spectral ratio. Bull Seism Soc Am 96, 914–925
4. Pitilakis K, Gazepis C, Anastasiadis A (2006) Design response spectra and soil classification for seismic code provisions. In: Proceedings, ETC-12 workshop
5. Hashash YMA, Musgrove MI, Harmon JA, Ilhan O, Groholski DR, Phillips CA, Park D (2017) DEEPSOIL 7.0, nonlinear and equivalent linear seismic site response of one-dimensional soil columns user manual

6. Bardet JP, Ichii K, Lin CH (2000) EERA: a computer program for equivalent-linear earthquake site response analyses of layered soil deposits. University of Southern California, Los Angeles
7. Schnabel PB, Lysmer J, Seed HB (1972) SHAKE: a computer program for earthquake response analysis of horizontally layered sites. Report n° EERC72-12, University of California at Berkeley
8. Idriss IM, Sun JI (1992) User's Manual for SHAKE91, A computer program for conducting equivalent linear seismic response analysis of horizontally layered soil deposits, Center for geotechnical modeling, Department of civil and environmental engineering, University of California, Davis
9. Ordonez GA (2002) SHAKE2000: a computer program for 1-D analysis of geotechnical earthquake engineering problems, User's manual, Geomotions, LLC, Lacey, Wash, 264. https://www.geomotions.com
10. EduPro civil system Inc. (2003) Proshake, version 1.12. Sammamish, Washington. https://proshake.com
11. Modaressi H, Forerster E (2000) CyberQuake. User's Manual, BRGM, France
12. Kirar B, Maheshwari B (2015) Soil amplification studies for Roorkee region. In: International geotechnical symposium on disaster mitigation in special geoenvironmental conditions Chennai, India, vol 6, 21–23
13. Anbazhagan P, Sitharam T, Divya C (2006) Soil amplification and liquefaction studies for Bangalore city, Indian geotechnical conference 2006. Channai, India, pp 14–16
14. Trivedi S (2011) Soil amplification study for Ahmadabad region. In: Institute of technology, International conference on current trends in technology, vol 2011, 382–481
15. IS: 1893:2016 (Part-I), Criteria for earthquake resistance design for structure: general provisions and building bureau of Indian standards (BIS), New Delhi
16. Centre for Advanced Engineering (2015) Report on geotechnical investigation for setting up of AIIMS at Kalyani West Bengal. 1–77. Centre for Advanced Engineering, Kolkata
17. IS 2131 (1981) Method for standard penetration test for soils. Bureau of Indian Standards (BIS) New Delhi
18. Kumar S, Srivastava T, Muley P (2019) Assessment of SPT-based liquefaction potential of kalyani region, Kolkata. In: Proceedings of the Indian geotechnical conference, SVNIT Surat
19. Andrus RD (2000) Stokoe, KH-II: liquefaction resistance of soils from shear-wave velocity. J Geotech Geoenviron Engg 126(11):1015–1025

Simulated Annealing Algorithm for Subsurface Shear Wave Velocity Investigation Using Ground Vibration Data

Anushka Joshi, Sateesh K. Peddoju, and Mohit Pandey

Abstract Subsurface shear wave velocity plays an important role in designing earthquake-resistant structures. Average shear wave velocity up to 30 m depth, known as V_{s30}, is used as a common design parameter. Ambient noise data that is generated by ground vibrations due to the passing of vehicles or other passive sources carry an important information about the subsurface shear wave velocity structure of the region. The seismic instrument records these disturbances along with three different directions i.e. two horizontal and one vertical. The ratio of the spectral amplitude of the horizontal to the vertical component is an important factor that is dependent on subsurface velocity structure of the investigating area. Ground vibration data from ambient noise recording system has been recorded at Bayasi site in the Garhwal Himalaya. A computer code DISHV in FORTRAN has been developed in this work for obtaining HV curve having a limited data from a large sampled HV curve having similar spectral property. The obtained output from DISH has been effectively used as an input to the HV-Inv software developed by Garcia et al. (Comput Geosci 97:67–78, 2016). The HV curve obtained from recorded data is modelled with a simulated annealing algorithm. The obtained velocity model from Simulated Annealing algorithm is further refined using forward modelling approach to match obtained and theoretical HV curve.

Keywords HVSR · Simulated annealing · Shear wave velocity

A. Joshi
Department of Computer Science and Engineering, Govind Ballabh Pant Institute of Engineering and Technology, Pauri, Uttarakhand, India
e-mail: anushkazinc@gmail.com

S. K. Peddoju
Department of Computer Science and Engineering, Indian Institute of Technology Roorkee, Roorkee, Uttarakhand, India
e-mail: drpskfec@iitr.ac.in

M. Pandey (✉)
Department of Earth Sciences, Indian Institute of Technology Roorkee, Roorkee, Uttarakhand, India
e-mail: pmohit42@gmail.com

T. G. Sitharam et al. (eds.), *Local Site Effects and Ground Failures*, Lecture Notes in Civil Engineering 117, https://doi.org/10.1007/978-981-15-9984-2_3

1 Introduction

Shear wave velocity is a valuable indicator of the dynamic behaviour of the soil due to its relationship with the shear modulus (i.e. $\mu = \rho \cdot V_s^2$), which depends on strain level. Therefore, it is the important parameter in construction purposes and engineering studies. Shear waves are the type of elastic waves, where particle motion is perpendicular to the direction of wave propagation. These waves are also called S-wave or Secondary wave because of its appearance on the seismogram immediate after the first arrival P-wave. Shear waves travel slowly in the rock as compared to the P-wave. The P-wave velocity is 1.37 times the S-wave velocity. Shear wave velocity is used to classify the soil types based on V_{s30} (S-wave velocity over top 30 m of depth) by characterizing the effect of sediment stiffness on the ground motion. There are many soil classifications based on V_{s30} like NEHRP, EUROCODE8, Uniform Building Code (UBC), etc. which is used for earthquake engineering design with the expectation that sites in the same class will respond similarly to a given earthquake. The shear wave velocity can be estimated by many geophysical methods. Out of which, Horizontal to Vertical Spectral Ratio (HVSR) technique is one of the geophysical method to estimate the shear wave velocity and for the study of ground motion amplification. This technique is non-destructive and cheapest. However, there is no evidence of the particular composition of the noise field at HV curve. The spectral ratio curve has been assumed to be dominated by the body waves around the peak of the *H*/*V* [6, 10, 15]. The HVSR ratio curve from ambient noise on other hand, has been related to the Rayleigh wave ellipticity by Lermo and Chávez-García [11], Malischewsky and Scherbaum [13]. When Rayleigh and Love waves come from various directions an ellipticity analysis becomes very complicated. However, successful inversion schemes based on surface wave ellipticity have been used [2] . The description of HV curve by taking into account the combined effect of complete seismic ambient-vibration wave field [21]. This is called diffuse field assumption. The purpose of this paper is to get the best shear wave velocity model by modelling Horizontal to Vertical Spectral Ratio curve by using Simulated Annealing Method based on software developed by García-Jerez et al. [9]. The input to this software require HV spectral ratio data in very limited samples. In the present work, a program has been developed to model the spectral ratio curve from ambient noise data and obtain a reliable limited sampled spectral ratio data.

2 Methodology

It has been verified by many workers that earthquake ground motion is strongly effected by local site conditions [1, 19]. Site effects is characterized by the means of spectral ratios of recorded motions with respect to reference rock site (e.g. Cadet et al. [7]). On the other hand, for more quiescent areas seismic noise severe as a major tool to provide useful information about spectral ratio contents of ground motion. Nakamura [14] was among initial workers who have related spectral ratio of horizontal to vertical component of recorded ground motion with the site effects. Lermo and Chávez-García [11] has obtained reliable estimates of the soil resonance frequencies from the spectral ratio of horizontal with vertical motions of microtremors. The HV curve has been studied to explain its strengths and limitations by different workers viz. [4, 6, 12, 17]. The HV ratio curve reveals the site dominant frequency f_0 but the amplitude of this HV curve is not well understood [17]. The recent work of Lunedei and Albarello [12] is a significant step towards the understanding of the amplitude as well as frequency part of HV spectrum.

In the present work, noise is assumed to be diffuse wavefield containing all types of elastic waves. Underlying this assumption is the connection between diffuse fields and the Green's function [8, 20, 23–25], which implies the proportionality between the average energy densities of a diffuse field and the imaginary part of Green's Function at the source [16, 20]. This can be summarised as:

$$\langle u_i(x_A, \omega) u_j^*(x_B, \omega) \rangle \propto \mathrm{Im}[G_{ij}(x_A, x_B; \omega)] \tag{1}$$

where $G_{ij}(x_A, x_B; \omega)$ is the Green's function defined as the displacement in the ith direction at point x_B due to the application of a unit harmonic point force in the jth direction applied at point x_A, $u_i(x_A, \omega)$ is the displacement field component in i direction at point x_A, $\omega = 2\pi f$ is the angular frequency. For coinciding source and receiver, the autocorrelation of both the components gives the directional energy density $E_i(x, \omega)$ at point x as:

$$E_i(x, \omega) = \rho \omega^2 \langle \| u_i(x, \omega) \|^2 \rangle \propto \mathrm{Im}[G_{ii}(x, x; \omega)] \tag{2}$$

In this expression ρ is the mass density. Therefore, the HV can be expressed as the square root of the ratio of the corresponding imaginary parts of Green's tensor components [2, 22]:

$$HV(x, \omega) = \sqrt{\frac{\mathrm{Im}[G_{11}(x, x; \omega)] + \mathrm{Im}[G_{22}(x, x; \omega)]}{\mathrm{Im}[G_{33}(x, x; \omega)]}} \tag{3}$$

In above expression the subscripts 1 and 2 represent the horizontal directions while the subscript 3 corresponds to the vertical one, respectively.

HV-Inv is a computer code for forward calculation and inversion of H/V spectral ratios of ambient noise (HVSRN) based on the diffuse field assumption (DFA) developed by García-Jerez et al. [9]. The software supports joint inversion of HVSRN and dispersion curves by using several local and global algorithms which include simulated annealing algorithm.

The definition of an optimal solution for any optimisation problems can be an extremely challenging task or some time it is practically difficult. This is because when a problem gets very large, we need to search a vast number of the possible solution to find the optimal solution. This type of problem can be solved using Simulated Annealing Technique (SAT). This is a probabilistic technique for approximating the global optimum of a given task. It is efficacious and commonly used form of optimization problems, which is used for solving the unconstrained and bound-constrained optimization problems. Simulated annealing is not affected by the initial guess, as it is a global optimization technique and gives the global optima instead of local optima, which is an optimal solution [5]. This algorithm was originally motivated by the process of annealing of metal. "Annealing" discusses the process of thermodynamics, which involves the cooling and heating of the material to modify its physical properties due to change in its internal structure [18]. The algorithm is a hill-climbing except instead of picking the best move, it picks a random move. At each iteration of the simulated annealing technique, a new point is randomly generated. If the selected iteration improves the solution, then it is accepted. Otherwise, the algorithm makes the iteration with some probability less than one. The difficulty with this approach is that while it rapidly finds a local minimum, it cannot get from there to the global minimum. There are so many uses of Simulated Annealing (SA technique in different domains. It has a broad range of application that is still being explored. In the present work the spectral content of HVSR curve is mainly dependent on the P and S wave velocity profile of region and density of the subsurface layers. Initial range of these parameters is selected from the heuristic guess based on geophysical and geological observations. The iterative computation of HVSR curve is based on Simulated Annealing algorithm. In the present work, HV-Inv has been used to study the site specific case study of the Bayasi region in the Garhwal Himalaya.

The ground vibration created by any disturbance at the surface of the earth can be recorded by three-component recorders. These recorders record ground vibrations in two horizontally and one vertically oriented sensor. Nakamura [14] has shown that the ratio of the amplitude spectrum of the horizontal component to the vertical component is directly related to the shear wave velocity profile of the subsurface of the earth. Normally in this technique, the ground disturbance created by passing vehicle is recorded for a specific duration of time in three-component sensors i.e. two horizontal components along north–south and east–west direction ($a_{ns}(t)$, $a_{ew}(t)$) and one vertical component ($a_v(t)$) as shown in Fig. 1. Fourier transform of these components are as follows:

$$\mathrm{a_{ns}(t)} \longrightarrow \mathrm{A_{ns}(f)} \tag{4}$$

$$\mathrm{a_{ew}(t)} \longrightarrow \mathrm{A_{ew}(f)} \tag{5}$$

$$\mathrm{a_{v}(t)} \longrightarrow \mathrm{A_{v}(f)} \tag{6}$$

Amplitude spectrum of each component is calculated as:

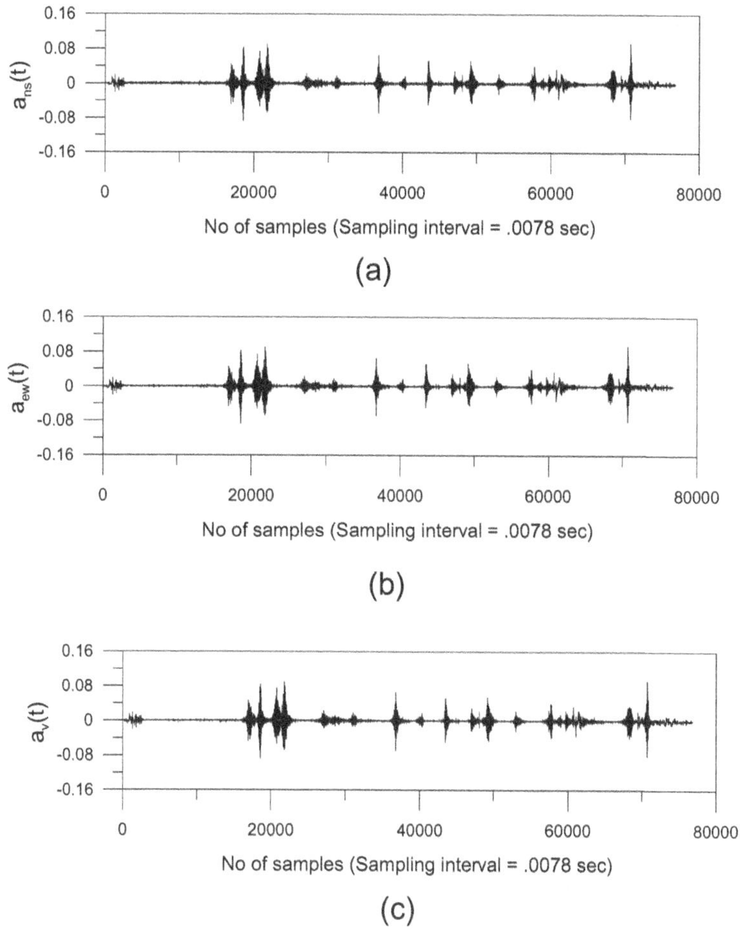

Fig. 1 Time series record of the **a** North South '$a_{ns}(t)$', **b** East West '$a_{ew}(t)$' and **c** vertical '$a_v(t)$' component of ground vibrations created by passing vehicle

$$\mathrm{Amp_{ns}(f)} = [\mathrm{Re}(\mathrm{A_{ns}(f)})^2 + \mathrm{Im}(\mathrm{A_{ns}(f)})^2]^{1/2} \tag{7}$$

$$\mathrm{Amp_{ew}(f)} = [\mathrm{Re}(\mathrm{A_{ew}(f)})^2 + \mathrm{Im}(\mathrm{A_{ew}(f)})^2]^{1/2} \tag{8}$$

$$\mathrm{Amp_{v}(f)} = [\mathrm{Re}(\mathrm{A_{v}(f)})^2 + \mathrm{Im}(\mathrm{A_{v}(f)})^2]^{1/2} \tag{9}$$

As we have two horizontal and one vertical component, so there are two amplitude spectrums of NS and EW component. The amplitude spectrum of $a_{ns}(t)$, $a_{ew}(t)$ and $a_v(t)$ is shown in Fig. 2.

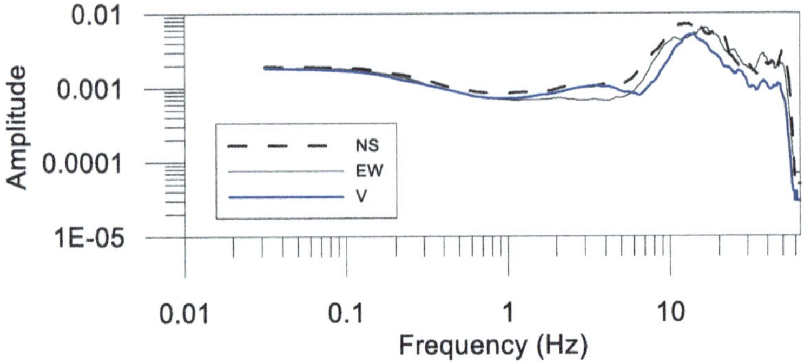

Fig. 2 Amplitude spectrum of time series shown in Fig. 1 along North South, East west and Vertical component of ground vibration

We can calculate the average of horizontal components either by geometrical or root mean square amplitude of horizontal components by the following formula:

$$\text{Root mean square} = \left[\left\{\left(\text{Amp}_{\text{NS}}(\text{f})\right)^2 + \left(\text{Amp}_{\text{EW}}(\text{f})\right)^2\right\}/2\right]^{1/2} \quad (10)$$

or

$$\text{Geometrical mean} = \left[\text{Amp}_{\text{NS}}(\text{f}) \times \text{Amp}_{\text{EW}}(\text{f})\right]^{1/2} \quad (11)$$

In the present work, Geometrical mean has been used for obtaining the H/V curve at any site. H/V curve can be obtained by the following expression and is shown in Fig. 3.

$$\text{HV}(\text{f}) = \left[\text{Amp}_{\text{NS}}(\text{f}) \times \text{Amp}_{\text{EW}}(\text{f})\right]^{\frac{1}{2}}/\text{Amp}_{\text{V}}(\text{f}) \quad (12)$$

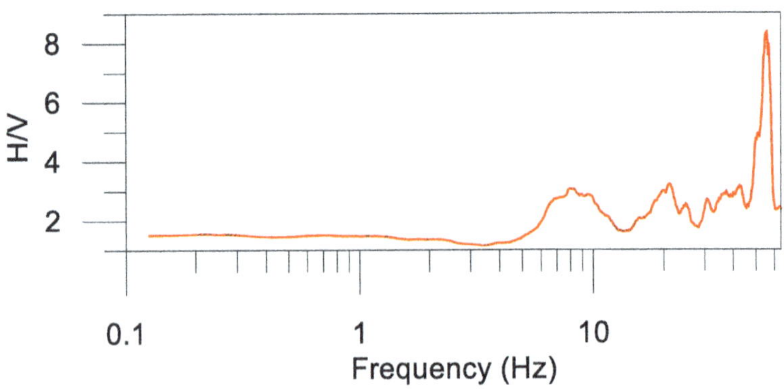

Fig. 3 Ratio of amplitude spectrum of horizontal by vertical component of record shown in Fig. 1

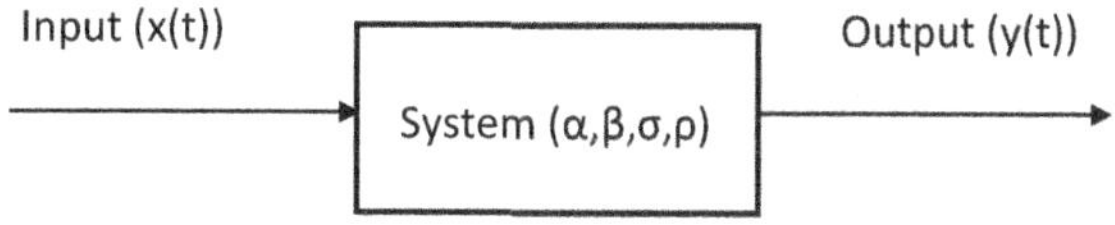

Fig. 4 Flow graph of forward modelling approach

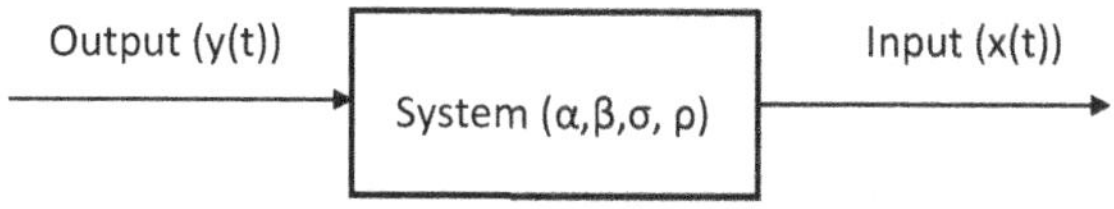

Fig. 5 Flow graph of inverse modelling approach

According to system theory, the output 'y(t)' is obtained when any input 'x(t)' is passed through a system as shown in Fig. 4.

In any mathematical approach, when input is known and system response is known and we are calculating the output, then such system is termed as the forward approach. Similarly, when the output of the system is known and we are interested to calculate input from the system, then this approach is termed as inverse approach. In both approaches, we know the system response. Inverse approach is shown in Fig. 5.

The obtained HVSR curve at any site is theoretically represented by the formulation given which is mainly dependent on parameters of the earth model. In the formulation, system response mainly depends on P wave velocity (α), Shear wave velocity (β), the density of the medium (ρ) and Poisson ratio (σ). The purpose of present work is to select final earth parameters using Simulated Annealing algorithm developed by García-Jerez et al. [9]. Different authors have attempted to explain the H/V phenomenology in terms of SH waves [10, 14], of Rayleigh waves [3, 11, 13], and by adding the effects of Love. Recent studies have considered the role of all waves, known as a total field [21, 22]. It was found that results differ not only as a function of the type of waves but also as a function of the temporal and spatial distribution of the sources and their strength. The inversion of the H/V curve requires not only the knowledge of the specific sources acting at the site but also of several other soil parameters (e.g., Poisson's ratio, damping of each layer, 2D effects, etc.) which cannot be easily determined. Under these condition various algorithms helps in selecting most preferred model out of various possible range of modelling parameters.

Modelling of HV curve can be made by Simulated Annealing algorithm in HV-Inv software. This software requires HV curve as an input function which is modelled by using appropriate parameter space that can be estimated using several algorithms supported in the software. In the present work the acquired data consist of a time series of 12 min length that has sampling frequency 128 Hz. Therefore, the data of ground vibration consist of 92,160 samples along three different mutually perpendicular directions. This gives a similar number of samples in HV curve in frequency domain. However, software HV-Inv needs very less number of data point in the HV curve as an input to the algorithm. Therefore, the challenging task before us is to re discretize 92,160 samples into very less samples retaining the spectral property of HV curve. We need to discretize it so that all peaks should remain preserved in an HV curve of fewer samples. For this purpose, a software DISHV is prepared in FORTRAN that prepares an HV curve of small length i.e. of nearly 30 samples in frequency range 0.1–100. Input to this program is a time series in NS, EW and Vertical component of ground vibrations. The program calculates FFT of these input ground vibration in various small windows which can be decided by the user. Presently 38,000 samples at sampling frequency 128 Hz has been used as input of ground vibration along with NS, EW and Vertical direction. Figure 6 shows the time series along with NS, EW and vertical direction. This entire ground vibration is divided into 10 windows and HV curve in each window is calculated. Mean and Standard deviation of all HV curve is calculated which is used as input to the algorithm. Figure 7a shows the HV curve when window length consists of 3800 samples. The software DISHV further check the number of points in each logarithmic window and discretize uniformly 10 samples in each window. The discretize HV curve having 39 points is shown in

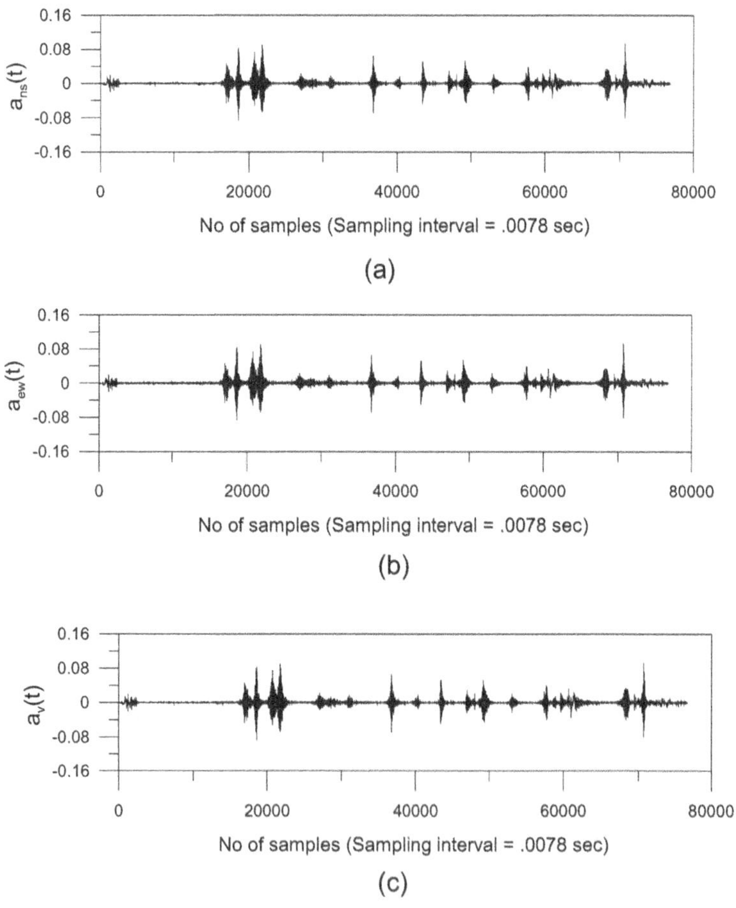

Fig. 6 **a** NS, **b** EW and **c** vertical component of ground vibration

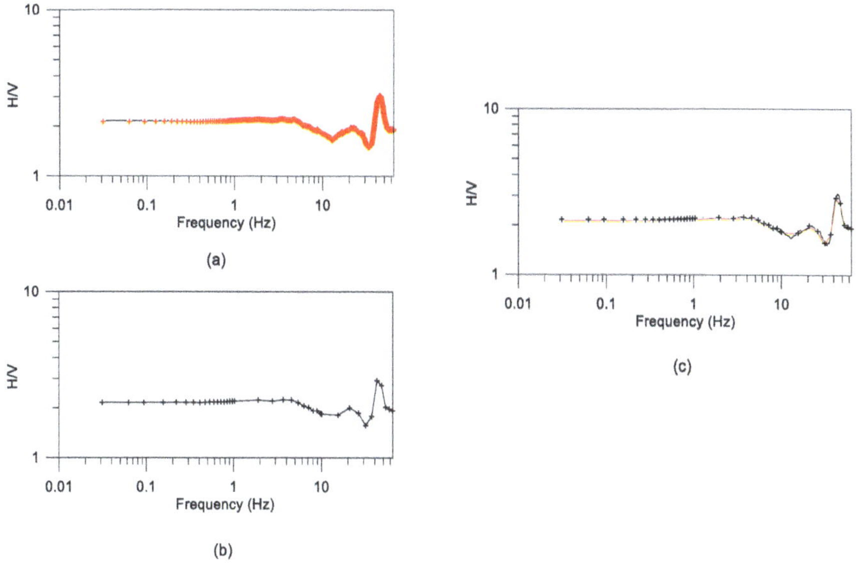

Fig. 7 **a** HV curve obtained from time series containing 3800 samples, **b** recreating HV curve from obtained curve using 39 samples and **c** comparison of HV curve having 3800 and 39 samples respectively

Fig. 7b. Comparison of HV curve that is having 3800 samples and 39 points clearly shows that peak are preserved and both curves has characteristic similarity shown in Fig. 7c. This reduces approximately 100 times the sample present in actual data, thereby retaining the main spectral contents of the HV curve. The output from DISHV program is further used as an input to HV-Inv program for modelling of HV curve using Simulated Annealing algorithm.

3 Case Study for the Bayasi Site, Uttarakhand

The ambient noise data has been recorded at the Bayasi site, Uttarakhand of the Garhwal Himalaya in two horizontal and one vertical direction shown in Fig. 8. The amplitude spectrum of time series of recorded noise data is shown in Fig. 9a along North–South, East–west and vertical component of ground vibration. The H/V curve obtained by the ratio of averaged horizontal amplitude spectrum to that of the vertical spectrum is shown in Fig. 9b. The obtained H/V curve is then used as input to the DISHV program to discretize the HV curve for less number of samples. The discretize HV curve having 36 samples is shown in Fig. 9c.

The parameter ranges of P-wave velocity, S-wave velocity and density are chosen for the inversion of HV curve using simulated annealing algorithm is shown in Fig. 10. The average velocity profile and best fit model of V_P and V_S obtained after

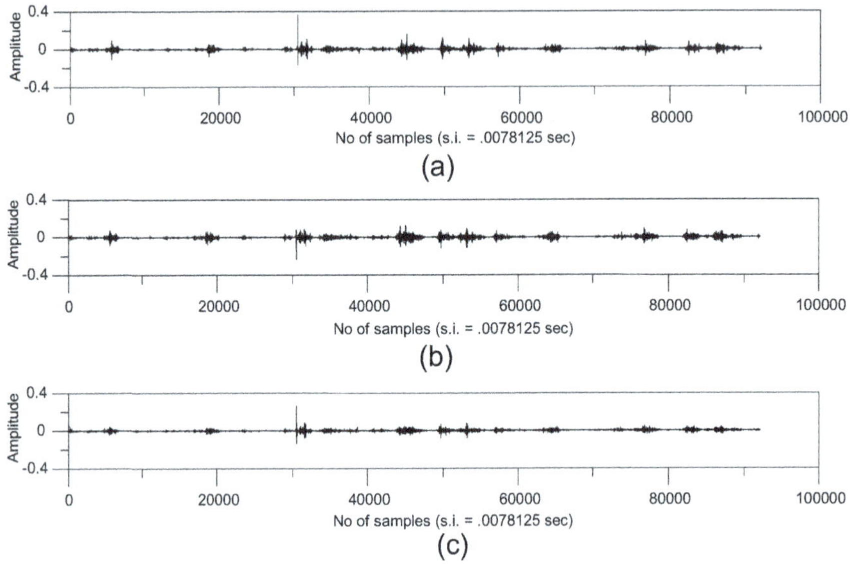

Fig. 8 Time series record of the **a** North South '$a_{ns}(t)$', **b** East west '$a_{ew}(t)$' and **c** vertical '$a_v(t)$' component at Bayasi site

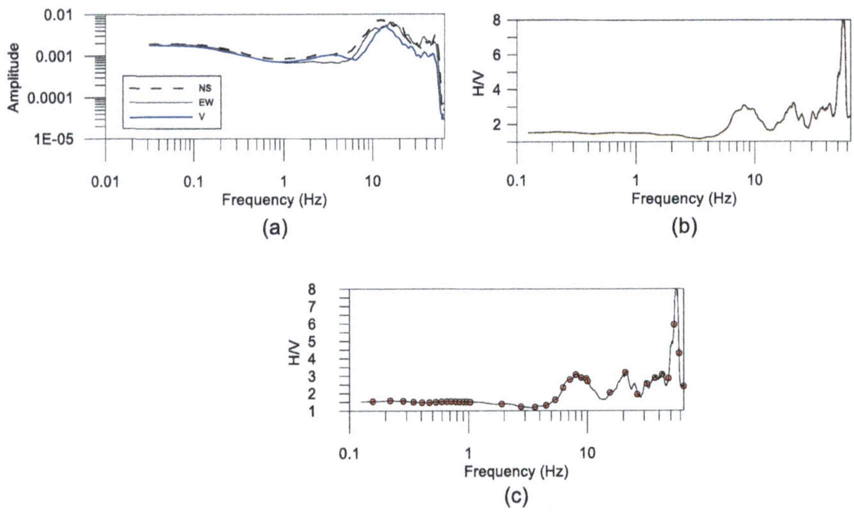

Fig. 9 **a** Amplitude spectrum of time series, **b** ratio of amplitude spectrum of horizontal by vertical component of record, **c** HV curve from 36 samples

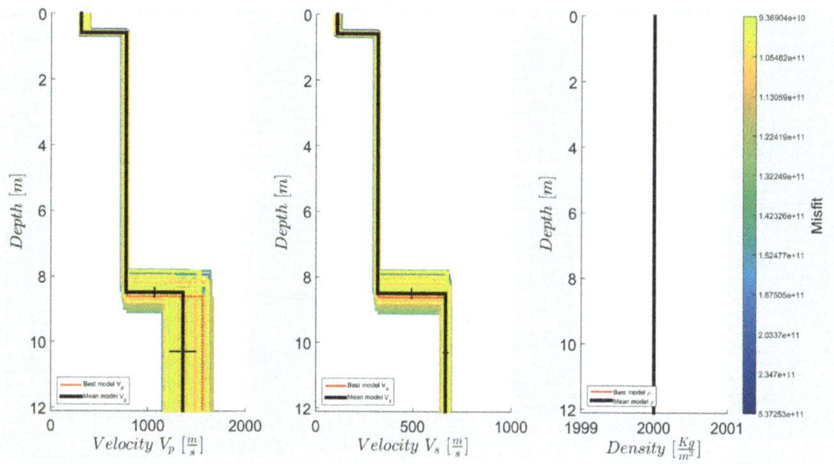

Fig. 10 Set of tested models (blue and yellow lines), average and best-fit models showing with black and red colour line

the inversion process is shown in Fig. 10 with black colour line and red colour line respectively. Figure 11 is showing the best-modelled HV curve in red colour after inversion and the yellow and blue lines showing the set of forward HV curves for the different test model. The output obtained from Hv-Inv software by using input spectral ratio at 36 points using Simulated Annealing algorithm is shown in Fig. 12.

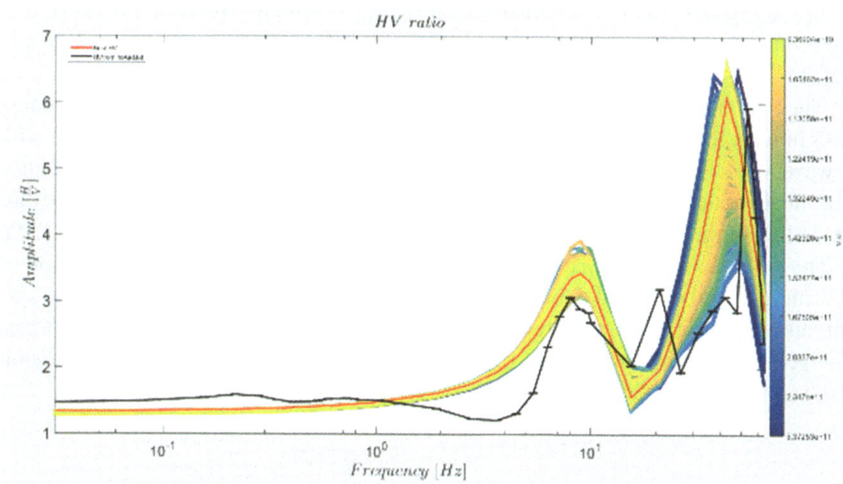

Fig. 11 Best modelled HV curve for the best model after inversion shown in red colour line

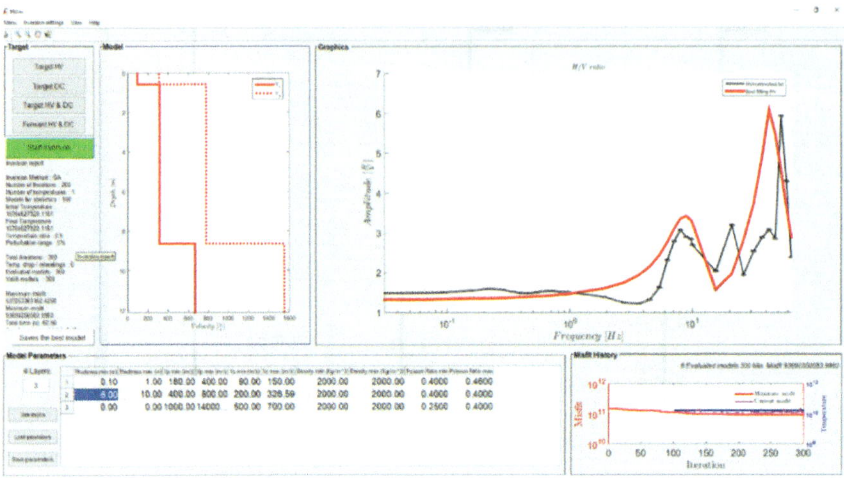

Fig. 12 HVInv program window of the inversion of HV curve using HV-Inv

4 Results and Discussion

The best-fit S-wave velocity model obtained from the simulated annealing algorithm is further refined by iterative modelling of HV curve using forward modelling approach. The minute change in the thickness and velocity according to the user gives the best model after the refinement. The modelled HV curve after the refinement is shown in Fig. 13. The one dimensional Shear wave velocity structure obtained after the refinement of model obtained from Simulated Annealing algorithm is shown in Fig. 14.

The final model obtained after refinement shows that the first layer is of thickness 0.45 m with S-wave velocity 100.0 m/s and the second layer of thickness 9 m with S-wave velocity about 320.1 m/s and the half-space velocity is about 645.2 m/s. The final model obtained after refinement in the model through iterative forward modelling of the model selected from Simulated Aneling algorithm is shown in Table 1. The work done in this paper shows that the HV-Inv program for the inversion of HV curve effectively model the HV curve by a global Simulated Annealing technique and allow the user to further refine the model obtained from the inversion process. The geological input of the area obtained from the field investigation of the region also support this observation.

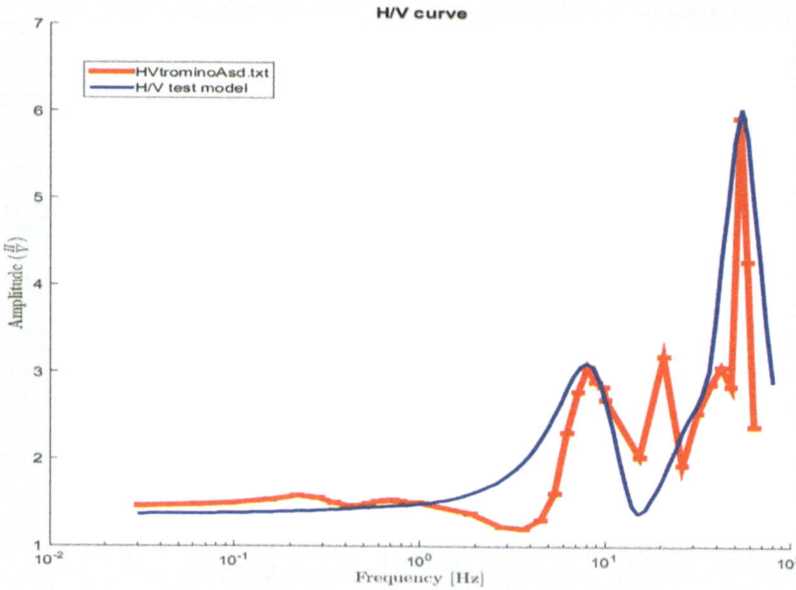

Fig. 13 Final HV curve obtained after the refinement

5 Conclusion

The ratio of the amplitude of horizontal to the vertical component of ground vibration created by an ambient noise is dependent on the subsurface velocity structure of the investigating area. Ground vibration data from ambient noise recording system has been recorded at the Bayasi site in the Garhwal Himalaya. A computer code DISHV in FORTRAN has been developed in this work has been used for obtaining HV curve with a limited number of data points that can be effectively used as an input to the HV-Inv software developed by García-Jerez et al. [9]. The HV curve obtained from DISHV is further used to obtained theoretical HV curve corresponding to a model of earth by using simulated annealing algorithm. The obtained velocity model from simulated annealing is utilized for the forward modelling approach that further refines the obtained shear wave velocity model by comparing the simulated and the observed HV curves.

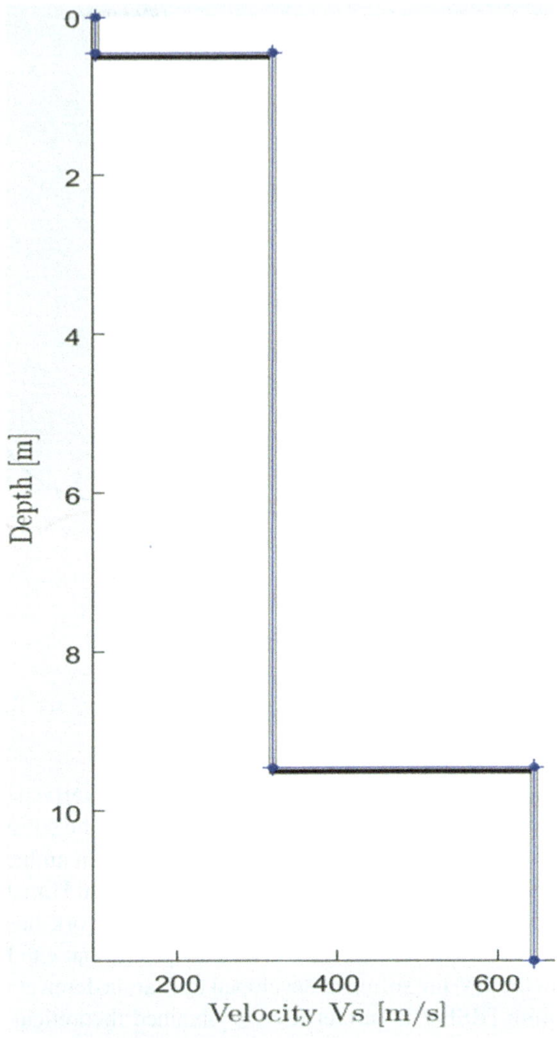

Fig. 14 Final S-wave velocity model after the further refinement in the model

Table 1 Final model parameter after the refinement

Thickness (m)	P-wave velocity (m/s)	S-wave velocity (m/s)	Density (Kg/m^3)	Poisson ratio
0.5	350.7	100.0	2000.0	0.45
9.0	792.8	320.1	2000.0	0.40
0.0	1415.9	645.3	2000.0	0.36

References

1. Aki K (1988) Separation of source and site effects in acceleration power spectral of California earthquakes. In: Proceedings of 11th world conference on earthquake engineering, vol 8, pp 163–167
2. Arai H, Tokimatsu K (2004) S-wave velocity profiling by inversion of microtremor H/V spectrum. Bull Seismol Soc Am 94(1):53–63
3. Arai H, Tokimatsu K (2005) S-wave velocity profiling by joint inversion of microtremor dispersion curve and horizontal-to-vertical (H/V) spectrum. Bull Seismol Soc Am 95(5):1766–1778
4. Bard PY (2008) The H/V technique: capabilities and limitations based on the results of the SESAME project. Bull Earthq Eng 6(1):1–2
5. Billings SD (1994) Simulated annealing for earthquake location. Geophys J Int 118(3):680–692
6. Bonnefoy-Claudet S, Köhler A, Cornou C, Wathelet M, Bard PY (2008) Effects of Love waves on microtremor H/V ratio. Bull Seismol Soc Am 98(1):288–300
7. Cadet H, Bard PY, Rodriguez-Marek A (2010) Defining a standard rock site: propositions based on the KiK-net database. Bull Seismol Soc Am 100(1):172–195
8. Campillo M, Paul A (2003) Long-range correlations in the diffuse seismic coda. Science 299(5606):547–549
9. García-Jerez A, Piña-Flores J, Sánchez-Sesma FJ, Luzón F, Perton M (2016) A computer code for forward calculation and inversion of the H/V spectral ratio under the diffuse field assumption. Comput Geosci 97:67–78
10. Herak M (2008) ModelHVSR—a Matlab® tool to model horizontal-to-vertical spectral ratio of ambient noise. Comput Geosci 34(11):1514–1526
11. Lermo J, Chávez-García FJ (1994) Are microtremors useful in site response evaluation? Bull Seismol Soc Am 84(5):1350–1364
12. Lunedei E, Albarello D (2010) Theoretical HVSR curves from full wavefield modelling of ambient vibrations in a weakly dissipative layered Earth. Geophys J Int 181(2):1093–1108
13. Malischewsky PG, Scherbaum F (2004) Love's formula and H/V-ratio (ellipticity) of Rayleigh waves. Wave Motion 40(1):57–67
14. Nakamura Y (1989) A method for dynamic characteristics estimation of subsurface using microtremor on the ground surface. Railway Tech Res Instit Quart Rep 30(1)
15. Nakamura Y (2000) Clear identification of fundamental idea of Nakamura's technique and its applications. In: Proceedings of the 12th world conference on earthquake engineering, vol. 2656. New Zealand, Auckland
16. Perton M, Sánchez-Sesma FJ, Rodríguez-Castellanos A, Campillo M, Weaver RL (2009) Two perspectives on equipartition in diffuse elastic fields in three dimensions. J Acoust Soc Am 126(3):1125–1130
17. Pilz M, Parolai S, Leyton F, Campos J, Zschau J (2009) A comparison of site response techniques using earthquake data and ambient seismic noise analysis in the large urban areas of Santiago de Chile. Geophys J Int 178(2):713–728
18. Press W, Teukolsky S, Vetterling W, Flannery B (1992) Numerical recipes in C, 2nd edn
19. Sanchez-Sesma FJ (1987) Site effects on strong ground motion. Soil Dyn Earthquake Eng 6(2):124–132
20. Sánchez-Sesma FJ, Pérez-Ruiz JA, Luzon F, Campillo M, Rodríguez-Castellanos A (2008) Diffuse fields in dynamic elasticity. Wave Motion 45(5):641–654
21. Sánchez-Sesma FJ, Rodríguez M, Iturrarán-Viveros U, Luzón F, Campillo M, Margerin L, García-Jerez A, Suarez M, Santoyo MA, Rodriguez-Castellanos A (2011) A theory for microtremor H/V spectral ratio: application for a layered medium. Geophys J Int 186(1):221–225
22. Sánchez-Sesma FJ, Rodrıguez M, Iturrarán-Viveros U, Rodrıguez-Castellanos A, Suarez M, Santoyo MA, Garcıa-Jerez A, Luzón F (2010) Site effects assessment using seismic noise. In: Proceedings of the 9th international workshop on seismic microzoning and risk reduction

23. Sato H (2010) Retrieval of Green's function having coda waves from the cross-correlation function in a scattering medium illuminated by a randomly homogeneous distribution of noise sources on the basis of the first-order Born approximation. Geophys J Int 180(2):759–764
24. Weaver RL, Lobkis OI (2004) Diffuse fields in open systems and the emergence of the Green's function (L). J Acoust Soc Am 116(5):2731–2734
25. Yokoi T, Margaryan S (2008) Consistency of the spatial autocorrelation method with seismic interferometry and its consequence. Geophys Prospect 56(3):435–451

Nonlinear Soil Amplification Models for a Moderately Active Seismic Zone in India

C. Shreyasvi and K. Venkataramana

Abstract The dynamic stiffness and damping of the soil material, depth of the soil profile, impedance between the soil and the underlying bedrock and soil nonlinearity are the factors influencing the local site response. The important parameter in assessing the site response is the "amplification factor," which is usually correlated to Shear Wave velocity in the top 30 m ($Vs_{(30)}$). Though using $V_{S(30)}$ as an index for amplification is simple and robust, it is not recommended for site-specific applications. In the present study, two distinct soil types i.e. "Sand" and "Clay" with the same value of $V_{S(30)}$ demonstrated variable amplification characteristics. Hence, distinct site amplification models were derived for the two soil types considering the intensity of the input bedrock motion as the primary independent variable. The borehole data from nearly 50 locations in North Kerala, an intraplate region in the Southern part of India was collected. The ground response was simulated in 1-dimension considering equivalent linear behavior of soils on the SHAKE 2000 platform. The ground motions used in the simulation were scaled to the target spectrum obtained from the regional seismic hazard assessment. The average spectral amplification observed is 5 for "Clay" and 3.5 for "Sand" in the study region. The soil profiles categorized as "sand" exhibits nonlinear behavior. "Clay" deposits reveal sustained amplification at longer periods and hence, can significantly influence ground response during longer duration ground shaking. The empirical amplification equations developed from the study can be used to modify the generic ground motion prediction models to region-specific applications.

Keywords Spectral matching · Amplification function · North Kerala · Ground response simulation

C. Shreyasvi (✉)
Department of Earth Sciences, IIT Gandhinagar, Palaj 382355, Gujarat, India
e-mail: shreyasvic@gmail.com

K. Venkataramana
Department of Civil Engineering, NITK Surathkal, Mangalore 575025, India

T. G. Sitharam et al. (eds.), *Local Site Effects and Ground Failures*, Lecture Notes in Civil Engineering 117, https://doi.org/10.1007/978-981-15-9984-2_4

1 Introduction

The seismic source, wave propagation, and site effects are the fundamental factors that need to be accounted for when estimating ground motion at a point of interest on the surface. The seismic wave originating from the source traces multiple paths undergoing amplification and in some cases attenuation depending on the dynamic properties of the constituent medium. The top few meters of the soil deposits have the potential to alter the transmitted seismic wave both in terms of magnitude as well as the direction of propagation. The impact of local site effects has been amply demonstrated during past earthquakes such as Kutch (1819), Michoacan (1985), Bhuj (2001), Sendai (2015) and many more. The local site effect is a broad term collectively describing various phenomena such as local ground response, basin effects, and topographic effects. The present study addresses the local ground response aspect for a region that is treated as a moderately active seismic zone. The local ground response assesses the impact of shallow sediments on the ground motion by considering nonlinear behavior, impedance, and possible resonance effects of the soils. The objective of ground/site response studies is to determine the amplification of the seismic waves and the shape of the resulting response spectra in order to improve the seismic resistance of the structures.

Earlier attempts by Raghukanth and Iyengar [1] presented site amplification as a function of shear wave velocity in the top 30 m ($V_{S(30)}$) for India. A similar study by Choi and Stewart [2] presented nonlinear site amplification as a function of $V_{S(30)}$ for the USA. In the present study, it was observed that multiple factors are responsible for site amplification such as density, soil type, cementation, depth to bedrock, and natural period of the soil column. Most of the site response investigations in the study region undermine the influence of soil nonlinearity in ground motion estimation and has not received much importance. The present study is an attempt to capture the nonlinear soil amplification has a function of soil type in contrast to the conventional $V_{S(30)}$ approach. The nonlinear behavior of soils has been estimated using an equivalent linear analysis as it is more efficient and requires lesser computation time when compared to nonlinear analysis.

The distinct amplification pattern of two soil types (clay and sand) have been presented under various time frames. Finally, a nonlinear amplification model has been proposed for each soil type as a function of input bedrock motion. The amplification model has been developed to accommodate various earthquake scenarios and to provide an overall rational estimate of site amplification. These amplification models can be directly incorporated into the existing ground motion prediction models to provide surface-level estimates of ground motion parameters such as PGA, PGV and S_a (at 5% damping). Further, the proposed models can be updated with supplemental field data in the study region.

2 Study Region

The study region belongs to the Northern part of Kerala stretching from Kasargod to Calicut. The data corresponding to soil strata was obtained in the form of borelogs. The topographical map of the study area, as well as the location of boreholes, has been shown in Fig. 1. As evident from Fig. 1, the topography of the study area has higher elevation indicating hilly terrain with gradually sloping toward the beach. The study area is outlined by coastal line on one side and hills on the other side. These borelogs consist of penetration resistance value, i.e. SPT "N" value, unit weight (γ, kN/m^3), location of groundwater table, index and engineering properties, and soil description. The statistics of the collected field data have been compiled and presented in the form of a histogram as shown in Fig. 2. Most of the soil samples belong to categories III (N < 15) and I (N > 15) highlighting the existence of competent as well as weaker soil material.

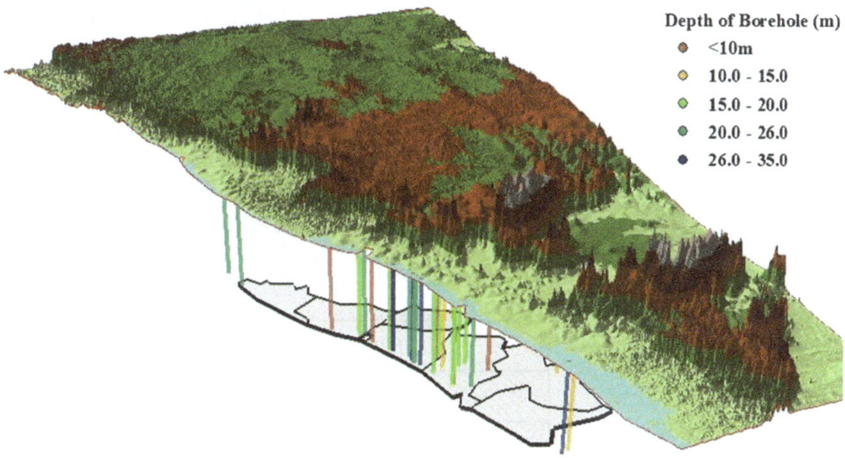

Fig. 1 Topographical map in the vicinity of borehole locations with the borehole depths dropped as a shadow [3]

Fig. 2 Histogram of the soil bore logs collected for the study

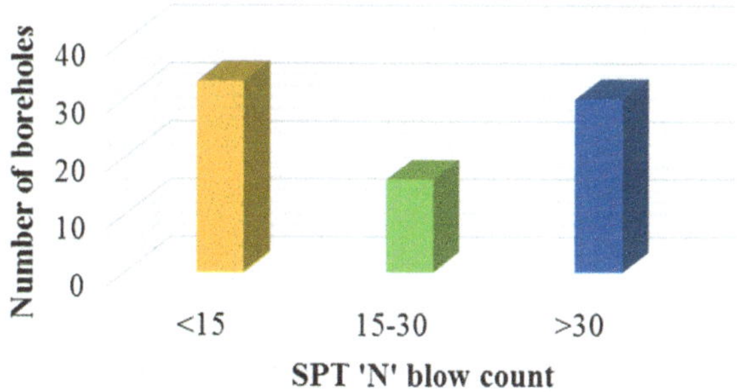

The collected soil data has been further classified into two soil types, i.e. sand and clay, depending on the grain size distribution and index properties. The representative stratigraphy of the two soil types is shown in Fig. 3. The soil formations typically vary from red fine loamy to clayey soils apart from red laterite with large variation in overburden thickness. Laterite rich in Iron and aluminum is found near to coastal plains. These laterites can be observed at higher elevations in the form of plateaus as well. Due to the absence of field measurements of V_S, an indirect method of

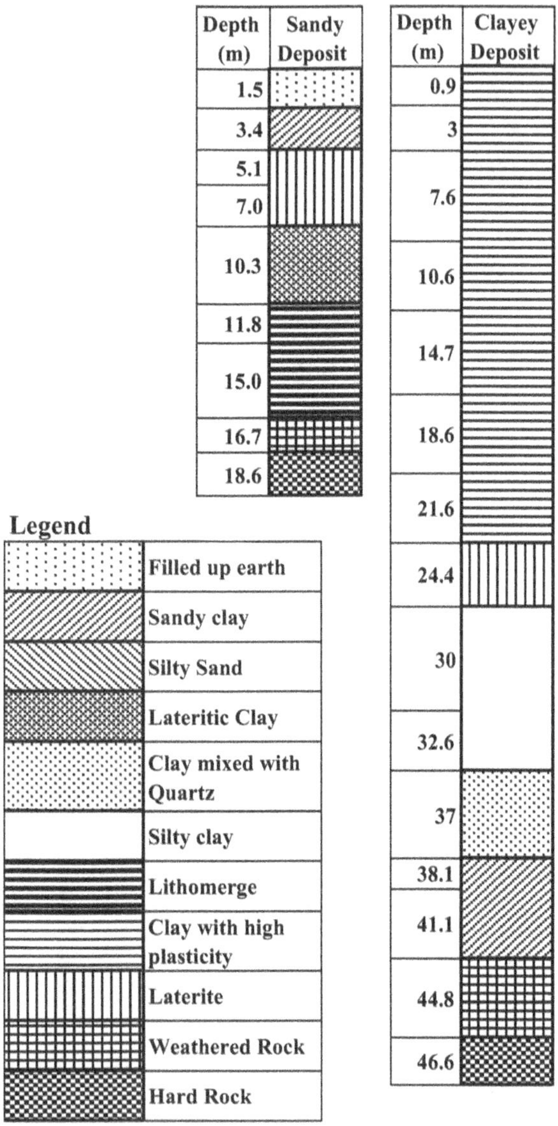

Fig. 3 Stratigraphy of the two representative soil types used in the study [3]

estimation of a V_S profile for different soil columns was adopted. The SPT "N" value has been correlated to V_S by many researchers as shown in Table 1. These correlations have been developed for similar topographical conditions as our study region. Further, the use of multiple correlations aids in reducing the epistemic uncertainty in the estimation of V_S.

A statistical comparison of estimated V_S profiles from various correlations for a regional soil profile revealed that few correlations such as HR, KMM, TK estimated extreme values and had to be discarded while finalizing the V_S profile for a given soil column. Finally, a total of 6 correlations were used for estimating the soil profile and a sample of this estimation has been shown in Fig. 4.

3 Soil Modeling and Ground Motion Records

The dynamic characteristics of the soil material need to be estimated and modeled to estimate the local site response. The modulus reduction (M-R) curves, i.e. shear modulus reduction (G/Gmax) versus cyclic shear strain (γc) and damping (D) curves (i.e. damping versus γc) are essential in modeling the dynamic behavior of soils. Based on experimental investigation, many classical M-R and damping curves have been proposed in the past. In the present study, the standard empirical relations of G/Gmax (Fig. 5) and D as a function of cyclic shear strain (γc) for sand and clay [19–22] have been used to model the dynamic soil response.

The second phase deals with the selection of ground motion records compatible with the target spectrum. The third phase involves the modification of the chosen records with respect to the target spectrum. The practice of selecting ground motion varies widely and no definitive guidelines or strict procedures exist. However, the codal provision provides a general guideline to consider a minimum of five recorded or simulated rock outcrop horizontal ground motion records from events with magnitude and distance range consistent with those controlling Maximum Considered Earthquake (MCE) ground motion [14].

The uniform hazard spectrum (UHS) for the study region has been adapted from Shreyasvi et al. [15]. The objective of the study is to quantify the seismic site amplification and the UHS resulting from a single hazard level may not be sufficient for developing the amplification equation. Hence, multiple hazard levels were considered. The design ground motions specified in the Indian seismic code correspond to 10% POE in 50 years, i.e. Design Basis Earthquake (DBE). With DBE as the standard design condition, two more hazard levels, i.e. 10 * DBE and 0.25 * DBE were defined based on the recommendations from Stewart and Afshari [16]. The DBE (abbreviated as HM) represents the median hazard level whereas 10 * DBE (HL) and 0.25 * DBE (HH) represent lower and higher hazard levels of the considered range. HL corresponds to 65% POE and HH corresponds to 2.5% POE in 50 years. The accelerograms selected for the study have been presented in Table 2 corresponding to DBE. The selected accelerograms are spectrally matched to the UHS of the study region in order to generate spectrum compatible ground motion records

Table 1 Vs–N Correlations investigated in the study [3]

Authors	Correlation	Soil type	Region
Maheshwari et al. (2010)[a]—**UM1** [4]	$V_s = 89.31 * N^{0.358}$ $V_s = 100.53 * N^{0.265}$ $V_s = 95.64 * N^{0.301}$	Clay Sand All soil	Chennai
Hanumanthrao and Ramana (2008)—**HR** [5]	$V_s = 79 * N^{0.434}$ $V_s = 82.6 * N^{0.43}$ $V_s = 86 * N^{0.42}$	Sand All soil Silty sand/sand silt	Delhi
Unal Dikmen (2009)—**UD** [6]	$V_s = 58 * N^{0.39}$ $V_s = 73 * N^{0.33}$ $V_s = 60 * N^{0.36}$ $V_s = 44 * N^{0.48}$	All soil Sand Silt Clay	Turkey
Chatterjee and Choudhury (2013)[a]—**CC1** [7]	$V_s = 78.21 * N^{0.38}$ $V_s = 77.11 * N^{0.39}$ $V_s = 54.82 * N^{0.53}$ $V_s = 58.02 * N^{0.46}$	All Soil Clay Silty Sand Silt	Kolkata
Kirar, Maheshwari et al. (2016)—**KMM** [8]	$V_s = 100.31 * N^{0.348}$ $V_s = 94.4 * N^{0.379}$ $V_s = 99.5 * N^{0.345}$	Sand Clay All Soil	Roorkee
Maheshwari et al. (2010)[a]—**UM2** [4]	$V_s = 90.75 * N^{0.304}$ $V_s = 96.29 * N^{0.266}$ $V_s = 83.27 * N^{0.365}$	All Soil Sand Clay	Chennai
Hasaneebi and Ulusay (2007)[a]—**NR** [9]	$V_s = 90 * N^{0.309}$ $V_s = 90.8 * N^{0.319}$ $V_s = 97.9 * N^{0.269}$	All Soil Sand Clay	Turkey
Anbazhagan et al. (2012)—**AZ** [10]	$V_s = 68.96 * N^{0.51}$ $V_s = 60.17 * N^{0.56}$ $V_s = 106.63 * N^{0.39}$	All Soil Sand Clay	Lucknow
Sil and Haloi (2017)[a]—**SH** [11]	$V_s = 75.478 * N^{0.3799}$ $V_s = 79.217 * N^{0.3699}$ $V_s = 99.708 * N^{0.3358}$	All Soil Sand Clay	Any region
Mhaske and Choudhury (2011)[a]—**MC** [12]	$V_s = 72 * N^{0.4}$	All soil	Mumbai
Thokchom et al. (2017)[a]—**TK** [13]	$V_s = 2.641 * N + 189.6$ $V_s = 3.925 * N + 143.1$ $V_s = 3.395 * N + 156.8$ $V_s = 3.311 * N + 160.5$	Sand Silt Clay All soils	Dholera, Western India
Chatterjee and Choudhury (2013)[a]—**CC2** [7]	$V_s = 78.63 * N^{0.37}$ $V_s = 78.03 * N^{0.38}$ $V_s = 58.62 * N^{0.45}$ $V_s = 56.44 * N^{0.51}$	All soil Clay Silt Silty sand	Kolkata

[a]Correlations finalized for estimating the V_S profile. Nomenclature ending with numbers 1 and 2 represents correlation derived based on uncorrected and corrected SPT ‘N’ values respectively

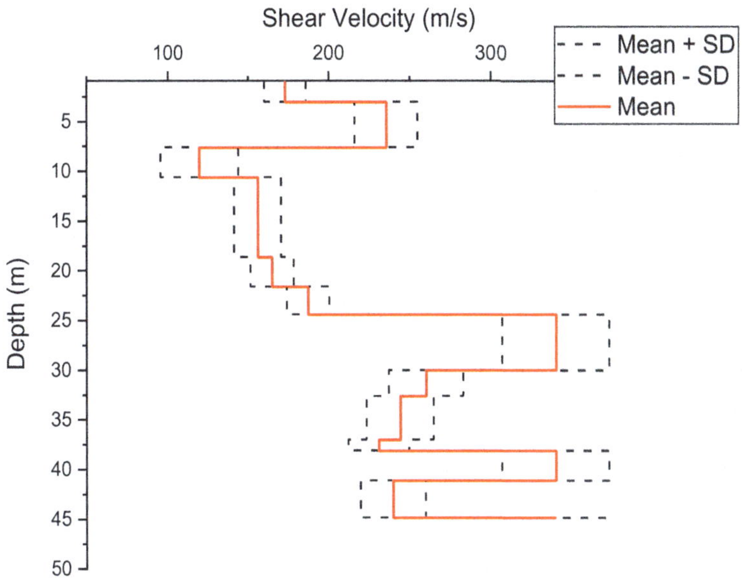

Fig. 4 Shear velocity profile of a typical clay deposit

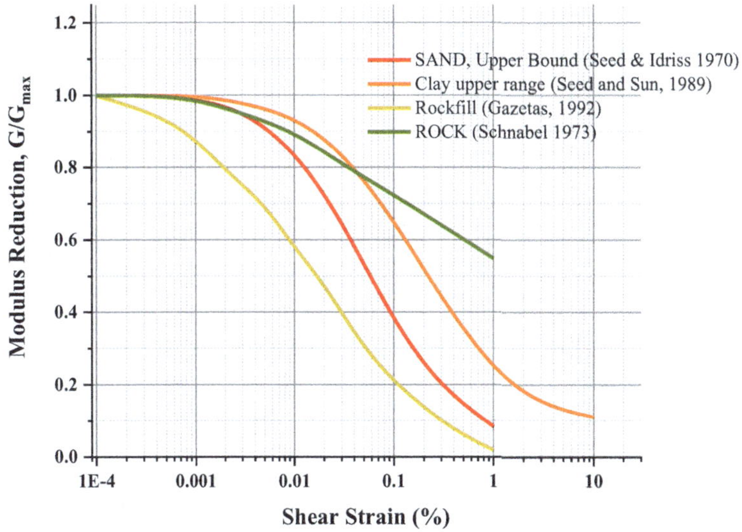

Fig. 5 Shear modulus reduction curves as a function of cyclic strain for different soil types

Table 2 Details of ground motions used in the study [3]

Earthquake name	Acronym	Date	Mw	R_{rup} (km)	PGA (g)	Scaled PGA (g)	Hazard level
RiviereDuLoup	RSN1688	06-03-2005	4.65	19.05	0.045	0.100	Median hazard level
RiviereDuLoup	RSN1771	06-03-2005	4.65	41.75	0.070	0.124	
Greenbrier	RSN6934	28-02-2011	4.68	6.27	0.030	0.135	
Greenbrier	RSN7052	28-02-2011	4.68	54.07	0.001	0.126	
ValDesBois	RSN4027	23-06-2010	5.1	52.94	0.018	0.102	
Sicily	IT.NOT	13-12-1990	5.6	48.30	0.090	0.111	
Central Italy	3A.MZ14	26-10-2016	5.9	36.60	0.515	0.108	
Central Italy	3A.MZ19	26-10-2016	5.9	30.40	0.096	0.118	
Central Italy	3A.MZ21	26-10-2016	5.9	30.70	0.183	0.104	
Central Italy	3A.MZ19_30	30-10-2016	6.5	22.60	0.363	0.114	
Central Italy	3A.MZ29	30-10-2016	6.5	26.90	0.689	0.131	

for the site response analysis. Individually matched accelerograms, original UHS termed as Target Spectrum and the mean of the matched accelerograms are plotted in Fig. 6.

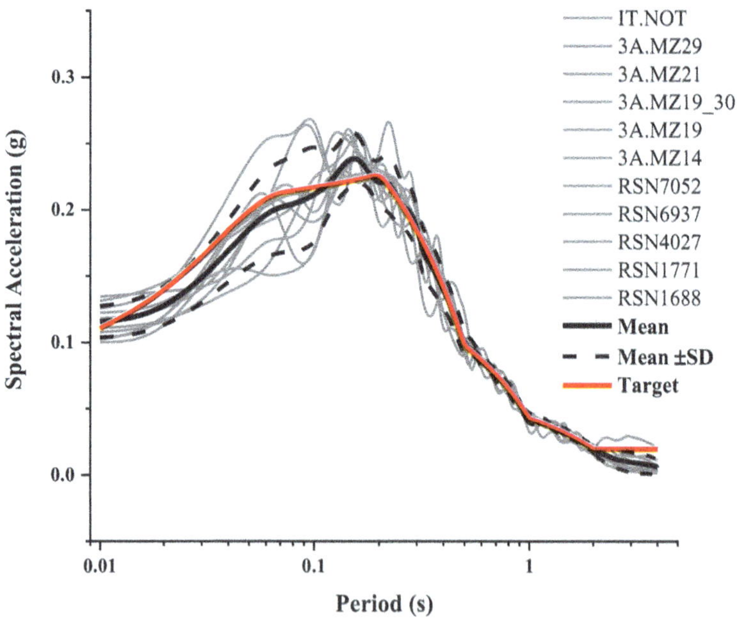

Fig. 6 Plot of 5% damped rock acceleration spectrum of ground motions scaled to HM [3]

4 Site Response Analysis

The dynamic simulation of the ground motion propagated through the layered soil medium was realized through the SHAKE2000 program. The compiled borehole information along with the selected and scaled input motions served as input to equivalent linear analysis. The dynamic characteristics of the local soil are explained through different output parameters such as amplification and surface response spectrum. Due to space constraints, the results of all the simulated soil columns are not presented. It was observed that the two soil profiles of nearly the same value of V_S demonstrated a distinct amplification pattern and hence, an alternate predictor variable for site amplification was mandatory.

Figures 7 and 8 represent the median amplification function along with the standard deviation and 95% confidence interval as a function of rock spectral acceleration assessed at different spectral periods. The amplification function varying with the period (AF(f)) for different input motion acceleration values (PGA and Sa (f)) was compiled for all the numerically modeled soil deposits belonging to clay and sand site category. Each soil profile generated 33 data points from the input ground motions for each period window. The data points from all the simulated soil columns have been presented in Figs. 7 and 8. Nonlinear regression was performed on these data points as shown.

The observed amplification data points are sparse and widely distributed for Sand. However, they were sufficient to draw a nonlinear amplification function for different

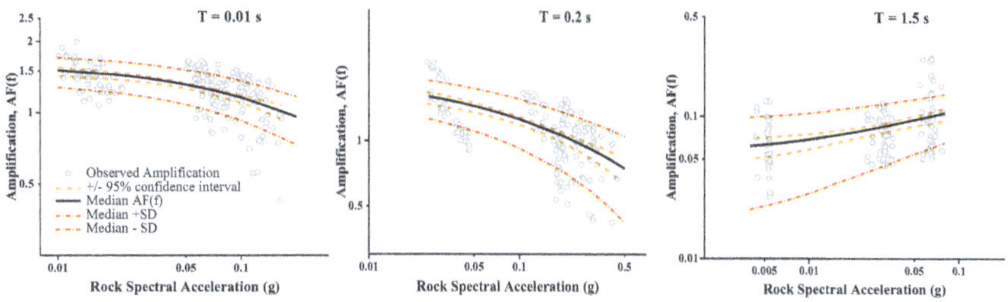

Fig. 7 Amplification factors regressed against rock spectral acceleration for 'sand' sites [3]

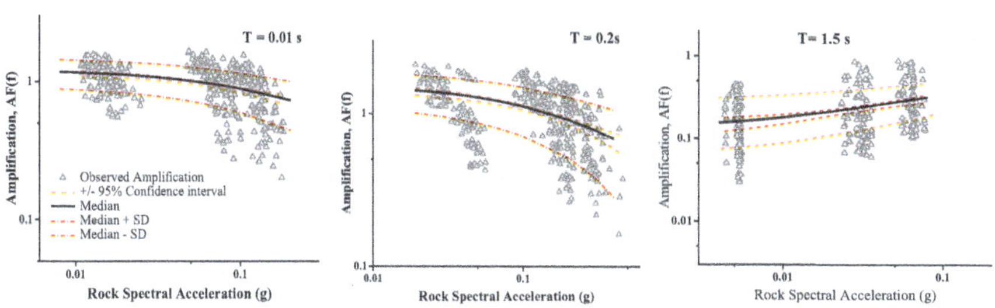

Fig. 8 Amplification factors regressed against rock spectral acceleration for 'clay' sites [3]

spectral periods. The natural period of the soil profiles lies in the range of 0.31–0.38 s. The amplification for the depicted soil type, i.e. sand tends to be highly nonlinear especially at periods below the average site period. Significant variability is visible for higher spectral acceleration values at T = 0.2 s, demonstrating even larger variation at T = 1.5 s. This is mainly due to the large differences in the intensity of the selected input ground motions. An upward shift was observed at intermediate and longer period range (T = 1.5 s). The inability of the equivalent linear methodology to converge to a solution for high-intensity records is the reason behind this upward shift [17]. Additionally, the equivalent linear method estimates higher amplification at the smaller period range and overestimates resonant responses when the soil becomes nonlinear [18]. Further, the amplification reduces drastically for periods after the resonant vibration period of the sites. Another interesting observation made during the study is the shift in the natural period depending on the strain induced by various ground motions.

The amplification observed in clay at T = 0.01 and 0.2 s is less when compared with sand. However, clay demonstrates sustained amplification at longer periods (T = 1.5 s). While the majority of the soil deposits have their natural period around 0.4 s, there are a few deposits with a period as high as 0.7–0.9 s displaying a diverse range. However, the shift in the natural period depending on the induced strain was noticed similar to the "sand" site. Due to this phenomenon, resonance is attained by soil profiles at the elongated site period (T = 1.5 s) driving the regression toward a positive correlation. As a result, the upward shift in the fitted median amplification curve was observed at these periods. The "clay" site demonstrated consistently higher AF(f) compared to the other soil categories implying amplification of longer period ground motions. Overall, "clay" sites demonstrate slower stiffness degradation and hence, less nonlinear when compared to "sand" sites. The $V_{S(30)}$ is widely accepted as an index for soil amplification.

However, the present study suggests that the soil characteristics affect the amplification to a greater extent. In order to validate this, two soil profiles (Sand & Clay) of the same Vs(30) were considered for plotting mean amplification in Fig. 9. As evident from Fig. 9, the sand exhibits higher amplification at lower input values but decreases gradually with the increase in PGA demonstrating its nonlinear behavior. For the "clay" site, amplification reduces as the intensity increases and de-amplification can be noticed for PGA values as small as 0.005 g. However, there is still a very clear amplification at longer periods as shown in Fig. 8. An important observation made during the analysis of "clay" deposits is that as the soil plasticity index increases, the behavior of the soils became less nonlinear.

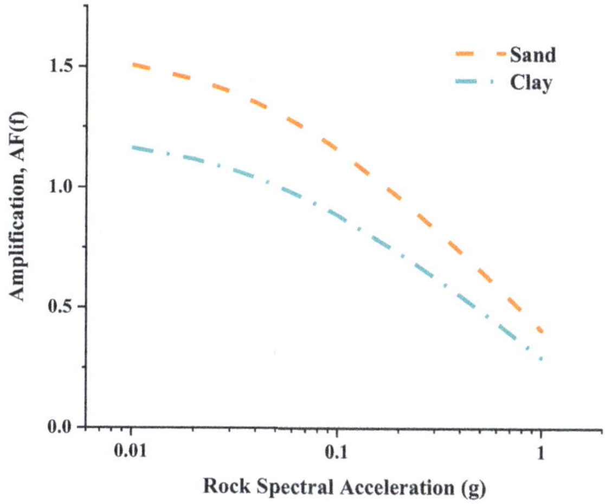

Fig. 9 Mean site amplification of the two soil types at T = 0.01 s [3]

5 Conclusion

The findings of the study are summarized below.

1. The soil profiles modeled in the study belongs to NEHRP "C" (360–760 ms^{-1}) and "D" (180–360 ms^{-1}) site categories and the study region belongs to seismic zone III (moderate level shaking).
2. The average spectral amplification observed 5 for "clay" and 3.5 for "sand" of the study region.
3. "Sand" sites exhibit nonlinear behavior by undergoing large amplification for smaller intensity measure but reduce substantially as the spectral acceleration (S_a) values exceed 0.1 g.
4. "Clay" sites exhibit amplification even at longer periods but become less nonlinear with the increase in plasticity index. Hence, the "clay" site plays a major role in the event of long-period seismic waves.
5. Three soil profiles of $V_{S(30)}$ in the similar range demonstrated distinct amplification characteristics. The "sand" site amplifies 29% more than the "clay" site for lower input S_a values.
6. The amplification characteristics observed in various spectral period frames suggest that PGA offers an unbiased and better prediction of amplification function. The same parameters studied at different spectral periods suffer from resonance (site-specific effect) and demerits of the EQL method (computational capacity).

References

1. Kanth SR, Iyengar RN (2007) Estimation of seismic spectral acceleration in peninsular India. J Earth Syst Sci 116(3):199–214
2. Choi Y, Stewart JP (2005) Nonlinear site amplification as function of 30 m shear wave velocity. Earthq Spectra
3. Shreyasvi C, Venkataramana K, Chopra S (2019) Local site effect incorporation in probabilistic seismic hazard analysis—a case study from southern peninsular India, an intraplate region. Soil Dyn Earthq Eng 123:381–398
4. Maheswari RU, Boominathan A, Dodagoudar GR (2010) Use of surface waves in statistical correlations of shear wave velocity and penetration resistance of Chennai soils. Geotech Geol Eng 28(2):119–137
5. Hanumantharao C, Ramana GV (2008) Dynamic soil properties for microzonation of Delhi, India. J Earth Syst Sci 117(2):719–730
6. Dikmen Ü (2009) Statistical correlations of shear wave velocity and penetration resistance for soils. J Geophys Eng 6(1):61–72
7. Chatterjee K, Choudhury D (2013) Variations in shear wave velocity and soil site class in Kolkata city using regression and sensitivity analysis. Nat Hazards 69(3):2057–2082
8. Kirar B, Maheshwari BK, Muley P (2016) Correlation between shear wave velocity (vs) and SPT resistance (N) for Roorkee region. Int J Geosynth Ground Eng 2:9
9. Hasancebi N, Ulusay R (2007) Empirical correlations between shear wave velocity and penetration resistance for ground shaking assessments. Bull Eng Geol Env 66(2):203–213
10. Anbazhagan P, Kumar A, Sitharam TG (2013) Seismic site classification and correlation between standard penetration test N value and shear wave velocity for Lucknow City in Indo-Gangetic Basin. Pure Appl Geophys 170(3):299–318
11. Sil A, Haloi J (2017) Empirical correlations with Standard Penetration Test (SPT)-N for estimating shear wave velocity applicable to any region. Int J Geosynth Ground Eng 3:22
12. Mhaske SY, Choudhury D (2011) Geospatial contour mapping of shear wave velocity for Mumbai city. Nat Hazards 59(1):317–327
13. Thokchom S, Rastogi BK, Dogra NN, Pancholi V, Sairam B, Bhattacharya F, Patel V (2017) Empirical correlation of SPT blow counts versus shear wave velocity for different types of soils in Dholera, Western India. Nat Hazards 86(3):1291–1306
14. American Society of Civil Engineers (2017) Minimum design loads and associated criteria for buildings and other structures, ASCE/SEI 7–16. Reston, VA
15. Shreyasvi C, Venkataramana K, Chopra S, Rout MM (2019) Probabilistic seismic hazard assessment of Mangalore and its adjoining regions, a part of Indian Peninsular: an intraplate region. Pure Appl Geophys 176(6):2263–2297
16. Stewart JP, Afshari K, Hashash YMA (2014) Guidelines for performing hazard-consistent one-dimensional ground response analysis for ground motion prediction (October 2014), 1–141
17. Papaspiliou M, Kontoe S, Bommer JJ (2012) An exploration of incorporating site response into PSHA-part II: sensitivity of hazard estimates to site response approaches. Soil Dyn Earthq Eng 42:316–330
18. Kim B, Hashash YMA, Stewart JP, Rathje EM, Harmon JA, Musgrove MI, Silva WJ (2016) Relative differences between nonlinear and equivalent-linear 1-D site response analyses. Earthq Spectra
19. Seed HB, Idriss IM (1970) Soil moduli and damping factors for dynamic response analysis. Report no. EERC 70–10. University of California, Berkeley, California
20. Seed HB, Sun JH (1989) Implication of site effects in the Mexico City earthquake of September 19, 1985 for earthquake-resistance-design criteria in the San Francisco Bay Area of California. Report No. UCB/EERC-89/03. University of California, Berkeley, California
21. Gazetas G, Dakoulas P (1992) Seismic analysis and design of rockfill dams: state-of-the-art. Soil Dyn Earthq Eng 11(1):27–61
22. Schnabel PB (1973) Effects of local geology and distance from source on earthquake ground motions (Doctoral dissertation, University of California, Berkeley)

Prediction of Future Surface PGA in the States of Indo-Gangetic Basin Considering Site Specific Studies

P. Anbazhagan, Mohammad Rafiq Joo, Meer Mehran Rashid, and Ketan Bajaj

Abstract Indo-Gangetic Basin (IGB) is the major geographical region of India extending from Punjab to Bihar including Uttar Pradesh (UP) and Haryana. In this region, major earthquake damages have been reported in the past. Several attempts have been made for different site response studies and estimation of earthquake risk in the region. However, most of them consider only shallow soil information < 50 m and available ground motions. Present study attempts to analyze the spatial variation of peak ground acceleration (PGA) at surface for possible future scenario earthquakes in and around IGB. The earthquakes were identified based on past seismic gaps and studies whose magnitudes ranged from Mw 7.5 to Mw 9.0. The earthquakes were simulated for 270 sites with available shear wave velocity data throughout the IGB. Using proper input parameters of soil column profiles, shear wave velocity, depth of input motion, suitable shear modulus reduction and damping curves; the detailed analyses were carried out using DEEPSOIL. This paper arrives at spatial variation of PGA at surface due to individual earthquakes. The response (bedrock as well as surface PGA) of different states toward each earthquake has been tabulated. Sites in Bihar reflect average and maximum surface PGA 0.15 g and 0.68 g, Uttar Pradesh 0.10 g and 1.18 g, Punjab and Haryana 0.12 g and 0.62 g respectively. These values are indicative of the sensitiveness toward earthquake damages. Maps representing surface PGA for each scenario earthquake were plotted giving detailed information about the surface seismic hazard of the area associated with each earthquake.

Keywords PGA · Indo-Gangetic Basin · IGB · Site response · Bed rock

---

P. Anbazhagan (✉) · K. Bajaj
Department of Civil Engineering, Indian Institute of Science, Bangalore 560012, India
e-mail: anbazhagan@iisc.ac.in; anbzhagan2005@gmail.com

M. R. Joo · M. M. Rashid
Department of Civil Engineering, National Institute of Technology Srinagar, Srinagar 190006, India

T. G. Sitharam et al. (eds.), *Local Site Effects and Ground Failures*, Lecture Notes in Civil Engineering 117, https://doi.org/10.1007/978-981-15-9984-2_5

1 Introduction and Study Area

Indo-Gangetic Basin, consisting of thick alluvial deposits, lies between longitude 77 E and 88 E and latitude 24 N and 30 N. It covers an area of around 2,50,000 km^2 and encompasses densely populated Indian states viz. Bihar, Punjab, Haryana and Uttar Pradesh as shown in Fig. 1. Its closeness to Himalayan boundary, presence of thick sedimentary deposits and higher amplification of seismic waves makes the area seismically vulnerable.

Several researchers in India reported high seismic risk in the region. Ambraseys [1] reviewed the seismicity of North India during the early instrumentation period 1892–1915 and summarized evidences for 50 events. Khattri [2] also studied three seismic gaps in the Himalayan Plate boundary which can lead to potential future earthquakes.

In the recent past, researchers have made several advances in site response studies. Anbazhagan et al. [3] used borehole data and synthetic ground motions to study limited sites for 1999 Chamoli earthquake. Researchers [4–8] considered limited locally recorded ground motions for site response studies and Kumar et al. [9] used worldwide recorded ground motion for the site-response study irrespective of seismic background of IGB.

Literature survey reveals that site response studies are mostly limited to 30 m soil column depth using limited recorded motions. For the first time, our study has attempted to carry out site response studies for different possible futuristic earthquakes with suitable input soil parameters. In the present work, 16 possible earthquakes with varying parameters were identified and subsequently used to generate

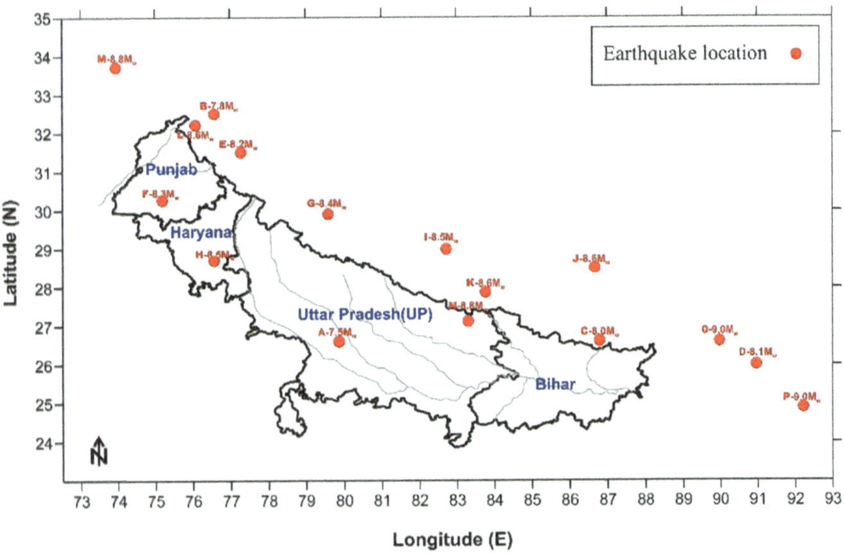

Fig. 1 Study area and earthquakes considered

synthetic ground motions for analysis. Contour maps illustrating variation of estimated surface PGA for each individual earthquake were developed. Wide range of PGA was observed corresponding to different earthquakes. Further each site exhibited different behavior toward different earthquakes. Sites in Bihar reflect average and maximum surface PGA 0.15 g and 0.68 g, Uttar Pradesh 0.10 g and 1.18 g, Punjab and Haryana 0.12 g and 0.62 g respectively. On an average, the percentage of sites hit by earthquakes in Bihar, Punjab and Haryana and UP are 38%, 35% and 37% respectively.

2 Simulation of Input Motions and Bedrock PGA

Synthetic ground motions are widely used for ground response studies, development of attenuation relationships, mapping of seismic hazards and amplification studies. To analyze 270 sites for identified seismic sources, ground motions at every site corresponding to each scenario earthquake were considered. This was done using EXSIM [10] taking into account seismotectonic parameters of the study area. The magnitudes of the identified earthquakes range between Mw 7.5 to 9.0, originating in and around IGB as depicted in Fig. 1. The detailed parameters of the earthquakes can be found from Ref. [11] and their locations and magnitudes can also be referred from Fig. 1. The ground motions used for analysis have varying parameters viz. frequency content, acceleration and duration. The bedrock PGA ranged between 0.0005 and 0.2996 g in Bihar, 0.0001 and 0.5771 in UP, 0.0003 and 0.6509 g in Punjab and Haryana. In whole IGB bedrock PGA varied from 0.0001 to 0.6509 g.

3 Input Soil Parameters

To analyze the sites with varying characteristics over IGB, different soil parameters viz nature of soil, density, shear wave velocity profiles and depth of input motion have been used as per provisions suggested by researchers. Shear wave velocity (Vs) as described by [12] while considering representative density from Vs of each layer [13], reliable depth level of input motion [14] and selected shear modulus reduction and damping curves as suggested by [15, 16]. The representative curves suggested have been therefore used to present site response studies of deep sites of IGB. In case of rock or hard layer, EPRI curve [17] and Zhang curve [18] have been used depending on whether Vs $\geq$ 800 m/s or Vs $<$ 800 m/s respectively for deposits of Quaternary type. For gravel sites with known particle size, Menq curve [19] has been used otherwise Zhang et al. [18] for deep gravel profiles. Zhang Curve [18] has also been used for deep sand deposits. For deep clay and silt sites, Darendeli curve [20] has been used.

4 Site Response Analysis and Surface PGA

For a given motion, PGA is the highest value of horizontal acceleration obtained from the accelerogram. Generally, horizontal accelerations describe ground motions because of their natural connection to inertial forces. It has been observed that ground motions with high peak accelerations are usually more damaging in nature. Site response analysis establishes the impact on ground surface motion due to the soils above the bedrock. Site response analysis finds its use in predicting ground surface motions for development of design response spectra, estimating liquefaction hazards and to measure earthquake induced forces. It requires the information of different field parameters like type, thickness and density of soil layers, shear wave velocity profiles, location of water table, depth of bedrock, shear modulus reduction and damping ratios. While designing any structure, it becomes important to know about the behavior of the soil column toward the incoming ground motions. The response influences the level of shaking which in turn governs the damage caused to the infrastructure in the area. To assess the site response, we simulated 16 earthquakes for 58 sites of Bihar, 136 sites of UP and 76 sites for Punjab and Haryana. Only those input motions which surpassed bedrock PGA 0.005 g were utilized in the study as the input motions and with PGA below 0.005 g were found ineffective in causing any significant damage to the infrastructure. 1D (one dimensional) nonlinear site response studies were carried out using DEEPSOIL V7 [21] and the results obtained are discussed in the subsequent section.

5 Results and Discussions

Seismic waves undergo amplification on traveling from bedrock to ground surface. Hence, parameters viz. peak ground acceleration, peak spectral acceleration, duration and frequency content associated with ground motions at surface differ from that at bedrock level or any hard layer. These modifications in seismic motions are site dependent and vary considerably from site to site. In present study, spatial variation of surface PGA corresponding to each earthquake has been shown. Surface PGA values due to each earthquake for UP, Bihar, Haryana and Punjab are summarized. These states showed varied levels of PGA and this behavior can be correlated to the magnitude, depth of fault, hypocentral distance, directivity, nature of soil deposits. Also, the PGA varied within the states giving rise to different hazard levels. The state wise analysis for future probable scenario earthquakes is briefly discussed below.

In Bihar, out of sixteen earthquakes considered, only nine have potential to generate bedrock PGA > 0.005 g and have therefore been used in site response analysis of 58 sites. The other seven earthquakes may be regarded having no or very little effect in the state. The highest bedrock PGA observed is 0.30 g in the state. The bedrock PGA is significantly amplified and goes as high as 0.68 g at surface. The PGA distribution for individual earthquakes is mapped in Figs. 2, 3, 4, 5, 6, 7,

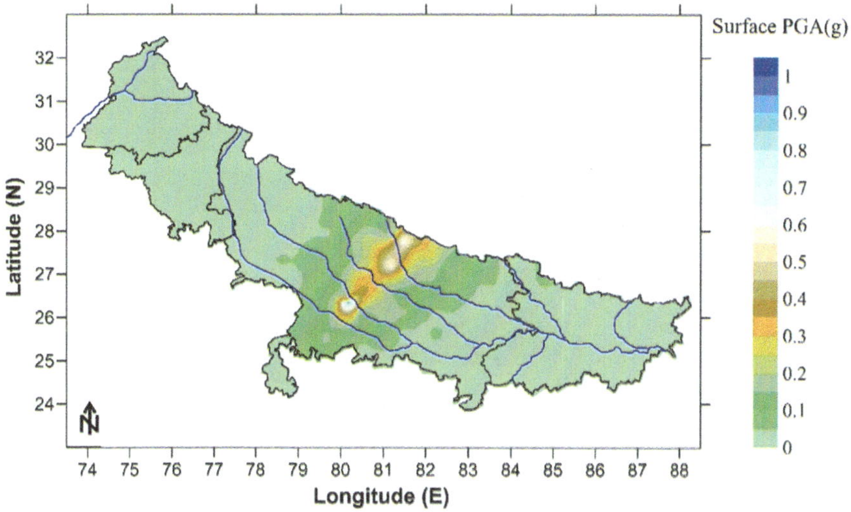

Fig. 2 Spatial variation of surface PGA for earthquake A

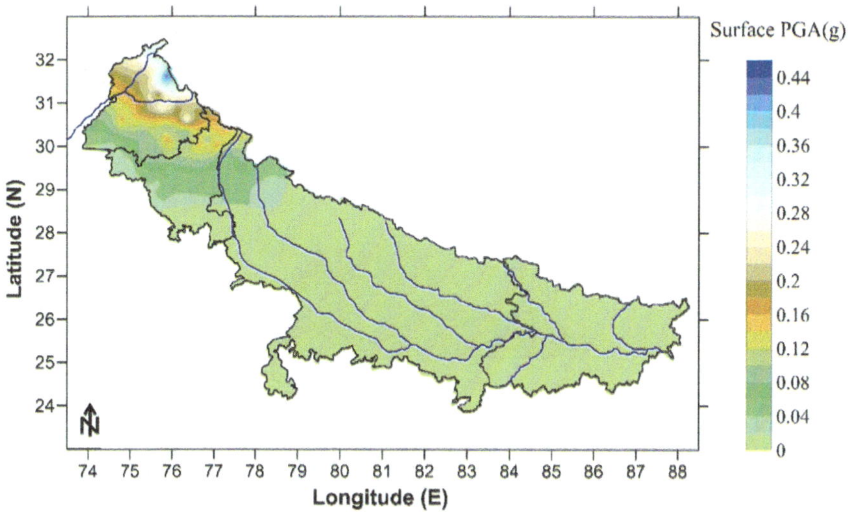

Fig. 3 Spatial variation of surface PGA for earthquake B

8, 9, 10, 11, 12, 13, 14, 15, 16 and 17. The surface PGA ranges from 0.007 to 0.68 g. Average surface PGA corresponding to each of 9 earthquakes are 0.03, 0.27, 0.1, 0.04, 0.17, 0.06, 0.30, 0.06 and 0.03 g. Earthquakes labeled as C and N affect 100% of the sites in Bihar and have a potential to generate surface PGA > 0.5 g at certain sites. Around 77% of analyzed sites in Bihar show average surface PGA > 0.1 g. The detailed information may be referred from Table 1.

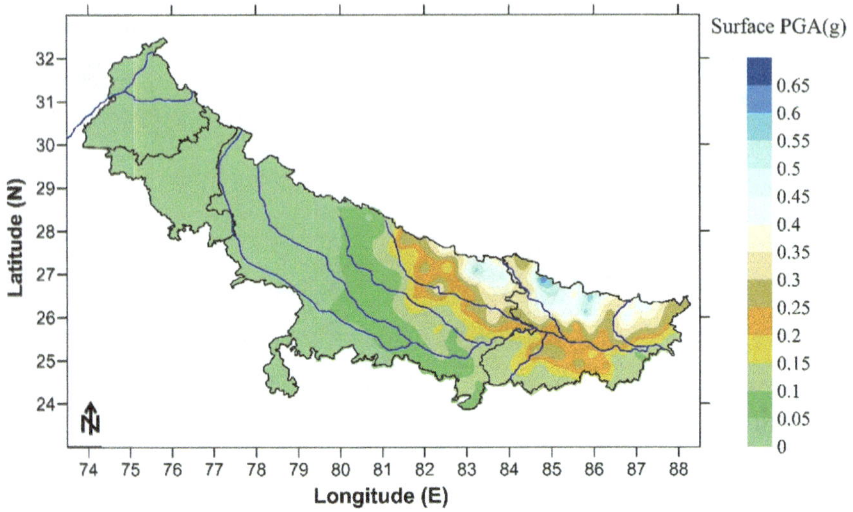

Fig. 4 Spatial variation of surface PGA for earthquake C

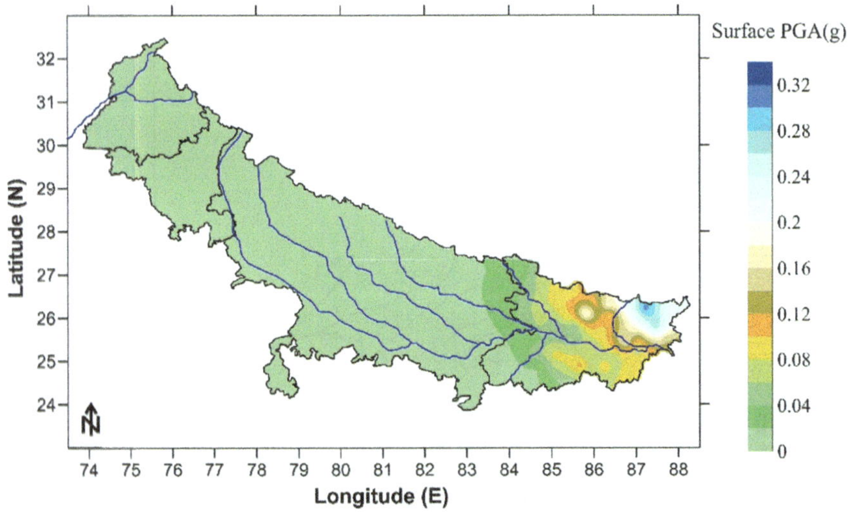

Fig. 5 Spatial variation of surface PGA for earthquake D

For Punjab and Haryana, eight earthquakes as listed in Table 2 were found to produce bedrock PGA > 0.005 g. The bedrock PGA varies between 0.0003 g and 0.6509 g. Ground motions (with PGA > 0.005 g) were inputted at sites in Punjab and Haryana to study site effects and arrive at distribution of surface PGA. Spatial variation due to 8 individual earthquakes can be referred from Figs. 2, 3, 4, 5, 6, 7, 8, 9, 10, 11, 12, 13, 14, 15, 16 and 17. Surface PGA also varies significantly for different

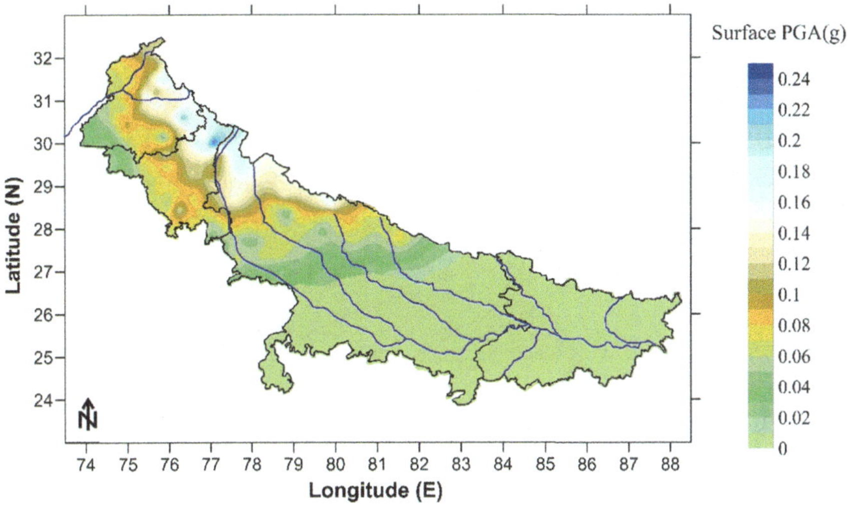

Fig. 6 Spatial variation of surface PGA for earthquake E

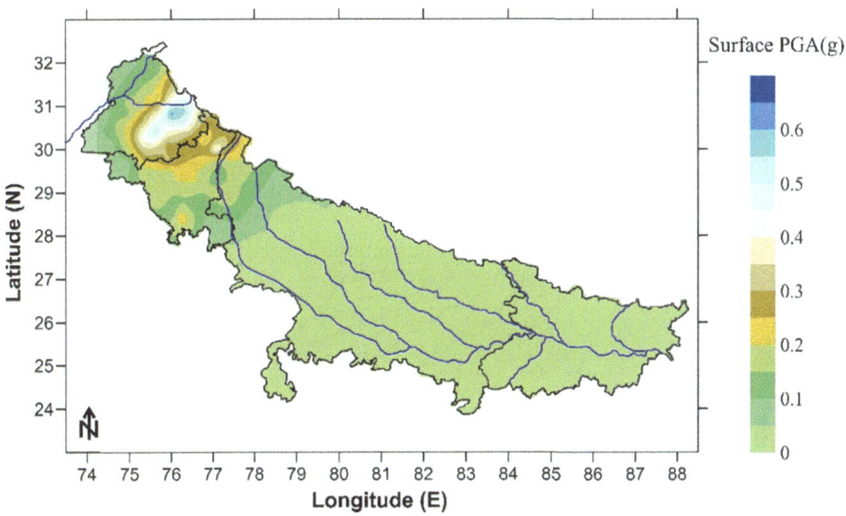

Fig. 7 Spatial variation of surface PGA for earthquake F

earthquakes and ranged in between 0.0011 g and 0.6238 g of surface PGA. The average values of surface PGA in the states due to selected earthquakes are 0.1285, 0.1030, 0.2174, 0.0319, 0.1254, 0.0018, 0.0769 and 0.0983 g. Earthquakes labeled as E, F and H respectively hit 97, 99 and 100% of sites in these states. Average PGA due to sources E, F and G are 0.1030, 0.2174 and 0.1254 g respectively. Structures in the area need specific provisions with respect to these seismic sources. About 65%

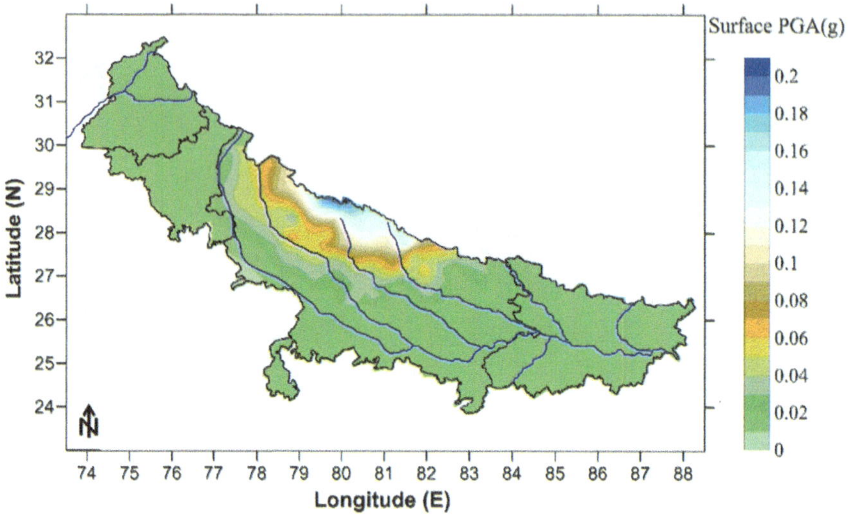

Fig. 8 Spatial variation of surface PGA for earthquake G

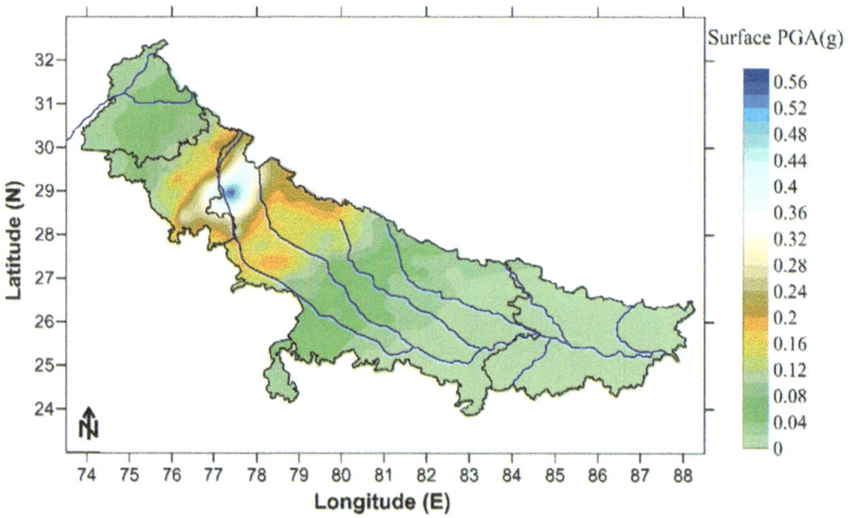

Fig. 9 Spatial variation of surface PGA for earthquake H

of sites analyzed in the region have average surface PGA > 0.1 g. Table 2 shows summary of results for Punjab and Haryana region.

Based on the same criteria, sites in Uttar Pradesh have been analyzed for thirteen seismic sources and surface PGA due to each considered earthquake has been evaluated. From the data obtained from DEEPSOIL program, spatial variation of surface PGA for each earthquake has been presented in Figs. 2, 3, 4, 5, 6, 7, 8,

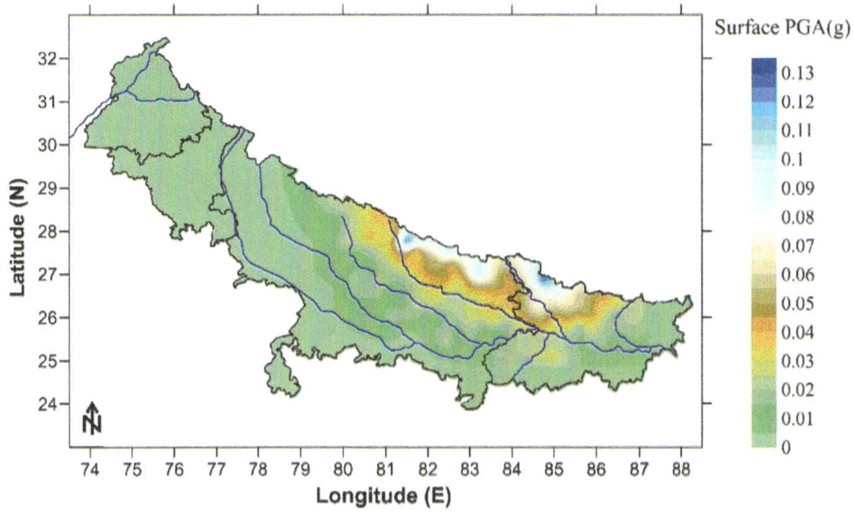

Fig. 10 Spatial variation of surface PGA for earthquake I

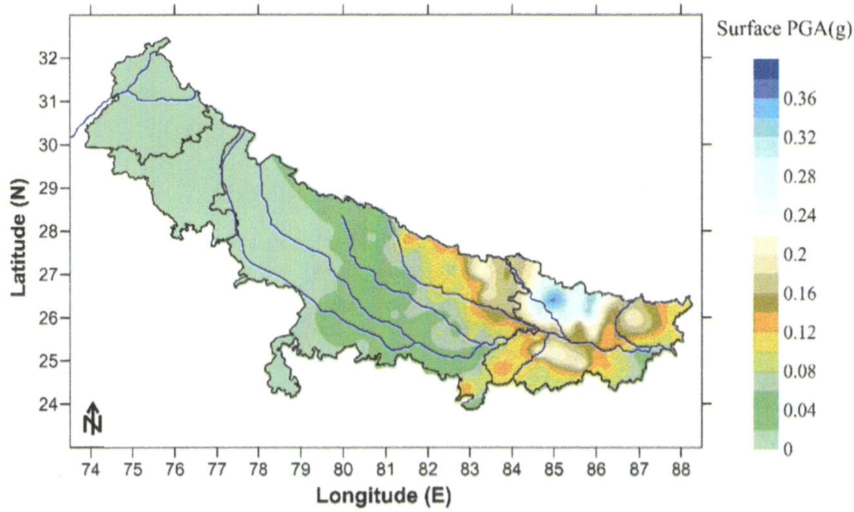

Fig. 11 Spatial variation of surface PGA for earthquake J

9, 10, 11, 12, 13, 14, 15, 16 and 17. A wide range of bedrock PGA is observed varying from 0.0001 g to 0.5771 g. Overall surface PGA for the state varies from 0.0018 g to 1.1827 g. The average values for each earthquake are 0.1096, 0.0420, 0.1504, 0.0205, 0.0669, 0.0556, 0.0578, 0.0924, 0.0260, 0.0724, 0.0208, 0.1912 and 0.1836 g as listed in Table 3. The sites in Uttar Pradesh show average PGA values lesser compared to Bihar, Punjab and Haryana and thus might be less prone to the

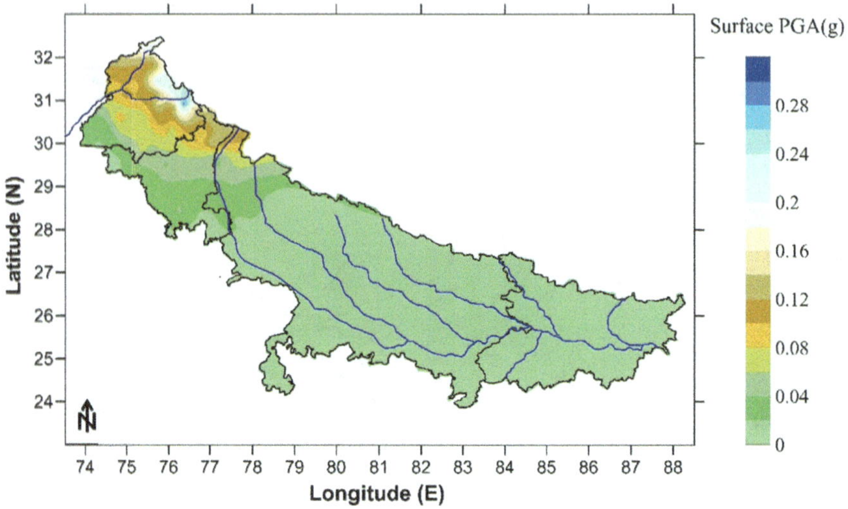

Fig. 12 Spatial variation of surface PGA for earthquake K

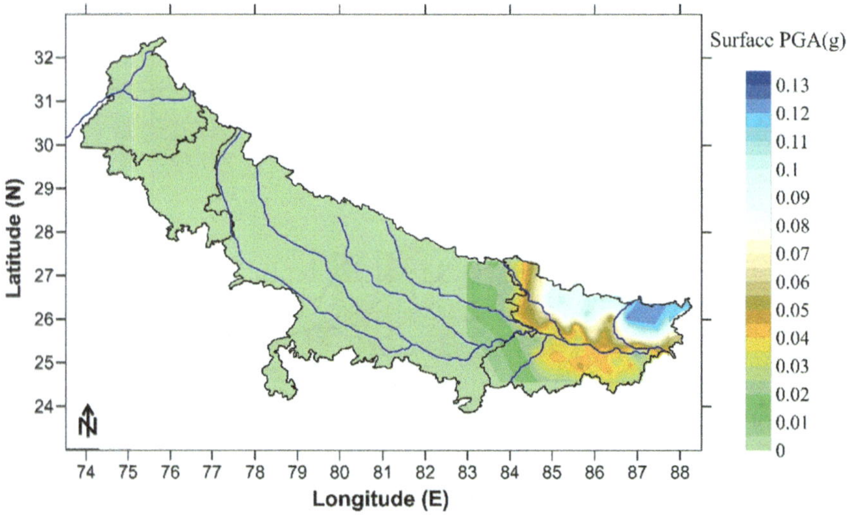

Fig. 13 Spatial variation of surface PGA for earthquake L

earthquake sources considered in this study. It is worth mentioning that surface PGA as high as > 1.0 g is observed for earthquakes 1 and 14 for few sites. Overall, 37% of analyzed sites show average surface values > 0.1 g.

Average surface PGA observed here is less as compared to PGA with respect to individual earthquakes. Overall average of 0.12 g is observed for the whole IGB which is quite different from state wise analysis. Average value of surface PGA

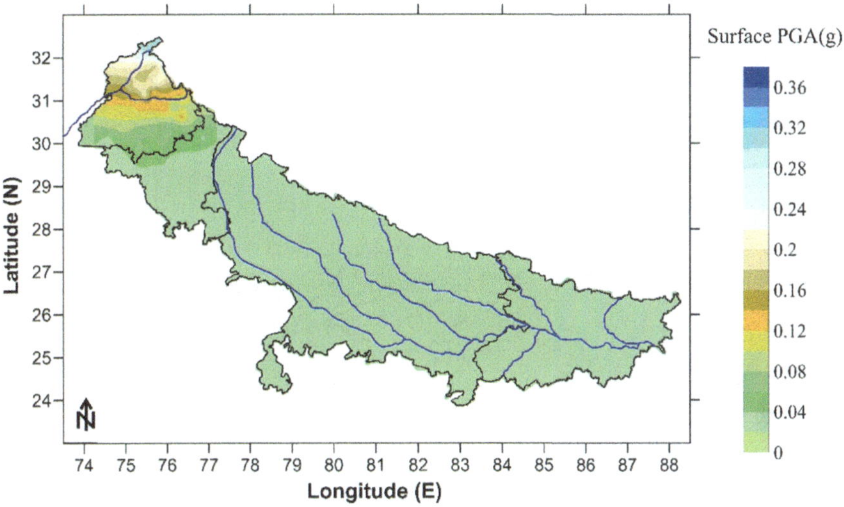

Fig. 14 Spatial variation of surface PGA for earthquake M

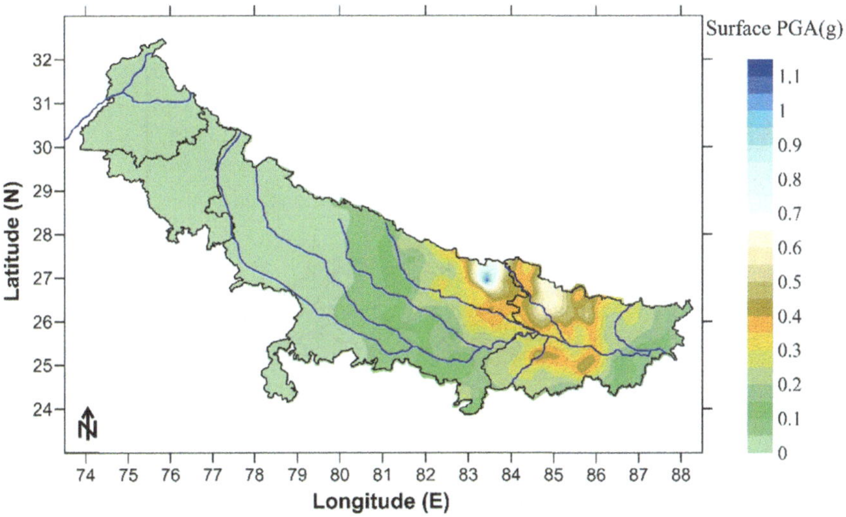

Fig. 15 Spatial variation of surface PGA for earthquake N

observed is as low as 0.014 g and as high as 0.32 g. An average value > 0.15 g is observed at a number of sites in IGB and therefore requires certain consideration in design of structures. The spatial variation may be referred from Fig. 18. Maximum surface PGA observed for IGB at any site is as high as 1.18 g and its spatial variation is shown in Fig. 19. A very smaller number of sites show maximum PGA > 0.7 g.

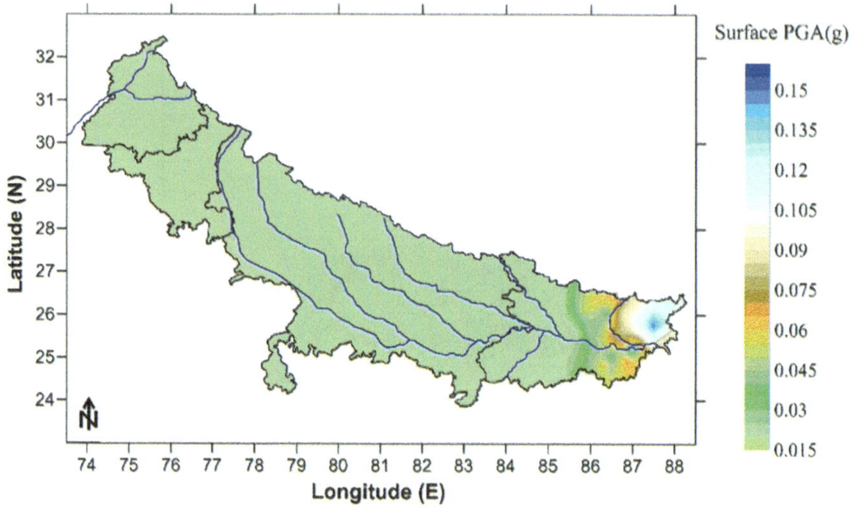

Fig. 16 Spatial variation of surface PGA for earthquake O

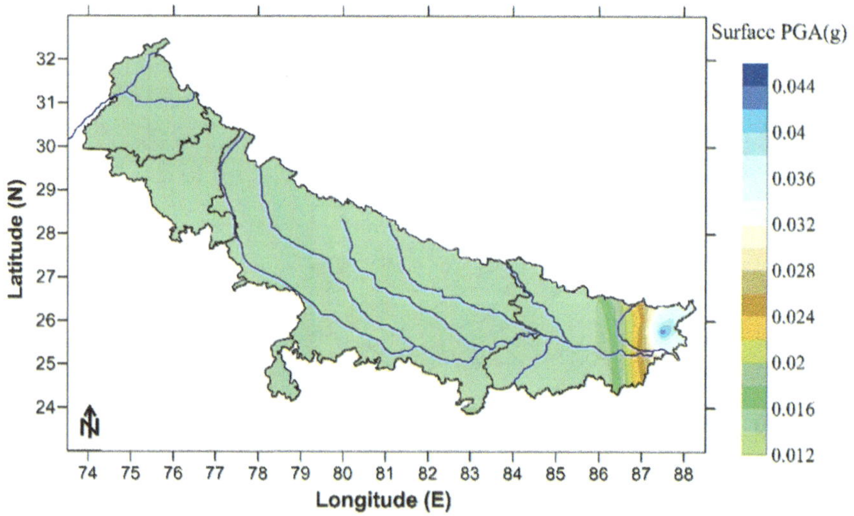

Fig. 17 Spatial variation of surface PGA for earthquake P

Majority of the sites show maximum PGA around 0.25 g. These maximum values may be referred for design of highly important structures, subjected to public use or national importance.

Table 1 Analysis of bedrock and surface PGA for Bihar

Earthquake	Bedrock PGA(g) range	Surface PGA(g) range	Average surface PGA(g)	Proportion (%) of sites
A	0.0005–0.0184	0.0093–0.0626	0.0316	21
C	0.0225–0.2996	0.0837–0.6546	0.2739	100
D	0.0021–0.1022	0.0215–0.3112	0.1005	89
I	0.0019–0.0468	0.0126–0.1310	0.0389	58
J	0.0267–0.1910	0.0311–0.3783	0.1671	98
L	0.0023–0.0463	0.0166–0.1266	0.0589	84
N	0.0289–0.3415	0.0599–0.6804	0.2985	100
O	0.0012–0.0428	0.0182–0.1518	0.0553	46
P	0.0008–0.0086	0.0122–0.0440	0.0280	12

Table 2 Analysis of bedrock and surface PGA for Punjab and Haryana

Earthquake	Bedrock PGA(g) range	Surface PGA(g) range	Average Surface PGA(g)	Proportion (%) of sites
B	0.0026–0.1383	0.0300–0.4319	0.1285	85
E	0.0102–0.0574	0.0297–0.2481	0.1030	97
F	0.0140–0.6509	0.0655–0.6238	0.2174	99
G	0.0004–0.0076	0.0229–0.0441	0.0319	8
H	0.0066–0.3302	0.0269–0.5875	0.1254	100
J	0.0003–0.0020	0.0011–0.0035	0.0018	5
K	0.0056–0.1040	0.0156–0.3087	0.0769	92
M	0.0012–0.1232	0.0208–0.3584	0.0983	67

6 Conclusion

1D nonlinear site response analyses for futuristic scenario earthquakes have been carried out at 270 sites covering UP, Punjab, Haryana and Bihar. Variation of estimated surface PGA for individual earthquakes has been deliberated using contour maps. Following are the main remarks concluded from the study:

- Wide range of bedrock PGA ranging from 0.0003 to 0.6509 g for Punjab and Haryana, 0.0005 to 0.2996 g for Bihar and 0.0001 to 0.5771 g for UP has been analyzed.
- Sites in Bihar reflect average and maximum surface PGA 0.15 and 0.68 g, Uttar Pradesh 0.10 and 1.18 g, Punjab and Haryana 0.12 and 0.62 g respectively.
- All the earthquakes hit varied proportion of sites in each state. On an average, the percentage of sites hit by earthquakes in Bihar, Punjab and Haryana and UP are 38, 35 and 37 respectively.

Table 3 Analysis of bedrock and surface PGA for Uttar Pradesh

Earthquake	Bedrock PGA(g) range	Surface PGA(g) range	Average surface PGA(g)	Proportion (%) of sites
A	0.0017–0.5771	0.0069–1.1593	0.1096	94
B	0.0001–0.0156	0.0266–0.0789	0.0420	4
C	0.0006–0.2534	0.0178–0.5983	0.1504	74
D	0.0001–0.0067	0.0029–0.0291	0.0205	3
E	0.0005–0.0522	0.0100–0.1919	0.0669	38
F	0.0002–0.0744	0.0148–0.1791	0.0556	19
G	0.0004–0.0466	0.0139–0.1846	0.0578	49
H	0.0010–0.3438	0.0093–0.4008	0.0924	67
I	0.0005–0.0561	0.0018–0.1319	0.0260	68
J	0.0013–0.1373	0.0116–0.2274	0.0724	78
K	0.0002–0.0322	0.0038–0.0684	0.0208	9
L	0.0002–0.0103	0.0137–0.0250	0.0192	8
N	0.0008–0.5105	0.0187–1.1827	0.1836	76

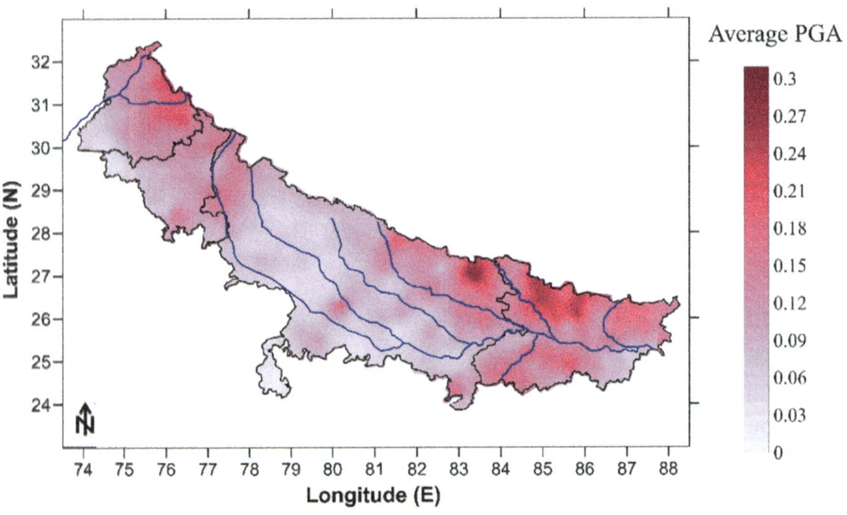

Fig. 18 Spatial variation of average surface PGA

- Earthquakes labeled as C, J and N have relatively higher impact on Bihar as they hit more than 90% of the sites, Punjab and Haryana show this response for earthquakes E, F, H and K while UP for A.
- Some of the earthquakes under consideration hit very little area (<10% of sites) viz. G and J for Punjab and Haryana, B, D, K and L for UP.

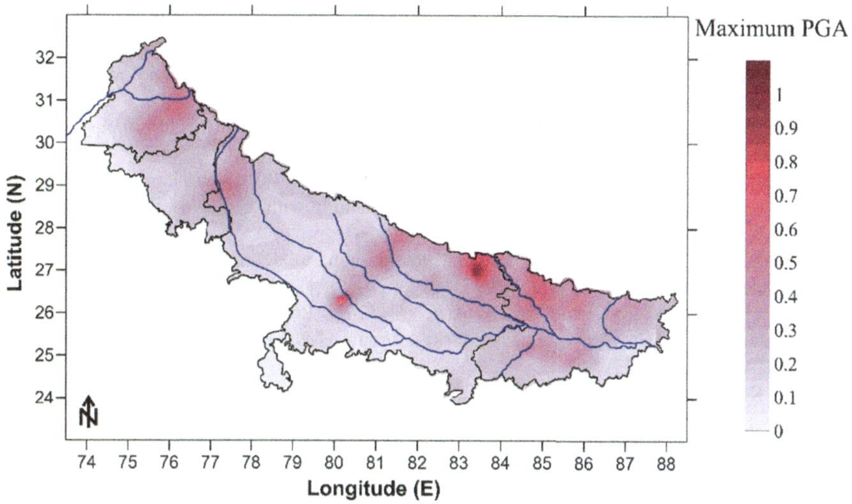

Fig. 19 Spatial variation of maximum surface PGA

- Higher surface PGA was observed for nearby seismic sources and have therefore higher expected damage level. These should be given due considerations while designing new structures in the area.
- PGA obtained from the study for futuristic earthquakes is much more than current seismic code of IS1893 [22].

Acknowledgements The authors are thankful for funding supported by the Science and Engineering Research Board (SERB), Department of Science and Technology [grant numbers SERB/F/162/2015-2016].

References

1. Ambraseys N (2000) Reappraisal of North-Indian earthquakes at the turn of the 20th century. Curr Sci 79:101–106
2. Khattri KN (1987) Great earthquakes, seismicity gaps and potential for earthquakes along the Himalayan plate boundary. Tectonophysics 38:79–92
3. Anbazhagan P, Kumar A, Sitharam TG (2010) Amplification factor from intensity map and site response analysis for the soil sites during 1999 Chamoli earthquake. In: Proceedings of the 3rd Indian young geotechnical engineers conference 2010. New Delhi, pp 311–316
4. Mahajan AK, Sporry RJ, Champati PK, Ranjan RR, Slob S, Van WS (2007) Methodology for site-response studies using multi-channel analysis of surface wave technique in Dehradun city. Curr Sci 92(7):945–955
5. Govindraju L, Bhattacharya S (2008) Site Response studies for Seismic hazard analysis for Kolkata city. In: Proceedings of the 12th international conference of the international association for computer methods and advances in geomechanics 2008, pp 2899–2907

6. Phanikanth VS, Choudhury D, Reddy GR (2011) Equivalent-linear seismic ground response analysis of some typical sites in Mumbai. Geotech Geol Eng 29(6):1109–1126
7. Kumar A, Anbazhagan P, Sitharam TG (2013) Seismic hazard analysis of Lucknow considering local and active seismic gaps. Nat Hazards 69:327
8. Jishnu RB, Naik SP, Patra NR, Malik JN (2013) Ground response analysis of Kanpur soil along the Indo-Gangetic Plains. Soil Dyn Earthquake Eng 51:47–57
9. Kumar A, Baro O, Harinarayan N (2016) Obtaining the surface PGA from site response analyses based on globally recorded ground motions and matching with the codal values. Nat Hazards 81:543–572
10. Motazedian D, Atkinson GM (2005) Stochastic finite-fault modeling based on a dynamic corner frequency. Bull Seismol Soc Am 95:995–1010
11. Anbazhagan P, Joo MR, Rashid MM, Al-Arifi NSN (2021) Prediction of different depth amplifications of deep soil sites for potential scenario earthquakes. Nat Hazards (Accepted, Submission Id: NHAZ-D-20-00073)
12. Bajaj K, Anbazhagan P (2019) Seismic site classification and correlation between Vs and SPT-N for deep soil sites in Indo-Gangetic Basin. J Appl Geophys 163:55–72
13. Anbazhagan P, Uday A, Moustafa SSR, Al-Arifi NSN (2016) Correlation of densities with shear wave velocities & SPT N values. J Geophys Eng 13:320–341
14. Bajaj K, Anbazhagan P (2019) Comprehensive amplification estimation of the Indo Gangetic Basin deep soil sites in the seismically active area. Soil Dyn Earthquake Eng 127:105855
15. Anbazhagan P, Prabhakaran A, Madhura H, Moustafa SSR, Al-Arifi NSN (2017) Selection of representative shear modulus reduction and damping curves for rock, gravel and sand sites from the KiK-Net downhole array. Nat Hazards 88(3):1741–1746
16. Bajaj K, Anbazhagan P (2019) Identification of shear modulus reduction and damping curve for deep and shallow sites: Kik-net data. J Earthquake Eng.https://doi.org/10.1080/13632469.2019.1643807
17. Electric Power Research Institute (EPRI) (1993) Guidelines for site specific ground motions. Palo Alto, California, November, TR-102293
18. Zhang J, Andrus R, Juang CH (2005) Normalized shear modulus and material damping ratio relationships. J Geotech Geoenviron Eng ASCE 131:453–464
19. Menq FY (2003) Dynamic properties of sandy and gravelly soils. PhD thesis, Department of Civil Engineering, University of Texas, Austin, TX
20. Darendeli MB (2001) Development of a new family of normalized modulus reduction and material damping curves. PhD dissertation, University of Texas at Austin, Austin, Texas
21. Hashash YMA, Musgrove MI, Harmon JA, Ilhan O, Groholski DR, Phillips CA, Park D (2017) DEEPSOIL 7.0, user manual
22. BIS 1893 (Part 1) (2016) Indian standard criteria for earthquake resistant design of structures. Bureau of Indian Standards, New Delhi, Sixth revision

Synthetic Ground Motion Simulation for Varanasi City

Manjari Singh, S. K. Duggal, V. P. Singh, and Keshav Kumar Sharma

Abstract Varanasi (latitude 25°28′ N and longitude 82°96′ E), the cultural capital of India, is presently clustered with a maze of ancient narrow lanes (Gullies) and old buildings. Being a sacred city, it is not well planned and structured, due to lack of adherence to earthquake-resistant building design philosophies and techniques. Consequently, even a smaller magnitude of an earthquake can cause a considerable loss. The city is also near to Faizabad ridge, which has been seismically sedentary for last 300 years. Due to unavailability of earthquake ground motion (G.M.) records in this region, it is necessary to simulate G.M. based on regional seismic data. In this regard, the stochastic approach has been adopted for the synthesis of G.M. at bedrock level. An EXSIM methodology has been used in this study for synthesis of strong G.M. for various identified faults (Allahabad Fault, Azamgarh Fault, Gorakhpur Fault, Deoria Fault, Lucknow Fault, Siwan Fault, Shajhanpur Fault and Great Boundary Fault) around the city. Various stress drops 70, 100, 125, 150, 175 and 200 bars have been taken for simulations to account for uncertainty in stress drop. Acceleration time histories due to various faults for Varanasi city has been simulated and plotted. The maximum PGA estimated was 0.078 g for Azamgarh Fault at 200 bar among all the faults around the city. Further Response Spectra has been plotted for stress drop (70–200 bar).

Keywords Varanasi city · Stochastic method · EXSIM · Finite fault · Stress drop

1 Introduction

During the movement of tectonic plates, accrued strain energy gets released in the form of seismic waves. All of the possible natural hazards, earthquakes are the most severe calamitous catastrophe for the society with respect to the loss of lives and

M. Singh (✉) · S. K. Duggal · V. P. Singh
MNNIT Allahabad, Prayagraj 211004, India
e-mail: majarisingh.2508@gmail.com

K. K. Sharma
NIT Jamshedpur, Jamshedpur 831014, India

T. G. Sitharam et al. (eds.), *Local Site Effects and Ground Failures*, Lecture Notes in Civil Engineering 117, https://doi.org/10.1007/978-981-15-9984-2_6

assets. Earthquakes are ineluctable, but their impact can be reduced appropriately by considering the soil properties, earthquake type, magnitude and area of impact during design of infrastructures. The Himalayas spans from J&K to Arunachal pradesh, Meghalaya, Manipur and Tripura and expanding to Nepal and China, which are the hub of the most severe earthquakes in the history. The most dreadful Gorkha earthquake 2015 that triggered in Nepal on April 25, 2015 and took the life of about 9000 people.

Improved structural design has a significant role in minimizing deaths and damage to property caused by earthquakes. For the design of earthquake-resistant structures, estimation of the level of ground shaking in terms of G.M. parameters: amplitude (Peak Ground Acceleration, PGA), frequency content (Fourier acceleration spectra, response spectra) and duration of motion is required, which reflects Characteristics of G.M.

Characteristics of G.M. are necessary to comprehend the peril arising from the event. For a specific site, design G.M. depends upon many factors which include site location with respect to potential seismic source, nature of rupture and seismicity of source, travel path between source and site as well as also the local site effects.

If an array of strong-motion networks would exist in the area, then the G.M. characteristics of the near field can be better understood. However, in those regions where is a paucity of strong G.M. data, as well as seismicity rates, are also less, earthquake simulations may then produce synthetic data based on regional seismic characteristics [1, 2].

G.M. simulation had begun by the pioneering work of Hartzell [3] and Irikura [4] about three decades ago. Afterwards, various researchers have come up with different simulation techniques which can be summarized in the following categories: Parameter-Prediction, Stochastic Process, Physics and Hybrid Methods.

Parameter prediction method requires a simple parameter to simulate G.M. (such as PGA, PGV and Sa (T)) at multiple periods [5–7]. It offers less computing time, relatively simple, but it is less useful in those regions where G.M. data is rarely available. Whereas, stochastic process-based, physics-based and hybrid methods simulate G.M. time history. Housner [8] was first to introduce stochastic process-based model by the use of white noise to represent earthquake G.M. Physics-based method encompasses various approaches: Finite fault method, Kinematic method and Dynamic method.

Finite fault method [9, 10] considers an earthquake as the sum of many smaller earthquakes by first assuming a rupture geometry and then modeling the earthquake as a distribution of smaller subevents on the rupture plane. Kinematic method [11, 12] consists of two steps one is a simulation of slip distribution, and another one is wave propagation. In the dynamic method [13, 14], pre-defined stress conditions and physics laws are simultaneously used to solve slip distribution and wave propagation. The dynamic method is comparatively more complex, and it is not generally adopted in the engineering field. On the other hand, hybrid method [15] is a combination of previously discussed methods which generate more realistic time history and convenient in the analysis of tall building that is excited at multiple frequencies.

In the present study, Extended Finite-Fault Simulation (EXSIM) software, based on stochastic finite fault approach to simulate G.M. database for Varanasi City has been used. In this city, there is not enough real G.M. data so, EXSIM is used to develop G.M. simulation for the city.

2 Seismotectonic Background and Seismicity of the Study Area

The Indo Gangetic Basin (IGB) is the largest newer alluvial sedimentary basin that comprises approx. 200 million population. IGB comprises the following major trending fault: NNE-SSW to NE-SW, NNW-SSE to NE-SE and EW. Among them, EW trending fault lies in Azamgarh-Gorakpur area and describes the MirganjGraben. The E-W elongated IGB was formed due to the uplift of Himalayas after the collision of Indian-Asian plates [16, 17]. The area includes several faults/ridges [18]. Several significant ridges in the IGB increase the seismic activity of the basements. The Faizabad and Munger ridge, near to Varanasi, is surrounded by many faults which are the extension of Bundelkhand and Satpura massif. The IGB is comparatively moderately seismic from the Himalayas [19].

Varanasi, one of the heritage city of India, belongs to zone III in the seismic map of India [20]. The city inevitably experienced shocks and had even suffered damages in past when strong earthquakes struck the neighboring vicinities of the Himalayan territory. Varanasi has also felt shocks of several past earthquakes that occurred in Uttar Pradesh such as 1925 Raibarielly earthquake M_w 6.0, 1956 Bulandshahar earthquake M_w 6.2, 1965 Gorakhpur earthquake M_w 5.7 and 1966 Moradabad earthquake M_w 5.8 earthquakes; Countrywide 1916 Dharchulla earthquake M_w 7.5, 1945 Uttarakhand earthquake M_w 6.5, 1991 Uttarkashi earthquake M_w 6.8 and 1997 Jabalpur earthquake M_w 6.0 and earthquakes like 1934 Nepal-Bihar earthquake M_w 8.4, 2015 Nepal earthquake M_w 7.8 from the neighboring country.

The main issue for the designer is the unavailability of systematic data of earlier earthquakes. Therefore the construction of seismic map for the region, considering the possibility of occurrence of major earthquake events in recent years, is required. It significantly includes sources like Main Central (MC) and Main Boundary (MB) Thrust that is 400 km distant from Varanasi city and are solitary earthquake-prone areas in the world.

Faizabad ridge, near to the city, is quiescent over more than 300 years as mentioned by the Disaster Risk Management Programme organized by the Home Affairs Ministry in collaboration with UN Development Programme (UNDP). Due to the huge seismic gap, the ridge is substantially stressed and has the capacity to produce higher magnitude of earthquake in upcoming years [21]. The Indian plate is subducted underneath the Asian plate by 5.25 m can produce an earthquake of M_w 8.0 stated by Earthquake Mitigation Department of Uttar Pradesh [21].

In order to identify the potential seismic sources, a vast literature survey has been carried out to observe the presence of linear, aerial and point sources. Only linear subsurface faults are observed as a potential seismic source around the study area and

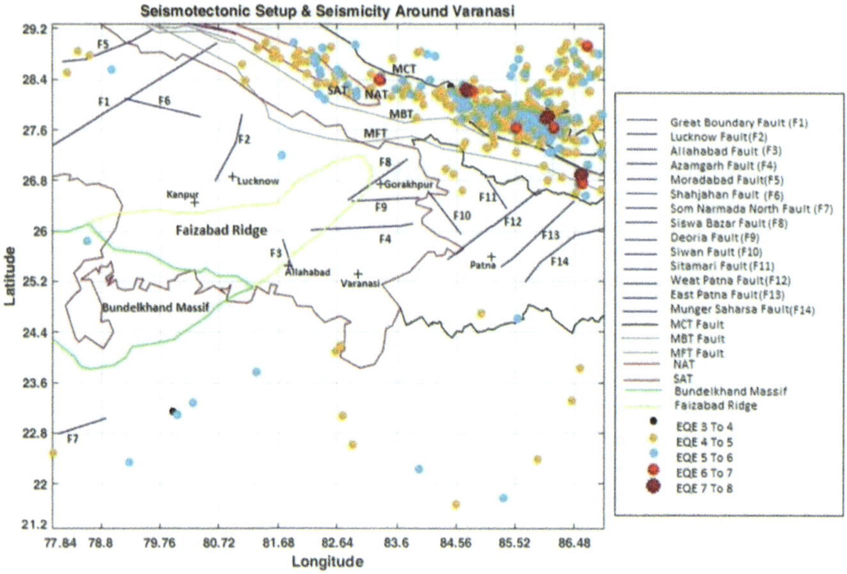

Fig. 1 Seismicity around the city

are defined in Seismotectonic Atlas of India. Presence of numerous faults around the study area increases the seismic hazard of Varanasi city. In the present study, faults around Varanasi between latitude 25.28° N–29.28° N and longitude 78.96° E–86.96° E (around 450 km) are plotted using MATLAB and are shown in Fig. 1. Further, the fault list of the area has been prepared, and various surface and sub-surface faults are listed and are tabulated in Table 1. The details of these faults like strike, dip, and mechanism (thrust/reverse, normal, and strike-slip) have been gathered from various publications like Dasgupta et al. [22], Singh et al. [23], and Kayal [24]. In this study, only faults having length more than 5.0 km are considered.

3 Methodology

The basic approach in G.M. simulation is to generate reliable synthetic data by modeling the seismic source mechanisms and properties of regional wave propagation. Hanks and McGuire [25] developed a basis of stochastic technique that observed high-frequency strong G.M. can be represented by finite duration band-limited white Gaussian noise. Boore [26] was the first who defined Band limited white-noise model in which Gaussian white noise of limited frequency band represents high-frequency motions of earthquakes.

Further, Beresnev and Atkinson [27] has used this approach to model the finite faults. The model has been discretized a rectangular fault plane into smaller sub

Table 1 Details of seismic sources

Faults names	Faults length (km)	Faults position	Strike (Deg.)	Dip (Deg.)	Latitude		Longitude		Mmax (as per WC 1994)
					Initial	Final	Initial	Final	
F1 (Allahabad Fault)	57	Subsurface	166	50	25.382	25.873	81.9	81.745	6.3
F2 (Azamgarh Fault)	158	Subsurface	87	50	26	26.074	82.19	83.778	6.9
F3 (Deoria Fault)	106	Subsurface	87	50	26.48	26.53	82.86	83.93	6.6
F4 (Gorakhpur Fault)	118	Subsurface	51	50	26.482	27.150	82.8	83.722	6.7
F5 (Lucknow Fault)	126	Subsurface	25	50	26.78	27.808	80.67	81.207	6.8
F6 (Siwan Fault)	94	Subsurface	64	60	26.636	25.945	84.09	84.64	6.6
F8 (Shajhanpur Fault)	130	Subsurface	106	50	28.11	27.788	79.17	80.445	6.8

faults and considered as a point source and summing the contributions of each sub fault. Motazedian and Atkinson [10] have used the term dynamic corner frequency (DCF).

In this concept, the total energy emitted from the fault has been conserved regardless of the selected sub fault size. In this paper, the EXSIM program (Extended Earthquake Fault Simulation Program) has been used for earthquake simulation, which is based on DCF using stochastic finite fault algorithm [10]. It is an updated form of Boore's (1983) point source stochastic model [26].

In this methodology, the Fourier amplitude spectrum of G.M. has been obtained from point source seismological model. Acceleration Fourier amplitude spectrum $A_i(f)$ of the ith sub fault is expressed as follows (Eq. 1):

$$A_i(f) = \left(\frac{M_{0i} H_i (2\pi f)^2}{1 + (f/f_{0i}(t))^2}\right)\left(\frac{\sqrt{2}\langle R_{\theta\varnothing}\rangle}{4\pi\rho V_s^3}\right)\left(Ge^{-\frac{\pi f R_i}{V_s Q}}\right)\left(F(f)e^{-\pi f k_0}\right) \tag{1}$$

where G denotes the geometric attenuation, M_i is the seismic moment of ith sub fault, V_s is shear wave velocity, R_i is the distance from sub fault to the site and Q is the quality factor. H_i is a scaling factor and $f_{0i}(t)$ defines DCF [10]. $F(f)$ denotes site amplification due to earthquake wave propagation from the source region to the surface. The term $e^{-\pi f k_0}$ is a high cut filter which takes care of the rapid spectral decay at high frequencies [28]. $\langle R_{\theta\varnothing}\rangle$ refers to radiation coefficient averaged over an appropriate range of azimuths and take-off angles and ρ is crust density at focal-depth. Moment of the ith sub fault is determined from slip distribution is defined as (Eq. 2):

$$M_{0i} = \frac{M_0 D_i}{\sum_{i=1}^{N} D_i} \tag{2}$$

where, M_0-total seismic moment of all sub faults, D_i-average final slip acting on the ith sub fault. The DCF $f_{0i}(t)$, can be defined from the following equation (Eq. 3) which relates the terms: the seismic-moment M_0 and the stress drop $\Delta\sigma$:

$$f_{0i}(t) = 4.9 \times 10^6 (N_R(t))^{-(1/3)} N^{-1/3} V_s \left(\frac{\Delta\sigma}{M_0}\right)^{(1/3)} \tag{3}$$

where, $N_R(t)$-a cumulative number of ruptured sub faults at (t) time. Scaling factor (H_i), can be expressed as (Eq. 4).

$$H_i = \left(N \frac{\sum_f \left(\frac{f^2}{1+(f/f_0)^2}\right)^2}{\sum_f \left(\frac{f^2}{1+(f/f_{0i}(t))^2}\right)^2}\right)^{(1/2)} \tag{4}$$

where, f_0 defines the corner frequency at the end of the rupture, and can be obtained by putting $N_R(t) = N$ in Eq. 3. In Eq. (4), the summation is done from 0 to the highest frequency exist in the signal.

In this study, the seismic model has been imposed in the time range by using numerical simulation for which Boore (1983) [26] recommended the following three steps:

(1) Simulated strong motion duration would be considered equal to the Gaussian stationary white noise sample of length [1]. In Eq. 5. the terms f_c and r denotes the corner frequency and hypocentral distance, respectively.

$$T = 1/f_c + 0.05r \tag{5}$$

(2) A non-stationary modulating function recommended by Saragoni [29] is multiplied by this sample. After the simulation, the sample is Fourier transformed into the frequency domain. Further, it is normalized using rootmean squares value and then multiplied with Eq. 1. Afterwards, it is converted back into time range to acquire a sample of sub-surface acceleration time histories.
(3) Summing up all simulated time histories for each sub faults of time delay (Δt_i) for generating final G.M. acceleration $A(t)$ and expressed as (in Eq. 6):

$$A(t) = \sum_{i=1}^{N} A_i(t + \Delta t_i) \tag{6}$$

Due to the modification between bedrock and A-type sites being a linear problem as well as 1-dimensional and hence these sites augmentation can be done by using the quarter wavelength method of Boore [30] directly. Database of kappa factors have been reported by Chandler et al. [31] and has given an equation (Eq. 7) for determining the kappa values (k_0) from the average shear wave in top 30 m of soil (V_{s30}).

$$k_0 = \frac{0.057}{V_{s30}^{0.8}} - 0.02 \tag{7}$$

For rock level G.M. simulation, kappa factor calculated as 0.057 for the soft soil in Varanasi. In this study, Table 2 shows the parameters used in the simulation.

The maximum magnitude of the earthquake M_w for all faults, as shown in Fig. 1 have been estimated by Wells and Coppersmith (WC) [35] relation (Eq. 8). Under the worldwide dataset, Mark [36] has suggested that the fault rupture length can be considered as (1/3–1/2) of the total fault length. Though taking such large sub-surface rupture length generates huge moment magnitude and doesn't fit with the historical earthquake records.

$$M_w = 4.38 + 1.49 * \log(RLD)(\text{Allrupturetypes}) \tag{8}$$

Table 2 Parameters used for simulation

Parameters	Value used in this study
V_s shear-wave velocity and ρ density at source	3500 m/s, 2.9 g/cc [32, 33]
$\Delta\sigma$ Stress drops	70, 100, 125, 150, 175, 200 bars [24]
$Q(f)$ quality factor	$142f^{1.04}$ [34]
G geometric attenuation	$1/r$ (*for* $r \leq 100$ km); $\sqrt{1/(100r)}$ (*for* $r > 100$ km) [32]
k_0 kappa factor	0.057 for deep profiles (Eq. 7)
$\Delta\sigma$ reference stress drop	100 bars
Pulsing percentage and Sub fault size	50% and 0.25 km × 0.25 km
Crustal amplification	ENA Hard rock amplification [1]

where M_w = moment magnitude; RLD = length of subsurface rupture (km).

3.1 Validation

2015 Nepal earthquake (also known as Gorkha earthquake) has occurred in the western part of Nepal with a magnitude M_w 7.8 toward North-West of Kathmandu [37] along the MFT active fault. The main shock of the earthquake was located at the Barpak village (Longitude 84.708° E and Latitude 28.147° N at shallow Depth 15 km) in the Gorkha district of Nepal. The earthquake epicenter was at Lamjung, around 80 km north-west of Kathmandu [37]. The characteristics of fault like strike, dip and depth have been taken from the USGS database [37]. Peak Ground Acceleration (PGA), event location and fault distance have been taken from the Center for Engineering Strong Motion Data CESMD (www.strongmotioncenter.org). The rupture length and width of the fault have been calculated by using empirical relationships given by WC [35]. The square grid has been used to categorize sub fault size configuration. The primary source parameters of Nepal earthquake, shown in Table 3.

The PGA values are recorded by five strong-motion accelerometers [CESMD] located at KTP, TVU, PTN, and THM operated by HU-TU (Hokkaido University and Tribhuvan University, Nepal). Another strong motion accelerometer is located at KATNP operated by USGS. The recorded PGA at respective stations is provided in Table 4.

Table 3 Source parameters of Nepal earthquake

Parameters	Value
M_w	7.8
M_o (Nm)	6.623×10^{20}
Strike (ϕ)	295°
Dip (δ)	10°
Rake (λ)	101°
Focal depth (km)	15
Length of rupture (L) (km)	221
Width of rupture (W) (km)	45
Sub fault length (*dl*) (km)	5
Sub fault width (*dw*) (km)	5
Rupture velocity (km/s)	0.8 β (β-fault Shearwave velocity)

Table 4 Recorded peak ground acceleration motion

Station number	Station name	North latitude (°)	East longitude (°)	PGA (g)
KATNP (USGS)	Kanti path, Kathmandu, Nepal	27.712	85.316	0.163
KTP (HOKU)	Municipality office, Kirtipur	27.682	85.272	0.260
TVU(HOKU)	Dep. Geology, Tribhuvanuniv., Kirtipur	27.681	85.288	0.234
PTN (HOKU)	Pulchowk, Tribhuvanuniv., Patan	27.681	85.318	0.154

Source www.strongmotioncenter.org

3.1.1 Model Input Parameters

The G.M. has been simulated from Stochastic Finite-Fault Model for A-type site (hard rock site) and estimated PGA from simulation compared with recorded PGA at KTP station. Site amplification has been ignored since the simulation is carried out for the A-type site. The V_s and density of hard rock site is taken as 2.97 km/s and 2600 kg/m^3 respectively [38]. Kappa factor $\boldsymbol{k}_0 = 0.005$s at hard rock level [38]. Stress drop and pulsing area were taken as 160 bar and 50% (www.nset.org.np/eq2015/intensity_maps.php). Geometrical spreading G is taken as $1/r$ (*for* $r \leq 100$ km); $\sqrt{1/(100r)}$ (*for* $r > 100$ km) [32]. Q = 500 $f^{1.04}$ is used as a quality factor [39]. The estimated PGA is 0.248 g as shown in Fig. 2 whereas recorded PGA at KTP station is 0.260 g.

For Azamgarh fault, moment magnitude M_w is determined using the WC [35] relation (Eq. 8) with sub-surface rupture length of fault as 158 km, it results in a magnitude of 6.8.

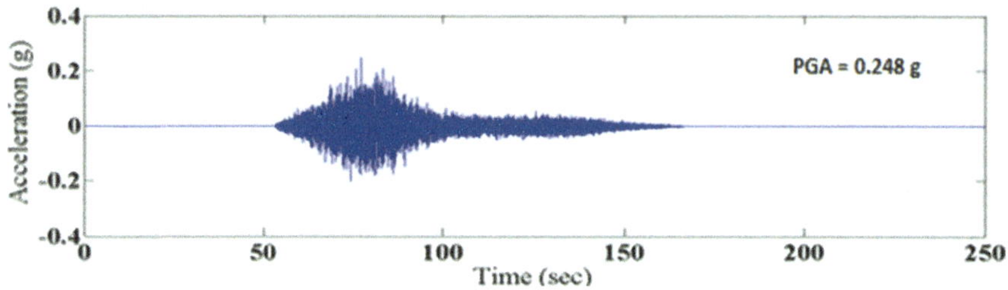

Fig. 2 Estimated PGA

Table 5 Simulated peak ground acceleration

Stress level (bar)	Peak ground acceleration (g)						
	Allahabad fault	Azamgarh fault	Deoria fault	Gorakhpur fault	Lucknow fault	Shahjahan fault	Siwan fault
70	0.0091	0.0254	0.0135	0.0172	0.0043	0.0008	0.0085
100	0.0111	0.0326	0.0157	0.0175	0.0047	0.0009	0.0084
125	0.0160	0.0576	0.0180	0.0157	0.0064	0.0014	0.0112
150	0.0168	0.0737	0.0078	0.0244	0.0059	0.0015	0.0079
175	0.0201	0.0506	0.0231	0.0297	0.0086	0.0021	0.0109
200	0.0211	0.0783	0.0275	0.0288	0.0077	0.0017	0.0136

4 Results and Discussions

Acceleration time histories due to all considered faults near Varanasi city have been simulated, for different stress drops (70–200 bars) and effective simulations are presented in Table 5. The table presents the maximum PGA values with corresponding stress drop for different faults. It was observed that Azamgarh fault gives the maximum PGA for all stress drop. The maximum PGA estimated was 0.078 g for Azamgarh Fault at 200 bar, and the minimum PGA was 0.0008 g at 70 bar for Shahjahan Fault, among all the faults around the city. Estimated acceleration time history for Azamgarh fault for all stress drop has been shown in Fig. 3. Also, for the same critical Azamgarh fault response spectra have been plotted and is shown in Fig. 4.

5 Conclusions

Varanasi city lies in IGB, which is close to the seismically active Himalayas, which is vulnerable to great earthquakes, surrounded by numerous inhomogeneities in the form of faults/ridges. The city is near to Faizabad ridge, which is greatly stressed due to nonoccurrence of any seismic activity for a prolonged duration. Each seismic

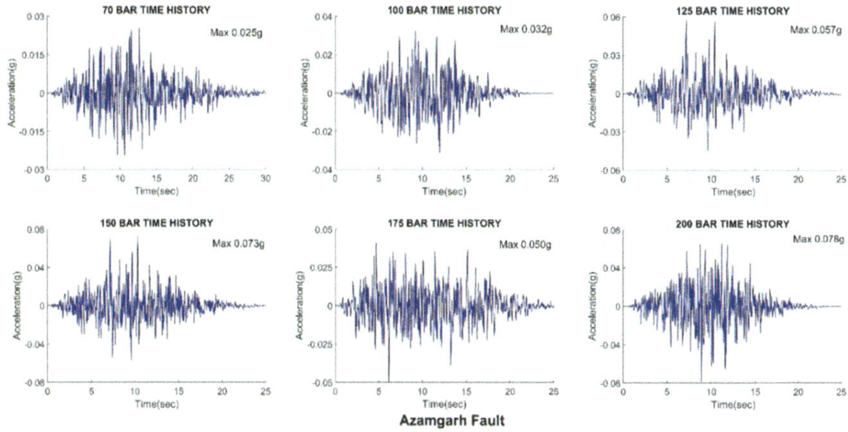

Fig. 3 Simulated acceleration time histories for Azamgarh fault

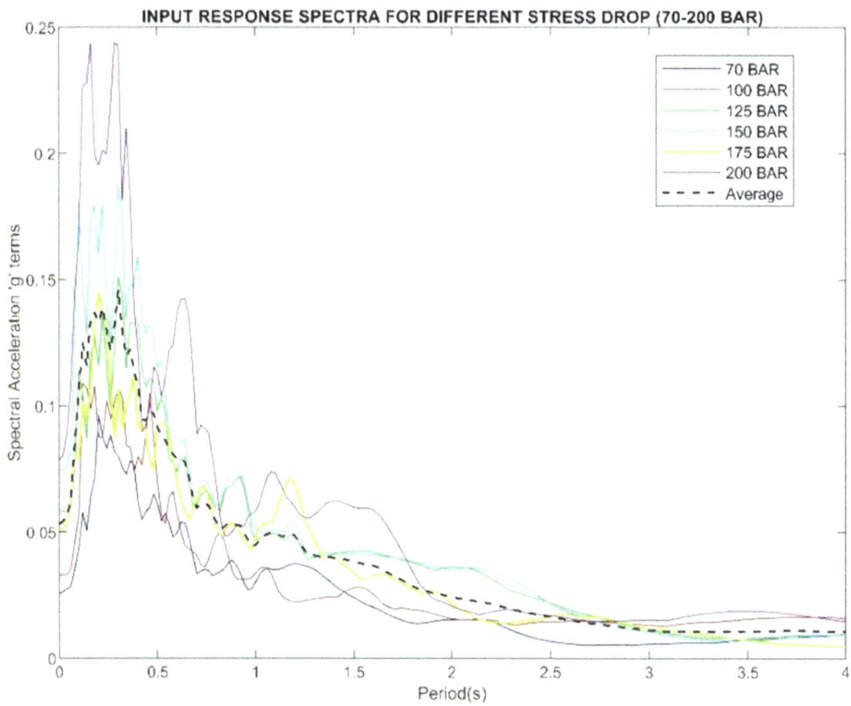

Fig. 4 Input response spectra for different stress drop (70–200 Bar)

source has a capacity to generate massive earthquakes in future. As the absence of recorded G.M. data of past earthquakes for the city, so G.M. has been simulated for all the identified faults within 450 km using stochastic finite-fault models for various stress drops (70–200 bar). Among all the observed faults, Azamgarh fault yielded maximum PGA (0.078 g), and Shahjahan Fault gives minimum PGA (0.0008 g) at all stress drops for Varanasi city. Furthermore, simulation procedure has been validated with recorded PGA of April 25, 2015 Nepal earthquake at KTP station.

References

1. Atkinson GM, Boore DM (2006) Earthquake ground-motion prediction equations for eastern North America. Bull Seismol Soc Am 96(6):2181–2205
2. Nath SK, Raj A, Thingbaijam KKS, Kumar A (2009) G.M. synthesis and seismic scenario in Guwahati city—a stochastic approach. Seismolo Res Lett 80(2):233–242
3. Hartzell SH (1978) Earthquake aftershocks as Green's functions. Geophys Res Lett 5(1):1–4
4. Irikura K (1986) Prediction of strong acceleration motion using empirical Green's function. In: Proceedings of 11th Japan earthquake engineering symposium, vol 151, pp 151–156
5. Esteva L, Rosenblueth E (1964) Espectros de temblores a distancias moderadas y grandes. Boletin Sociedad Mexicana de Ingenieria Sesmica 2:1–18 In Spanish
6. Abrahamson NA, Shedlock KM (1997) Overview. Seismol Res Lett 68(1):9–23
7. Baker JW, Jayaram N (2008) Correlation of spectral acceleration values from NGA G.M. models. Earthq Spectra 24(1):299–317
8. Housner GW (1947) Characteristics of strong-motion earthquakes. Bull Seismol Soc Am 37(1):19–31
9. Joyner WB, Boore DM (1986) On simulating large earthquakes by Green's-function addition of smaller earthquakes. Earthquake source mechanics. American Geophysical Union Monograph 37, pp 269—274
10. Motazedian D, Atkinson GM (2005) Stochastic finite-fault modeling based on a dynamic corner frequency. Bull Seismol Soc Am 95(3):995–1010
11. Brune JN (1970) Tectonic stress and the spectra of seismic shear waves from earthquakes. J Geophys Res 75(26):4997–5009
12. Hartzell S, Harmsen S, Frankel A, Larsen S (1999) Calculation of broadband time histories of G.M.: comparison of methods and validation using strong-G.M. from the 1994 Northridge earthquake. Bull Seismol Soc Am 89(6):1484–1504
13. Olsen KB, Madariaga R, Archuleta RJ (1997) Three-dimensional dynamic simulation of the 1992 Landers earthquake. Science 278(5339):834–838
14. Kozdon JE, Dunham EM (2013) Rupture to the trench: dynamic rupture simulations of the 11 March 2011 Tohoku earthquake. Bull Seismol Soc Am 103(2B):1275–1289
15. Mai PM, Imperatori W, Olsen KB (2010) Hybrid broadband ground-motion simulations: combining long-period deterministic synthetics with high-frequency multiple S-to-S backscattering. Bull Seismol Soc Am 100(5A):2124–2142
16. Dewey JF, Bird JM (1970) Mountain belts and the new global tectonics. J Geophys Res 75(14):2625–2647
17. Sinha R, Tandon SK, Gibling MR, Bhattacharjee PS, Dasgupta AS (2005) Late quaternary geology and alluvial stratigraphy of the Ganga basin. Himalayan Geol 26(1):223–240
18. Valdiya KS (1976) Himalayan transverse faults and their parallelism with subsurface structures of north Indian plains. Tectonophysics 32:352–386
19. Quittmeyer RC, Jacob KH (1979) Historical and modern seismicity of Pakistan, Afghanistan, northwestern India, and southeastern Iran. Bull Seismol Soc Am 69(3):773–823

20. IS 1893 (2016) Criteria for earthquake resistant design of structures, part 1—general provisions and buildings (Fifth Revision). Bureau of Indian Standards, New Delhi
21. Nadeshda TNN (2004) Lucknow is on earthquake list. Times of India. https://timesofindia.indiatimes.com/city/lucknow/Lucknow-isonearthquakelist/articleshow/679471.cms. Accessed 17 June 2018
22. Dasgupta et al (2000) Seismotectonic atlas of India and its environs. Geological Survey of India, Kolkata
23. Singh SK, Garcia D, Pacheco JF, Valenzuela R, Bansal BK, Dattatrayam RS (2004) Q of the Indian shield. Bull Seismol Soc Am 94(4):1564–1570
24. Kayal JR (2008) Microearthquake seismology and seismotectonics of South Asia, 1st edn. Springer
25. Hanks TC, McGuire RK (1981) The character of high-frequency strong G.M. Bull Seismol Soc Am 71(6):2071–2095
26. Boore DM (1983) Stochastic simulation of high-frequency G.M.s based on seismological models of the radiated spectra. Bull Seismol Soc Am 73(6A):1865–1894
27. Beresnev IA, Atkinson GM (1997) Modeling finite-fault radiation from the ω n spectrum. Bull Seismol Soc Am 87(1):67–84
28. Anderson JG, Hough SE (1984) A model for the shape of the Fourier amplitude spectrum of acceleration at high frequencies. Bull Seismol Soc Am 74(5):1969–1993
29. Rodolfo Saragoni G, Hart GC (1973) Simulation of artificial earthquakes. Earthq Eng Struct Dyn 2(3):249–267
30. Boore DM, Joyner WB, Fumal TE (1997) Equations for estimating horizontal response spectra and peak acceleration from western North American earthquakes: a summary of recent work. Seismol Res Lett 68(1):128–153
31. Chandler AM, Lam NTK, Tsang HH (2006) Near-surface attenuation modelling based on rock shear-wave velocity profile. Soil Dyn Earthq Eng 26(11):1004–1014
32. Singh SK, Dattatrayam RS, Shapiro NM, Mandal P, Pacheco JF, Midha RK (1999) Crustal and upper mantle structure of Peninsular India and source parameters of the 21 May 1997, Jabalpur earthquake (Mw=5.8): results from a new regional broadband network. Bull Seismol Soc Am 89(6):1631–1641
33. Mitra S, Priestley K, Bhattacharyya AK, Gaur VK (2005) Crustal structure and earthquake focal depths beneath northeastern India and southern Tibet. Geophys J Int 160(1):227–248
34. Mohanty WK, Prakash R, Suresh G, Shukla AK, Walling MY, Srivastava JP (2009) Estimation of coda wave attenuation for the National Capital region, Delhi, India using local earthquakes Pure. Appl Geophys 166:429–449
35. Wells DL, Coppersmith KJ (1994) New empirical relationships among magnitude, rupture length, rupture width, rupture area, and surface displacement. Bull Seismol Soc Am 84(4):974–1002
36. Mark RK (1977) Application of linear statistical models of earthquake magnitude versus fault length in estimating maximum expectable earthquakes. Geology 5(8):464–466

37. U.S. Geological Survey (2018) Earthquake facts and statistics. https://earthquake.usgs.gov/earthquakes/eventpage/us20002926#general_region. Accessed 25 Mar 2018
38. Khattri KN, Yu G, Anderson JG, Brune JN, Zeng Y (1994) Seismic hazard estimation using modelling of earthquake strong G.M.s: a brief analysis of 1991 Uttarkashi earthquake Himalaya and prognostication for a great earthquake in the region. Curr Sci 343–353
39. Mitra S, Paul H, Kumar A, Singh SK, Dey S, Powali D (2015) The 25 April 2015 Nepal earthquake and its aftershocks. Curr Sci 108(10):1938–1943

Dynamic Study of Existing Structure Influenced by Adjacent Deep Excavation

L. Geetha, M. N. Hegde, and M. Mohammed Ayaz

Abstract Rapid growth and expansion in metropolitan cities necessitate the construction of buildings very close to existing structures. Whenever there is an excavation made adjacent to structure, the soil below the footings gets disturbed due to the lateral escape/displacement of soil and subsidence of soil, and hence a decrease in the bearing capacity of the soil around it. It is therefore required to assess the impact of construction on the nearby existing structures. The main aim of the study is to identify the variation in the dynamic properties of an existing frame structure adjacent to deep excavation and to find the response of the structure under seismic excitation. The analysis of structure without adjacent excavation is carried out and is treated as reference for six other cases considered. For three cases, the distance of excavation from existing frame structure is kept constant (as 0.5 m) and the depth of excavation is varied (1.5, 3, and 4.5 m). In other three cases, the depth of excavation is kept constant as 3 m and the distance of excavation from the existing frame is varied as 1, 2, and 3 m. In all the cases the width of excavation is kept constant as 3 m. The finite element modeling and analyses are carried out using software. The parameters studied are displacement, stress distribution, frequency, mode shapes, and amplitudes.

Keywords Deep excavation · Adjacent excavation · Soil loosening due to excavation

---

L. Geetha (✉) · M. N. Hegde
Department of Civil Engineering, Dr. Ambedkar Institute of Technology, Bengaluru 560056, India
e-mail: itsgeethashiv@gmail.com

M. Mohammed Ayaz
Department of Civil Engineering, Dayananda Sagar College of Engineering, Bengaluru 560078, India

T. G. Sitharam et al. (eds.), *Local Site Effects and Ground Failures*, Lecture Notes in Civil Engineering 117, https://doi.org/10.1007/978-981-15-9984-2_7

1 Introduction

The constant growth and expansion of city infrastructures across the globe produce a vast number of engineering problems. In metropolitan areas, construction and renovation projects often occur within close proximity of existing structures due to space limitations. A major concern during the planning and execution of projects involving deep excavation is the impact of construction-related ground movements on adjacent structures. Excavation support systems are often designed to ensure the structural stability of adjacent structures. Although this approach produces a support system that adequately prevents structural failure, it may yield excessive deformations and ground movements. In response to these ground movements, adjacent structures will translate, rotate, deform, and possibly sustain damage.

Construction-related ground movements associated with excavation include movement in response to ground loss during excavation, and short-term and long-term movements in response: changes in the state of stress in the ground as load is transferred from the ground to the excavation support system; dewatering activities; ground treatment or modification; deep foundation installation and vibration caused by blasting, foundation installation, or general equipment operation. The excavation for the new structure induces stresses on the existing structure. It is therefore required to assess the impact of construction on existing structures. In general, the following should be examined:

1. Variation in the bearing capacity of the soil adjacent to the deep excavation.
2. The extent of displacements that may affect the structural or operational integrity of adjacent structures.
3. Response of the structure influenced by adjacent deep excavation under seismic load.

Ni et al. [1] studied numerical calculation model of the dynamic response of a tall building base foundation system. The structure was modeled and the response of the structure was found out for both horizontal and vertical seismic action using Time-History Seismic Analysis, with different elastic modulus for the foundation. The result showed that the structures on soft foundation take on stronger capacity of resisting vertical earthquake than that on rigid foundation, and the opposite is the case of resisting horizontal earthquake. The story drifts on soft foundation under both horizontal and vertical earthquake action were much less than that on rigid foundation. Che-Ani et al. [2] conducted a study on complaints received by the Local Authority in Selangor, Malaysia about the impact of adjacent new construction works on the conditions of existing buildings in the neighboring vicinity. For that a case study consisting of 28 units of two-storey terrace houses were considered. Inspection form was used as a main instrument along with a basic surveying tool with a digital caliper for measuring cracks. Crack-width data were collected and analyzed using the Statistical Package Social Sciences (SPSS) software. It showed that the existing building had an impact from the adjacent new construction works. Sebastian Bryson and Kotheimer [3] presented the results of three-dimensional finite element analyses of a building adjacent to an excavation, which were used to evaluate the magnitude of strain that developed in the interior walls in response to the excavation-related ground movements. The analyses showed that the initial cracking observed in selected infill wall panels could not have occurred solely in response to excavation-related deformations. Consequently, it was found that the wall panels cracked as a result of a combination of strains induced in the structure from self-weight settlement and excavation-induced displacement. Metwally et al. [4] a thorough analysis of the pipeline failure influence in different soils on adjacent buildings was investigated. ANSYS Civil FEM was used to perform numerical simulations. The variable parameters like, pipeline settlement, position of settlement, burial depth, soil stiffness, infiltration of sewage and groundwater were used to simulate the pipeline failure. Results were presented as vertical and horizontal displacements of ground beneath building.

A two bay four storey portal frame as shown in Fig. 1 is considered for the study. The width of beams and columns is taken as 230 mm and depth as 450 mm. M25 grade concrete is used for the structural elements and 30° is the angle of friction considered for soil. The analysis of structure without adjacent excavation is carried out and is treated as reference for six other cases considered. For three cases, the distance of excavation from existing frame structure is kept constant (as 0.5 m) and the depth of excavation is varied (1.5, 3, and 4.5 m). In other three cases, the depth of excavation is kept constant as 3 m and the distance of excavation from the existing frame is varied as 1, 2, and 3 m. In all the cases the width of excavation is kept constant as 3 m. In ANSYS, the element SOLID65 is used to represent the beams and columns and SOLID45 is used to represent soil structure, with linear-plastic properties used in that element. The mesh size used for meshing the frame is 0.2 and 0.6 m for soil. Meshed models of all cases considered are shown in Fig. 2.

Fig. 1 Model of frame considered

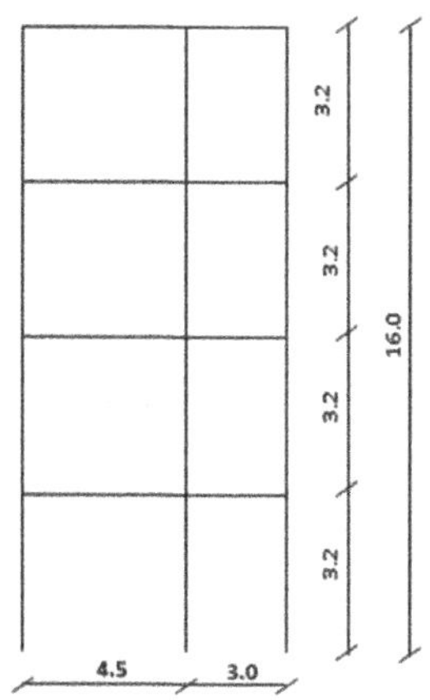

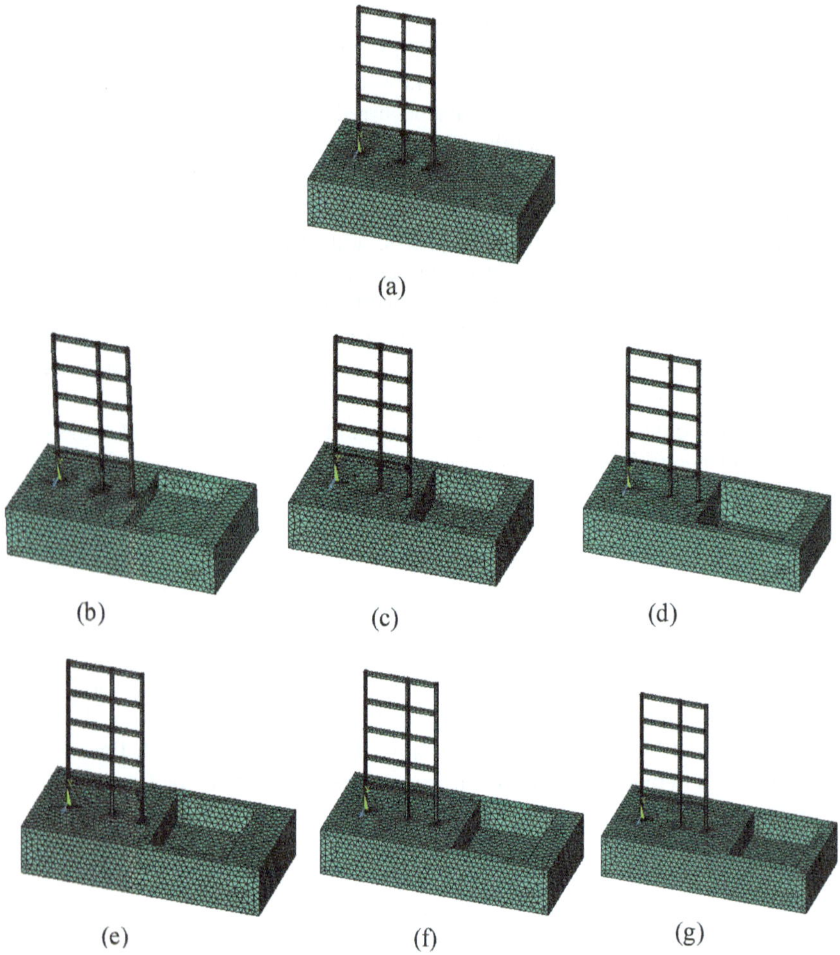

Fig. 2 Meshed models of different cases considered **a** without adjacent excavation **b**, **c**, and **d** 1.5 m deep, 3 m deep, and 4.5 m deep adjacent excavation, respectively. **e**, **f**, and **g** excavation 1 m away, 2 m away, and 3 m away from structure, respectively

2 Static Analysis

The displacement and stress variation in the structure under static load condition is tabulated in Tables 1 and 2. Contour plots of both parameters are from Figs. 3, 4, 5, 6 and 7. Results show that the displacement is more for structure with nearer and deeper adjacent excavation.

Table 1 Maximum displacement and maximum stress for different depth of excavation

Type of excavation	Maximum displacement (mm)	Maximum stress (MPa)
Without excavation	3.798	3
1.5 m depth	9.295	3.43
3 m depth	12.354	3.46
4.5 m depth	13.203	3.48

Table 2 Maximum displacement and maximum stress for different distances of excavation

Type of excavation	Maximum displacement (mm)	Maximum stress (MPa)
Without excavation	3.798	3
Excavation 1 m away	4.064	3.26
Excavation 2 m away	3.989	3.24
Excavation 3 m away	3.883	3.15

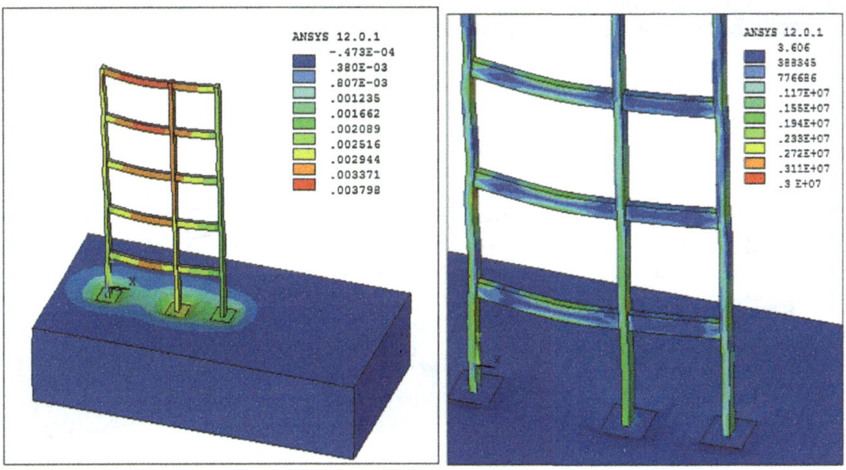

(a) Displacement (b) Stress distribution

Fig. 3 Contour plot of frame without excavation

3 Modal Analysis

With the increase in the depth of the excavation there is a decrease in the natural frequency of the structure because of the reduced stiffness of the soil near the excavated area as shown in Fig. 8. Similarly nearer the excavation, to the existing structure

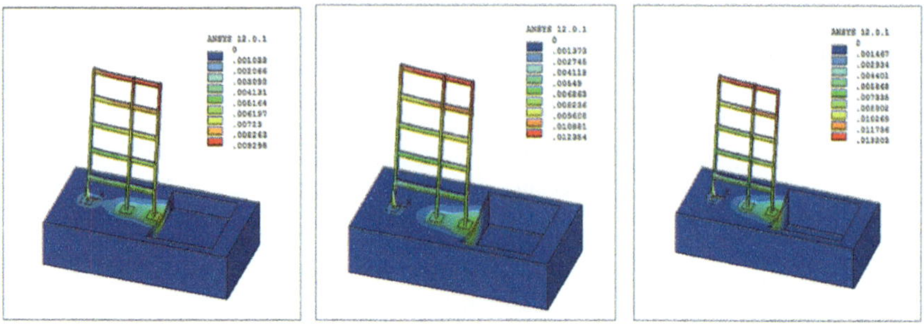

Fig. 4 Contour plot of displacement for frames with 1.5, 3, and 4.5 m deep excavation

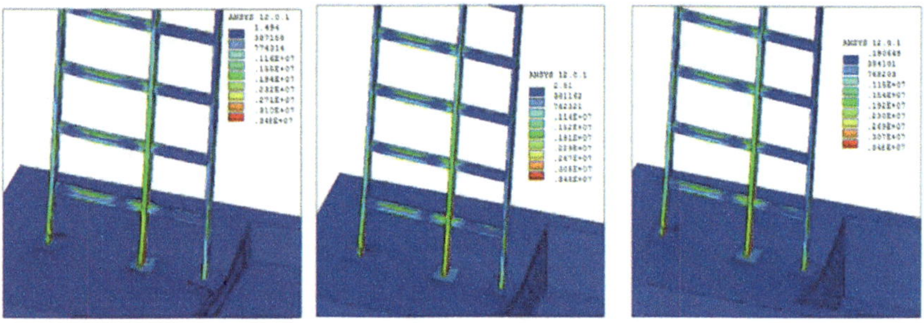

Fig. 5 Contour plot of stress variation for frames with 1.5, 3, and 4.5 m deep excavation

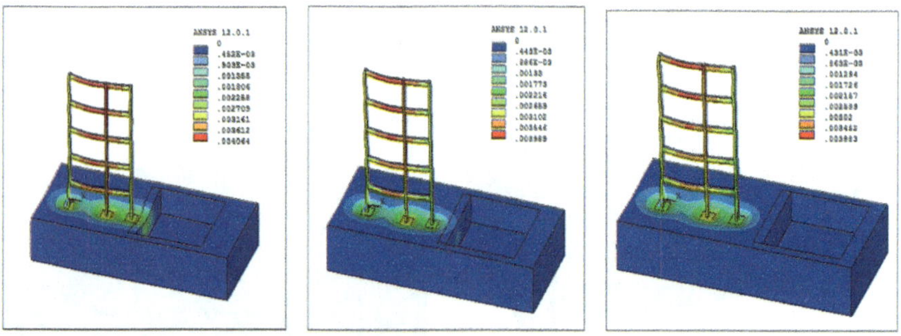

Fig. 6 Contour plot of displacement for frames with excavation 1, 2, and 3 m away

lesser is the frequency as shown in Fig. 9. This also affects the boundary condition of the adjacent column.

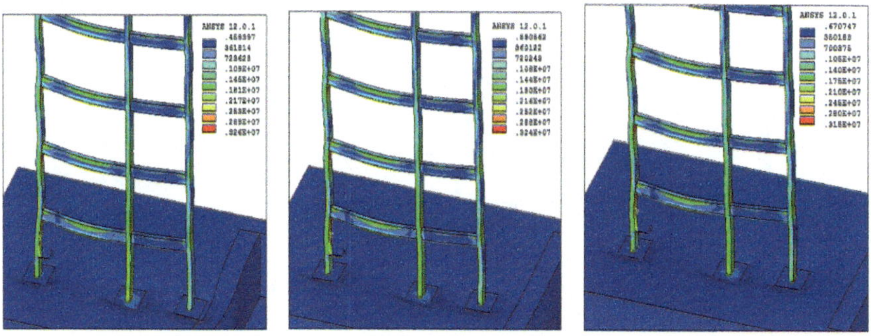

Fig. 7 Contour plot of stress variation for frames with excavation 1, 2, and 3 m away

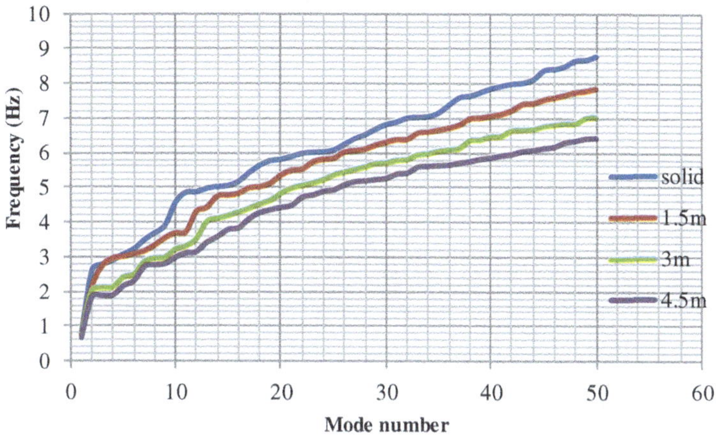

Fig. 8 Mode number versus natural frequency for different depths of excavations

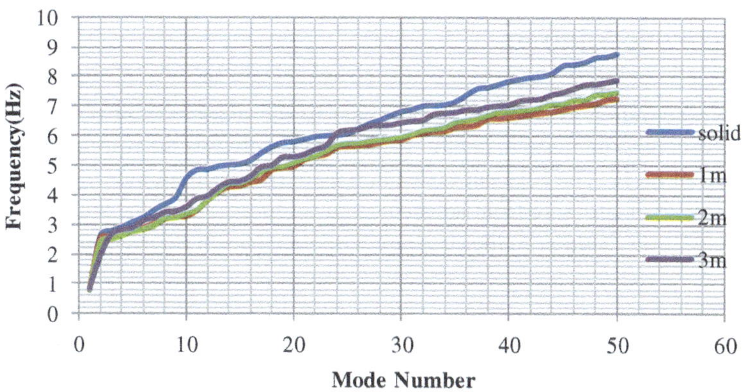

Fig. 9 Mode number versus natural frequency with excavation 1, 2, and 3 m away

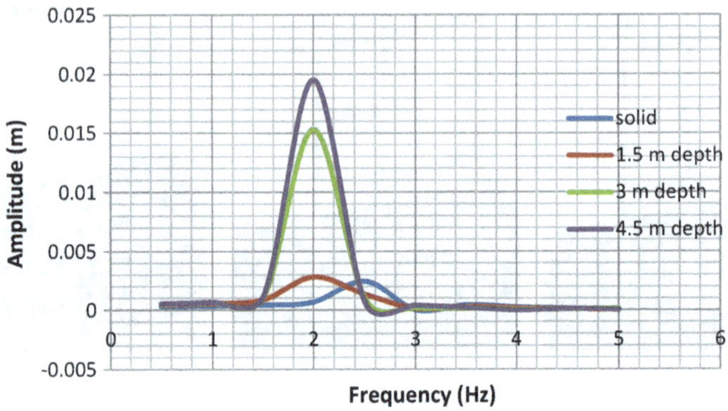

Fig. 10 Amplitude versus frequency for different depths of excavations

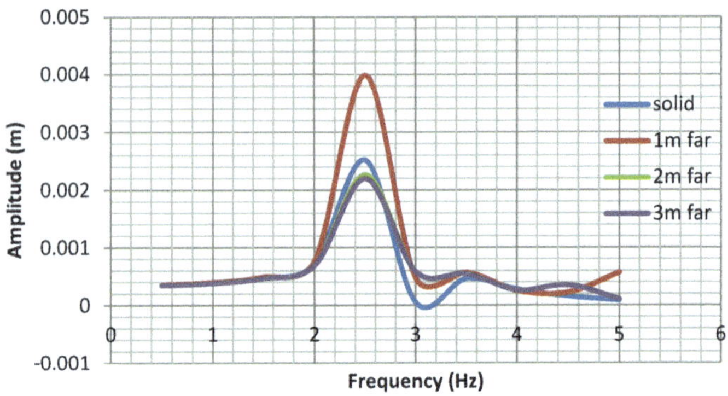

Fig. 11 Amplitude versus frequency for excavation 1, 2, and 3 m away

4 Harmonic Analysis

Depth of excavation is directly proportional to the amplitude of response of the existing structure. Increase in depth of excavation increases the response amplitude as shown in Fig. 10. The adjacent excavation increases the amplitude of the structure. If the distance of structure from adjacent excavation increases, the amplitude of the structure will vary similar to the structure without adjacent excavation as in Fig. 11.

5 Conclusions

1. It can be observed that the increase in depth of the excavation increases the maximum stress of the existing structure around 14–16% and displacements

can be increased almost 3 times displacement of structure without adjacent excavation.

2. As the distance of excavation from the existing footing increases the displacement and stress decreases.
3. The excavation made adjacent to a structure will induce a variation in the dynamic characteristics of the structure like Natural frequencies, amplitude, and Mode shapes.

References

1. Ni X, Zhang Y, Zhang S, Huang W (2011) Comparative analysis on seismic response of tall building on soft and rigid foundation. Adv Mater Res 250–253:2216–2220
2. Che-Ani A, Noor ZM, Pheng LS, Tawil NM, Tahir MM (2011) Building cracks: evidence on the impact of new construction works on existing buildings. Struct Surv 29(4):337–351
3. Sebastian Bryson L, Kotheimer MJ (2011) Cracking in walls of a building adjacent to a deep excavation. J Perform Constr Facil 25(6)
4. Metwally KG, Hussein MM, Akl AY (2011) Damage assessment of buildings due to different parameters of pipeline deterioration. Life Sci J 8(3):278–289

1D and 2D Dynamic Site Response of Landfill Site Through Numerical Analysis

Parul Rawat and Supriya Mohanty

Abstract Solid waste management and disposal become a challenge on global level. Due to this rapid growth of waste, we are losing low-lying areas and valuable lands. Therefore, it is a peak demand to utilize the waste-filled areas for future use. The objective of this study is to get the dynamic response of a landfill site by using already available data in past studies. The response study of landfill site helps to predict the deformation and amplification behavior of the MSW fill sites. So that required preventive measures and ground improvement techniques can be implemented according to the response of site. The study area used in this paper includes landfill sites of Delhi area, which are mostly designed for about 20 m but now they are overfilled with more than 50 m. The aim is to find the seismic response of the site under defined dynamic loading. The response of landfill site was numerically analyzed both in one and two dimensions with the help of Cyclic1D and Plaxis2D software, respectively. The analysis performed is based on finite element method under the earthquake of magnitude of 6.5 with duration of 39.9 s and PGA of 0.3 g.

Keywords Landfill · Seismic response · Cyclic1D · Plaxis2D

1 Introduction

India is a fast developing country with neglecting its side effects like generation of wastes. It generates about 62 million tons of municipal solid waste (MSW) per annum which is likely to increase by 165 million tons by the end of 2031 [1]. To fulfill this requirement we have only 59 constructed landfill sites and 376 are under planning [2]. The major part of landfill site consists of MSW, which is a composite mixture of household, commercial, industrial, medical waste, etc. This not only leads

P. Rawat (✉) · S. Mohanty
Department of Civil Engineering, Indian Institute of Technology (BHU) Varanasi, Varanasi 221005, Uttar Pradesh, India
e-mail: parulrawat.rs.civ18@itbhu.ac.in

S. Mohanty
e-mail: supriya.civ@iitbhu.ac.in

T. G. Sitharam et al. (eds.), *Local Site Effects and Ground Failures*, Lecture Notes in Civil Engineering 117, https://doi.org/10.1007/978-981-15-9984-2_8

to the piling of this waste in landfill sites but also hinders in our economy. The best solution could be to see its potential use and reduce the waste by converting it into energy; landfill mining can also be one of the solutions. However, generally sanitary landfills are designed in such a way that once they achieve designed capacity, they are covered and can be used for future use such as public parks or construction. For this particular study Delhi region is considered which comes under seismic zone IV as per the seismic hazard map of India [3]. The two major landfill sites of Delhi are Ghazipur and Okhla, which are already filled almost thrice there design capacity. The Ghazipur dumpsite covers an area of about 28 ha, which is about 680 m long and 360 m wide. At present, the site receives about 2600 Metric tons of solid waste per day. Slopes forming the periphery of dump varied between 37° and 75° with horizontal. The Okhla site spreads in an area of about 16.2 ha and is about 705 m long and 303 m wide. The site receives about 1100 Metric tons of solid waste per day. The periphery slope of the dump varied between 36° and 46° [4].

1.1 Past Study on Dynamic Analysis of MSW

There are various past as well as ongoing field and laboratory studies on MSW and landfill sites have been carried out by researchers. As stability of landfills in active seismic zone is a very important issue, which may lead to contamination of groundwater, slope failure as well as other environmental hazards. Therefore, to eliminate these effects, static and dynamic properties of waste are required to be understood properly before design of landfills. There are various methods to find the composition, unit weight, strength and dynamic parameters of landfill sites all over the world, but as MSW itself is a heterogeneous material, the properties differ from place to place. It can be seen that unit weight profile is more important parameter than average unit weight for seismic analysis [5]. Zekkos [6] from various laboratory and field data analyzed that unit weight is function of waste composition, compaction effort, depth-dependent confining pressure that is acting on landfill and gave a hyperbolic equation for unit weight of landfills. It was also investigated from large-scale cyclic triaxial test that the composition of waste has significant effects on dynamic properties of MSW [7]. Waste materials are generally soft and show nonlinear behavior but it shows significant shear resistance due to presence of fibrous material [8]. To understand the seismic behavior of MSW with single clay liner landfill founded on sand, a dynamic centrifuge testing was conducted at various earthquake intensities and frequency [9]. The site-specific one-dimensional seismic response study was conducted for Delhi Okhla landfill site. For this both field and laboratory data were used for seismic analysis. The amplification phenomenon observed from the study was useful for the design of cover system, assessment of permanent displacements and global stability for the dumpsite during closure [10]. Shear modulus is major parameter to evaluate the dynamic property of MSW and low strain shear modulus depends on its mass density and shear wave velocity. The study conducted on Bangalore landfill site shows dynamic shear modulus of the waste increases with increasing confining pressure and shear strains for both dry and dumped wastes [11].

1.2 Objective of Study

The objective of the study is to get the dynamic response of a landfill site by using already available data in past studies. The response study of landfill site has been carried out with the help of numerical technique (1D and 2D seismic analysis) to predict the deformation and amplification behavior of the MSW fill, so that required preventive measures and ground improvement techniques can be implemented according to the response of site.

2 Brief Discussion About Numerical Methods

2.1 Modeling Using Cyclic1D

Cyclic1D is a user interface finite element program for execution of the site response, which includes analysis of liquefaction-induced deformations [12]. The program operates in time domain with linear and nonlinear analysis for dry as well as saturated strata. In most of the cases, one-dimensional (1D) site response analysis is performed to assess the effect of soil conditions on ground shaking because vertically propagating and horizontally polarized shear waves dominate the earthquake ground motion wave field [13]. The number of elements in Cyclic1D can be chosen from 10 to 400. For the present study in 1D the maximum elements taken after convergence analysis was 360.

2.2 Modeling Using Plaxis2D

Plaxis2D is a popular two-dimensional (2D) finite element software for the analysis of deformation, stability and groundwater flow. The finite element model generated is based upon geometry model, which involves the composition of surface, lines and points. In FEM, the continuum is divided into number of elements, which consist of nodes, in Plaxis it can be 6-noded or 15-noded triangular element [14].

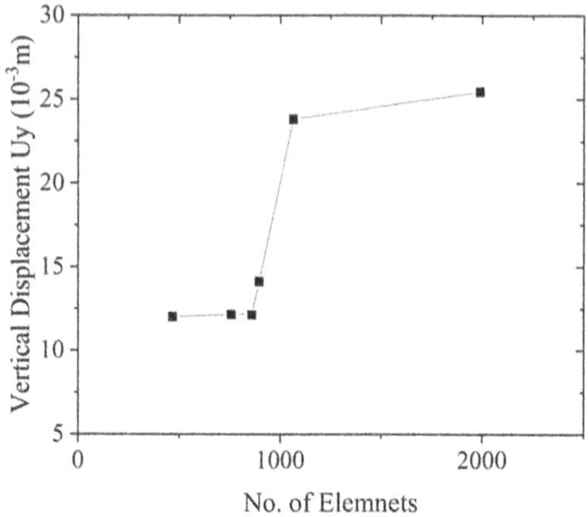

Fig. 1 Convergence analysis for Plaxis2D

2.2.1 Meshing

Once geometry and parameters are defined, the model was analyzed for convergence and mesh was generated. For this particular study, the model converged at 1989 number of elements with relative element size of 0.7 m (Fig. 1).

2.2.2 Stage Construction Phase

The landfill site was treated as embankment, so it is assumed to be constructed in 5 stages and consolidated for 4 years each, i.e., total of 20 year. The last stage is added for the input of seismic load.

3 Model Description

The two major landfill sites of Delhi area (Ghazipur and Okhla) were studied and a generalized profile was considered accordingly. The MSW landfill model was considered as an embankment with 50 m fill height and side slope of 45°. The model consists of two layers of subsoil (fine sand and silty sand) of 20 m each with average groundwater at 12 m below the ground level.

3.1 MSW Parameters

The MSW parameters of two major landfill sites of Delhi were taken from past studies [4, 10, 15]. Hardening soil model was considered for the study as it is superior to any linear elastic model and produce more realistic results as well as capable of modeling modulus reduction with increase in strain. The Young's modulus of MSW for this analysis was correlated from SPT data of the site [16].

3.2 Subsoil Parameters

The general site profile of Delhi was considered in study from past data [17, 18]. The parameters considered are shown in Table 1.

3.3 Input Earthquake Parameters

The input motion selected was Uttarkashi earthquake of magnitude 6.5 and total duration of 39.9 s (Fig. 2) with peak ground acceleration (PGA) value of 3.04 m/s^2.

Table 1 Model parameters of MSW landfill and subsoil

	MSW	Fine sand	Silty sand
Model	Hardening soil	Hardening soil	Hardening soil
Material behavior	Drained	Drained	Drained
γ_{Dry} (kN/m^3)	11	17	18
γ_{Sat} (kN/m^3)	14	20	21
e_{int}	0.8	0.5	0.5
E_{50}^{ref} (kN/m^2)	19122.96	23045.62	30890.94
E_{oed}^{ref} (kN/m^2)	19122.96	23045.62	30890.94
E_{ur}^{ref} (kN/m^2)	40.00E3	48.00E3	62.00E3
ν'	0.3	0.3	0.3
c'	17	0	20
Φ (°)	32	33	30
Data set	Standard	Standard	Standard
	Medium	Medium	Medium Fine
$k_x = k_y$ (m/day)	1.04544E−3	10E−2	10E−4
Damping ratio (%)	15	10	10
Damping parameters			
α	0.628319	0.418879	0.418879
β	0.0318310	0.0212207	0.0212207

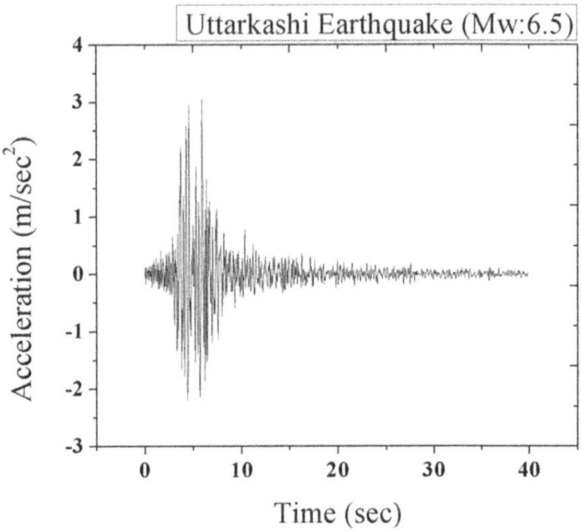

Fig. 2 Input motion used for the analysis (Uttarkashi Earthquake)

4 Results and Discussions

For 1D analysis using Cyclic1D, responses are presented along the depth of the MSW fill. For 2D analysis in Plaxis2D, seven sections at X = 20, 50, 150, 300, 450, 550 and 580 m all over the landfill are considered along which all the responses are presented below.

4.1 Displacements

1D Analysis: The results obtained from the analysis done using Cyclic1D show that vertical displacement is maximum for 50 m MSW landfill and it linearly reduces to zero at the bottom of the deposit (Fig. 3a). The low density of landfill material causes compression of the fill. Similarly, the horizontal displacement can also be seen maximum at the top layers (MSW fill layer) and reduces as the depth increases (Fig. 3b).

2D Analysis: The results obtained from the analysis conducted using Plaxis2D at 7 sections shows that vertical and horizontal displacements are more toward the end of the fill near the slopes (at section X = 20, 50, 550 and 580 m) at the end of dynamic phase (Fig. 4). The displacement reduces as the depth of fill increases. Similar kind of trends is shown in Cyclic1D results (Fig. 3). However, Cyclic1D analysis was done for single soil column, which shows less displacement as compared to Plaxis2D results.

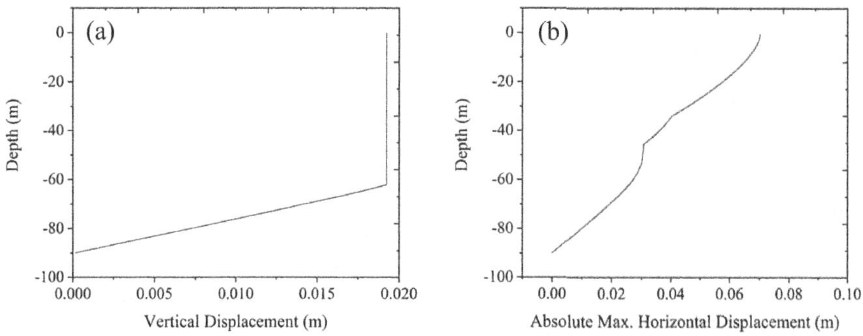

Fig. 3 **a** Vertical displacement **b** Absolute max. horizontal displacement variation with depth in 1D analysis

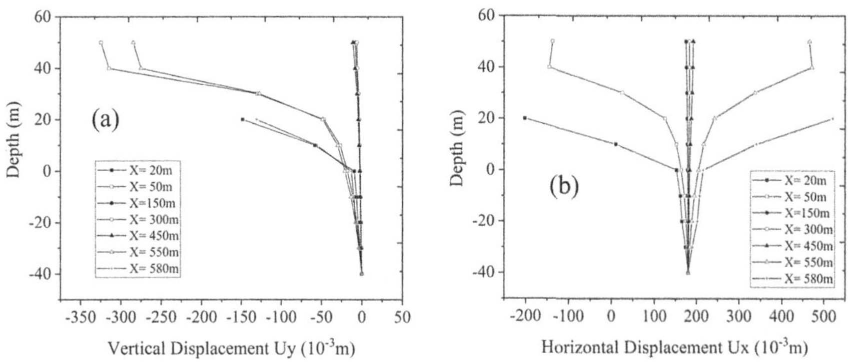

Fig. 4 **a** Vertical displacement variation with depth **b** Horizontal displacement variation with depth for different sections of landfill in 2D analysis

4.2 Stresses

1D Analysis: The drawback of 1D analysis was that only a single soil column can be considered, so the column was assumed to be at the center of the profile. The shear stress variation shows increment with depth and maximum shear stress concentration was noticed in lower layers (Fig. 5a). Similarly, effective confinement also shows increasing trend with depth (Fig. 5b).

2D Analysis: The vertical and horizontal stress increases with depth in dynamic phase as overburden pressure increase with depth (Fig. 6). The vertical stresses with depth at section X = 20 and 580 m are comparatively low, whereas shear stress increases as we move away from center of landfill toward the slope ends and low in the center sections due to confinements from both sides (Fig. 7).

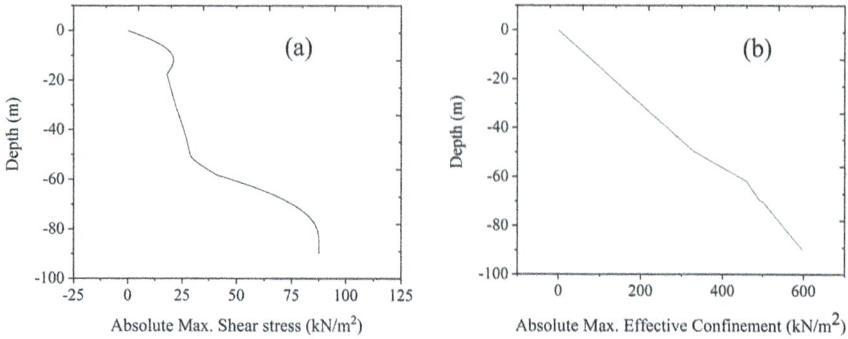

Fig. 5 **a** Absolute max. shear stress **b** Absolute max. effective confinement variation with depth in 1D analysis

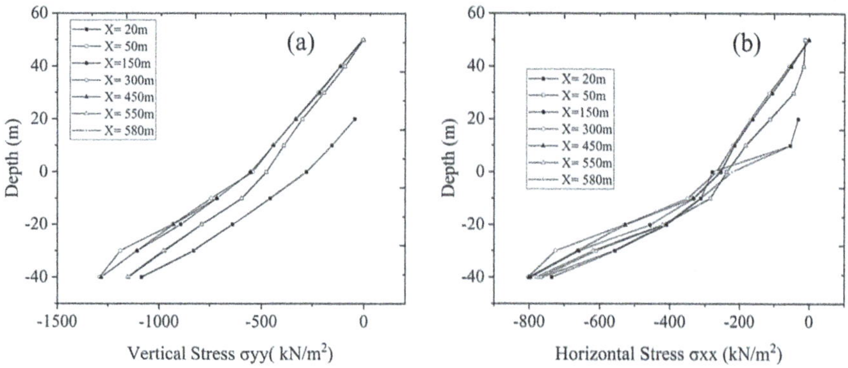

Fig. 6 **a** Vertical stress variation with depth **b** Horizontal stress variation with depth for different sections of landfill in 2D analysis

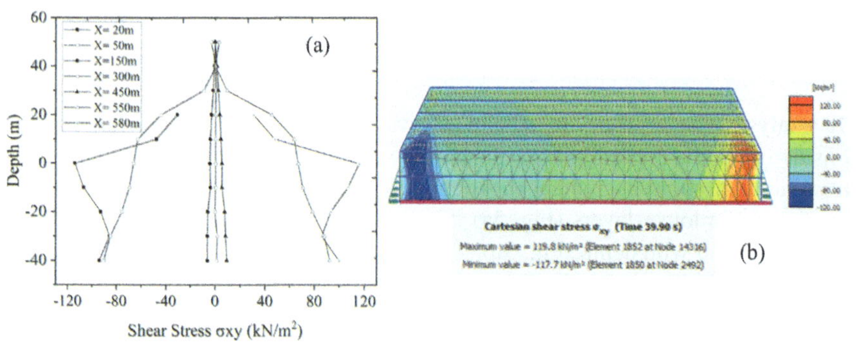

Fig. 7 **a** Shear stress variation with depth at different sections of landfill **b** Shear stress contour for 2D analysis

4.3 Acceleration

1D Analysis: The horizontal acceleration in 1D analysis can be seen amplified in top layer of fill and reduced to PGA value again at the bottom (Fig. 8). The amplification in top layer is because of low confinement, which reduces in middle and again amplifies at the bottom layer.

2D Analysis: The maximum horizontal acceleration at different sections (Fig. 9) coincide at bottom is about 2.9 m/s^2. The acceleration near the top surface of landfill can be seen more for all sections between 1.5 and 2 m/s^2, as at the top layer there is no vertical overloading of confinement.

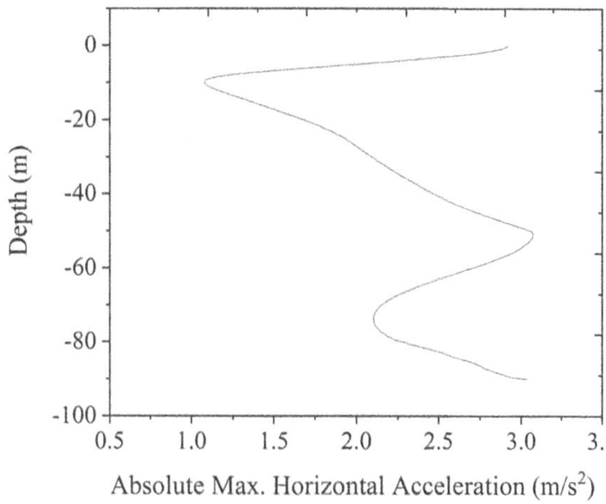

Fig. 8 Horizontal acceleration variation of landfill with depth in 1D analysis

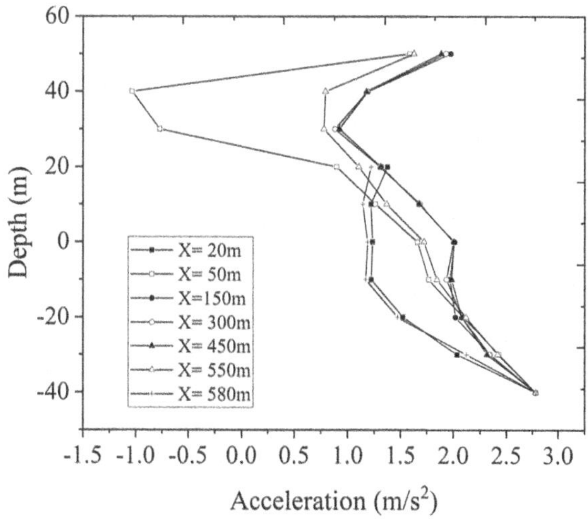

Fig. 9 Horizontal acceleration variation at different sections of landfill with depth in 2D analysis

4.4 Excess Pore Pressure

1D Analysis: The water table is considered at 12 m below the ground and in Fig. 10 for Cyclic1D, the water table starts from 62 m as 50 m is MSW landfill height and water table lies 12 m below that. The excess pore water pressure increases from water table to the bottom of the profile (Fig. 10).

2D Analysis: The excess pore pressure profile at different sections concludes that excess pore pressure increases with depth and achieves maximum values at bottom. The profile (Fig. 11) shows that excess pore pressure is more in the center portion of landfill. The maximum excess pore pressure can be seen for the middle section,

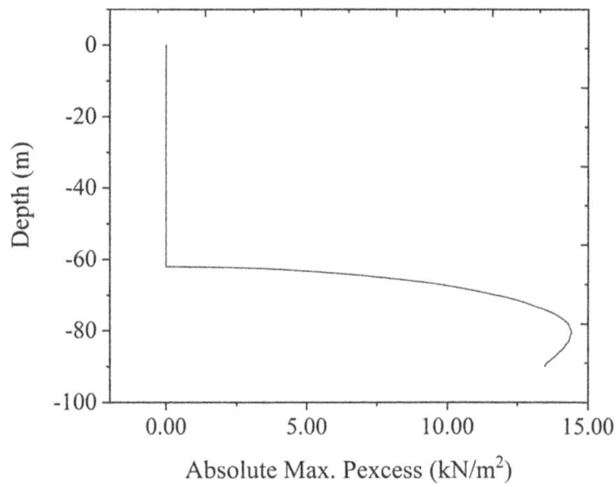

Fig. 10 Excess pore water pressure variation of landfill with depth in 1D analysis

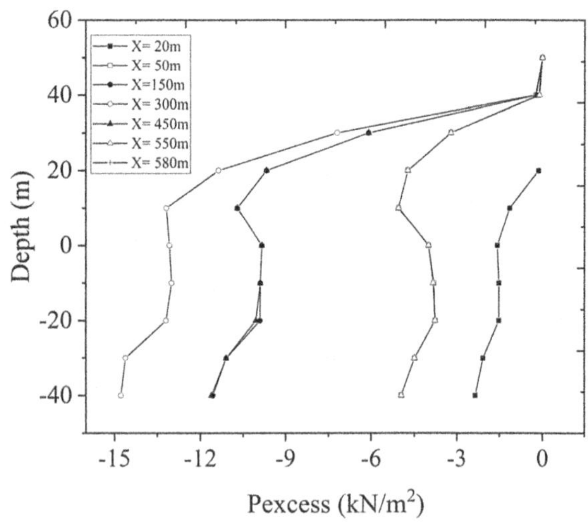

Fig. 11 Excess pore water pressure variation at different sections of landfill with depth in 2D analysis

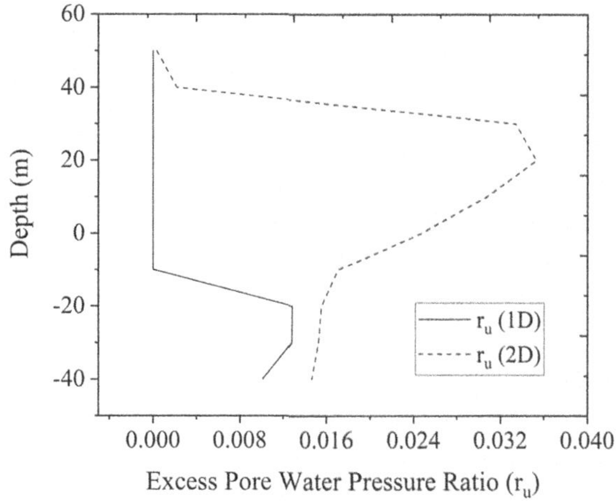

Fig. 12 Excess pore water pressure ratio variation with depth in 1D and 2D analysis

i.e., X = 300 m, as that section has more vertical as well as horizontal confinement, where as the side ends, i.e., section X = 20 and 580 m has lowest P_{excess} developed.

In Cyclic1D, the MSW landfill site was analyzed for liquefaction susceptibility by considering plasticity model into account. In Plaxis2D, hardening soil model was used which in contrast to elastic perfectly plastic model the yield surface of hardening plasticity model was not fixed in principal stress space, but it can be expanded due to plastic straining. The analysis in Cyclic1D was done in only one section or a column, which can be considered as middle section. Although the trends from both 1D and 2D analysis are same but shear stress values with depth are more for 1D analysis, if we compare with middle section of 2D analysis. The acceleration shows sudden change as material in layers changes at 50 and 70 m, whereas that sudden change can't be seen in 2D analysis.

Based on pore pressure-related criteria, soil liquefaction is defined as a state at which excess pore water pressure ratio (r_u) becomes equal to 1. This happens when pore water pressure increase becomes equal to initial effective overburden stress. Sometime this is also known as 'initial liquefaction' or 'partial liquefaction'. Figure 12 shows the variation of r_u with depth for 1D and 2D analysis (at section x = 300 m), which shows no signs of liquefaction of MSW fill as well as foundation soils.

5 Conclusions

The above results concluded the 1D and 2D seismic response of the Delhi landfill sites, which shows the amplification of acceleration in loose MSW fill and higher vertical and horizontal stresses in lower layers. The similar site response for Okhla Delhi site was shown by Ramaiah [19], variation of maximum horizontal equivalent acceleration (MHEA) with depth from 1D analysis for local earthquakes (Mw = 6.5) confirms the maximum horizontal acceleration in the top layers. The results from Cyclic1D are little conservative as, we can see the profile variations at only one section. The results from 2D analysis from Plaxis2D at middle section give almost similar results as in 1D analysis by Cyclic1D. For using the landfill sites for future use, it is required to focus on the slope stability of landfill as it is the zone of maximum shear stress. Although this modeling is not sufficient for study and design purpose, it just gives an idea about the site response during any earthquake.

References

1. Planning Commission Report (2014) Reports of the task force on waste to energy, vol I (in the context of Integrated MSW management)
2. CPCB Status report on municipal solid waste management (2013)
3. IS 1893 (2002) Criteria for earthquake resistant design of structures—Part 1: General provisions and buildings. Bureau of Indian Standards, New Delhi
4. Ramaiah BJ, Ramana GV, Datta M (2017) Mechanical characterization of municipal solid waste from two waste dumps at Delhi, India. Waste Manag 68:275–291
5. Kavazanjian Jr E, Matasovic N, Stokoe II KH, Bray JD (1996) Insitushear wave velocity of solid waste from surface wave measurements. In: Proceedings of the 2nd international congress on environmental geotechnics, vol 1, pp 97–102, Balkema, Rotterdam, Netherlands
6. Zekkos DP, Bray JD, Kavazanjian E, Matasovic N, Rathje E, Riemer M, Stokoe KH (2005) Framework for the estimation of MSW unit weight profile. In: 10th international waste management and landfill symposium, pp 3–7, Santa Margherita di Pula, Cagliari, Italy
7. Zekkos D, Bray JD, Riemer MF (2008) Shear modulus and material damping of municipal solid waste based on large-scale cyclic triaxial testing. Can Geotech J 45(1):45–58
8. Towhata I, Kawano Y, Yonai Y, Koelsh F (2004) Laboratory tests on dynamic properties of municipal wastes. In: 11th conference in soil dynamics and earthquake engineering and the 3rd international conference on earthquake geotechnical engineering, vol 1, pp 688–693
9. Thusyanthan NI, Madabhushi SPG, Singh S (2004) Modelling the seismic behaviour of municipal solid waste. In: 11th international conference on soil dynamics and earthquake engineering, Berkeley, USA
10. Ramaiah BJ, Ramana GV, Kavazanjian E Jr, Bansal BK (2016) Dynamic properties of municipal solid waste from a dump site in Delhi, India. Geo-Chicago 2016:121–130
11. Naveen BP, Sitharam TG, Sivapullaiah PV (2014) Evaluating the dynamic characteristics of municipal solid waste for geotechnical purpose. Curr Adv Civ Eng 2(1):28–34
12. Lu J, Elgamal AWM, Yang Z (2006) Cyclic1D: a computer program for seismic ground response. Department of Structural Engineering, University of California, San Diego
13. Kramer SL (1996) Geotechnical earthquake engineering. Prentice Hall, Inc., UpperSaddle River, New Jersey, p 653
14. Plaxis BV (2018) Plaxis 2D reference manual. The Netherlands
15. Ramaiah BJ, Ramana GV (2014) CPTu at a municipal solid waste site in Delhi, India. In: 3rd international symposium on cone penetration testing, pp 1083–1091, Las Vegas, Nevada, USA
16. Brahma P, Mukherjee S (2010) A realistic way to obtain equivalent Young's modulus of layered soil. In: Indian geotechnical conference, pp 305–308, Bombay, India (2010)

17. Inqualabi KQ, Garg RK, Rao KB (2017) Seismic vulnerability of urban bridges due to liquefaction using nonlinear pushover analysis and assessing parameters for damage detection. Procedia Eng 173:1739–1746
18. Datta M (2012) Geotechnology for environmental control at waste disposal sites. Indian Geotech J 42(1):1–36
19. Ramaiah BJ, Ramana GV, Bansal BK (2016) Site-specific seismic response analyses of a municipal solid waste dump site at Delhi, India. In: Proceedings of Geo-China, pp 191–198

A Study on Characteristics of Soil Profile of Guwahati City Against Different Ground Motions: 1D NonLinear Ground Response Analysis

Amar F. Siddique, Anusuya Acharjee, and Binu Sharma

Abstract One-dimensional nonlinear (NL) ground response analysis (GRA) is carried out for four locations viz. Dharapur and Azara area, VIP and Rani area, Jalukbari area and Assam Engineering College (AEC) area of Guwahati city. The analysis is done with the help of 23 boreholes using Indo-Burma earthquake, 1997 recorded at Nongpoh station and Indo-Burma earthquake, 1988 recorded at Nongstoin station with duration 52.94 s. These two input motions are further scaled up to 0.18 and 0.36 g and used for the GRA analysis. The response of the substrata obtained after the analysis has been presented in terms of peak spectral acceleration, maximum shear stress profile, strain distribution and peak ground acceleration (PGA) for all the locations irrespective of every motions. For most of the locations, soil layers with high SPT N value, generally stiffer soil, resulted in a lower value of surface PGA compared to soft soil layers. The PGA and overall energy content of the strong motion significantly affect the response of a multilayered soil profile, keeping all other parameters constant. Higher strain amplitudes have been observed at various sites and at various depths of the soil profiles which may undergo settlement or earthquake-induced liquefaction. The peak spectral acceleration obtained is much higher for 0.36 g scaled motion compared with the spectral acceleration of rocky or hard soil sites (IS: 1893–2002) for both the earthquakes which indicate sensitiveness of the locations to the induced motions.

Keywords Nonlinear ground response analysis · Peak ground acceleration · Peak spectral acceleration

---

A. F. Siddique (✉) · A. Acharjee · B. Sharma
Department of Civil Engineering, Assam Engineering College, Guwahati 781013, India
e-mail: amaribnesiddiq98@gmail.com

A. Acharjee
e-mail: anusuyaa798@gmail.com

B. Sharma
e-mail: binusharma78@gmail.com

T. G. Sitharam et al. (eds.), *Local Site Effects and Ground Failures*, Lecture Notes in Civil Engineering 117, https://doi.org/10.1007/978-981-15-9984-2_9

1 Introduction

The Guwahati city is situated beside the mighty river Brahmaputra in the Indian state of Assam which lies between coordinates of 26.183 N and 91.733E. According to IS 1893 [1], the city falls in the most elevated seismic zonal level (zone V). As per Nath et al. [4], the city had been hit by quakes of various sizes extending from 5 to 8.7 before. This has featured the need to comprehend the vulnerability of land to seismic risks caused by future quakes in Guwahati city. Due to the high seismicity level together with presence of alluvial soil in Guwahati city, one-dimensional (1D) ground response analysis (GRA) utilizing the nonlinear (NL) procedure was performed for four locations in Guwahati city viz. Dharapur and Azara, VIP and Rani, Jalukbari and Assam Engineering College (AEC) area, where bore logs up to 30 m depth are available with the required physical properties, using the software DEEPSOIL v6.0. NL analysis is done because it has more potential to simulate soil behavior accurately and is more realistic in nature.

2 Motivation and Objective(s)

Guwahati city soil consists of alluvial deposits with layers of both coarse and fine-grained soils. Hence, large modifications in earthquake waves can occur due to change in variation in soil properties near to the surface of earth. The main objective is therefore to study the response of the soil deposits of Guwahati city when subjected to strong ground motion. The aim is to determine the peak ground acceleration (PGA), maximum shear stress profile, strain distribution with depth and peak spectral acceleration of 23 boreholes spreading in four locations of Guwahati city. The boreholes(BH) considered for AEC area are BH01, BH03, BH23, BH136 and BH143; for Dharapur and Azara area the boreholes considered are BH07, BH08, BH15, BH16, BH17 and BH18; for VIP and Rani area BH09, BH10, BH11, BH12 and BH13 are considered and for Jalukbari area BH02, BH04, BH05, BH19, BH21 and BH26. Response spectrum is dealt in detail in this work to see the spectral acceleration behavior with respect to time. The response spectra obtained with the input motion from 1997 Indo-Burma earthquake and 1988 Indo-Burma earthquake in addition with the response spectra obtained after scaling the input motion are compared with the response spectra of rocky or hard soil sites (IS: 1893 [1]). Since response spectra of a site are more important to design buildings hence these findings will have considerable influence in building design and estimation of induced effects of earthquakes.

3 Methodology

The present study is accompanied with few steps which are utilized during the analysis. The steps include: (1) to gather information, (2) to model them for the software used, (3) to execute the program and (4) to interpret the outcomes. 1D ground response analysis can be done with the help of three approaches viz. Linear, Equivalent Linear (EL) and Nonlinear (NL). Linear analysis can be done in both frequency domain and time domain and it is one of the simplest processes out of all the analysis. But the main disadvantage in this approach is that the soil behavior is assumed linear in nature and the soil properties (shear modulus and damping ratio) are constant throughout the analysis. To overcome the limitation of linear analysis, equivalent linear method was developed to analyze the nonlinear response of soil using frequency domain with the aid of linear transfer functions. The nonlinear behavior of the soil is modeled in terms of equivalent linear properties corresponding to effective shear strain using iterative procedure (Kramer [3]). However, this technique is computationally helpful and gives sensible outcomes but it is unfit to speak to the adjustment in soil stiffness that really happens amid the seismic tremor. Henceforth, nonlinear ground response analysis is performed to obtain the realistic stress–strain behavior of soil analyzed in time domain with the help of numerical integration. The current study adopts the nonlinear analysis with the help of economically accessible program, i.e.DEEPSOIL v6 software. The modulus reduction curve and damping curves are absent for the site, therefore, for sandy soil, Seed and Idriss (mean limit) curve and for clay, Vucetic and Dobry [6] curve is used for the nonlinear GRA analysis. The curves are fitted with the help of MRDF procedure. Kodner and Zelasko [2] stress–strain model is used for the analysis including the Mashing criteria. The input parameter shear wave velocity can be used instead of shear stiffness to represent the stiffness of the soil layer during the analysis in the DEEPSOIL v6 software. The correlation given in Eq. (1) by Sharma and Rahman [5] is used to determine the shear wave velocity for all types of soil.

$$V_s = 74.639 * N^{0.3876} \tag{1}$$

4 Soil Study

The study of soil of a site is considered important to understand the type of soil present and to know the index properties of the soil, depth of water table and many other characteristics. For the present study, information of twenty three (23) boreholes up to 30 m depth were taken from a project sponsored by the Department of Science and Technology, India. The borehole data consist of SPT N values, depth of water table and other engineering properties of the soils. Laboratory experiments with the help of undisturbed and disturbed samples led to determination of engineering properties of the soil. The fine grained portion for the most part comprises of soils of arrangement classification CL, CI and CH as indicated by the Indian standard soil classification. In a couple of areas inorganic soil of characterization ML and CL-ML and nonplastic inorganic soil were additionally experienced. The coarse grained fraction is for the most part of grouping SP, SW, SC, SM, SP-SC.

5 Input Motion

The recorded acceleration time histories of Indo-Burma earthquake, 1997 and Indo-Burma earthquake, 1988 recorded at rocky sites have been selected as the input motions for the ground response analysis which are further scaled to peak bedrock acceleration value of 0.18 g and 0.36 g. The motions that are selected have a very low value of bedrock peak horizontal acceleration (weak motions). These weak motions do not lead to any harmful effect to the structures. In order to see the vulnerability and risk of the structures the motions are scaled up to get a strong motion. Table 1 depicts the input data used for the current study. After the input motions are scaled up, the energy content increases and all other parameters remain constant.

6 Results and Discussion

One-dimensional nonlinear ground response analysis (GRA) has been done for Dharapur and Azara area, VIP and Rani area, Jalukbari area and Assam Engineering College (AEC) area. The results after the analysis are obtained in terms of PGA with depth, maximum shear stress profile, strain (%) profile and response spectra. The PGA value gets amplified at the ground surface. The amplification factor for all the locations are also reported which is defined as the ratio of surface PGA to the input bedrock PGA. Table 2 represents the summary of the range of surface PGA, amplification factor, strain values, maximum shear stress and the maximum spectral acceleration for all the input motions applied to the four locations. It is found that for the input motion 0.04 g and scaled up motion 0.18 and 0.36 g recorded at Nongpoh (NGPH) station for Indo-Burma Earthquake (IND-BM EQR), 1997, the surface PGA

Table 1 Details of input motion

Earthquake	Recording station	Description
1997 Indo-Burma	Nongpoh	M_W 6, depth 90 km,
		PGA 0.04 g, duration 47.64 s, site class A
1988 Indo-Burma	Nongstoin	M_W 7.2, depth 34 km,
		PGA 0.05 g, duration 52.94 s, site class A

Table 2 Results of ground response analysis

Earthquake	Input Motion	Surface PGA (g)	Amplification Factor	Maximum Strain (%)	Stress Ratio	Spectral Acceleration(g)
AEC area						
1997 IND-BM	0.04 g	0.10–0.18	2.60–4.52	0.00–0.04	0.10–0.28	1.17
	0.18 g	0.25–0.45	1.39–2.50	0.00–0.16	0.25–0.95	3.11
	0.36 g	0.47–0.80	1.31–2.24	0.01–0.33	0.46–1.35	4.63
	0.05 g	0.08–0.14	1.78–2.98	0.00–0.05	0.08–0.25	0.83
1988 IND-BM	0.18 g	0.28–0.43	1.58–2.43	0.00–0.18	0.36–0.76	2.69
	0.36 g	0.45–0.74	1.25–2.08	0.01–7.44	0.49–1.30	4.47
Dharapur and Azara area						
1997 IND-BM	0.04 g	0.07–0.15	1.75–3.75	0.00–0.05	0.07–0.15	1.00
	0.18 g	0.22–0.37	1.27- 2.11	0.01–0.81	0.17–0.38	2.49
	0.36 g	0.37–0.52	1.05–1.44	0.01–1.74	0.23–0.67	3.80
	0.05 g	0.09–0.16	1.86–3.32	0.00–0.05	0.09–0.21	0.88
1988 IND-BM	0.18 g	0.22–0.45	1.27–2.50	0.01–4.22	0.23–0.44	2.53
	0.36 g	0.39–0.62	1.11–1.72	0.02–30.0	0.18–0.65	3.48
VIP and Rani area						
1997 IND-BM	0.04 g	0.08–0.18	2.17–4.65	0.00–0.05	0.08–0.20	0.97
	0.18 g	0.31–0.63	1.74–3.52	0.00–0.33	0.32–0.63	3.76
	0.36 g	0.37–0.76	1.05–2.11	0.01–1.68	0.38–0.91	4.28
	0.05 g	0.08–0.15	1.78–3.12	0.00–0.04	0.09- 0.18	0.66
1988 IND-BM	0.18 g	0.24–0.46	1.37–2.60	0.00–1.69	0.26–0.59	2.05
	0.36 g	0.38–0.53	1.08–1.49	0.01–9.19	0.31–0.83	3.28
Jalukbari area						
1997 IND-BM	0.04 g	0.07–0.18	1.97–4.65	0.00–0.06	0.08–0.34	1.10
	0.18 g	0.22–0.43	1.23–2.41	0.00–0.24	0.18–0.67	2.61
	0.36 g	0.38–0.57	1.08–1.59	0.01–0.61	0.28–0.88	3.64
	0.05 g	0.08–0.15	1.70–3.08	0.00–0.13	0.11–0.28	0.95
1988 IND-BM	0.18 g	0.20–0.44	1.16–2.48	0.01–9.60	0.22–0.60	2.17
	0.36 g	0.43–0.65	1.22–1.72	0.02–48.5	0.30–0.86	3.32

values vary from 0.10 to 0.80 g for AEC area, 0.07–0.52 g for Dharapur and Azara area, 0.08–0.76 g for VIP and Rani area, 0.07–0.57 g for Jalukbari area. Similarly, for the input motion 0.05 g and scaled up motion 0.18 g and 0.36 g recorded at Nongstoin(NGSTN) station for Indo-Burma Earthquake (IND-BM EQR), 1988, the surface PGA values vary from 0.08 to 0.74 g for AEC area, 0.09–0.62 g for Dharapur and Azara area, 0.08–0.53 g for VIP and Rani area, 0.08–0.65 g for Jalukbari area. Also, the amplification factor shows the range correspondingly as per the variation of the surface PGA. Moreover, as per the input motion strain value and maximum stress ratio (the ratio of shear stress to the effective vertical stress of the soil) value also showed variation and it is found that for the former earthquake, the strain value ranges from 0.00 to 0.33% and stress ratio value ranges from 0.10% to 1.35% for AEC area, 0.00–1.74% and 0.07–0.67% for Dharapur and Azara area, 0.00–1.68% and 0.08–0.91% for VIP and Rani area, 0.00–0.61% and 0.08–0.88% for Jalukbari area. For the second earthquake, the strain value ranges from 0.00% to 7.44% and maximum stress ratio value ranges from 0.88 to 1.33% for AEC area, 0.00–30% and 0.09–0.65% for Dharapur and Azara area, 0.00–9.19% and 0.09–0.83% for VIP and Rani area, 0.00–48.5% and 0.11–0.86% for Jalukbari area.

Of the four locations, the result of AEC area is presented in this paper with the help of graphical representations and similar results are obtained for all other locations.

For AEC area, Fig. 1 shows the comparison of peak horizontal acceleration with respect to depth for the input motion of Indo-Burma Earthquake, 1997 and Fig. 2 represents the comparison of peak horizontal acceleration with respect to depth for the input motion of Indo-Burma Earthquake, 1988. It is found that soft soil layer shows higher surface PGA in comparison to stiff soil layer. Thus soft soil will amplify the motion because soft soil has a very low SPT N value (N < 10) and low shear wave velocity than stiff soil. As the value of peak bedrock acceleration (PBRA) increases after the input motion is scaled up, the value of surface PGA also increases because the energy content increases keeping the other parameters constant. From Figs. 1 and 2, it is observed that lower surface PGA and amplification factors are obtained for Indo-Burma Earthquake, 1988 recorded at Nongstoin station than Indo-Burma Earthquake, 1997 recorded at Nongpoh station. This is due to the reason

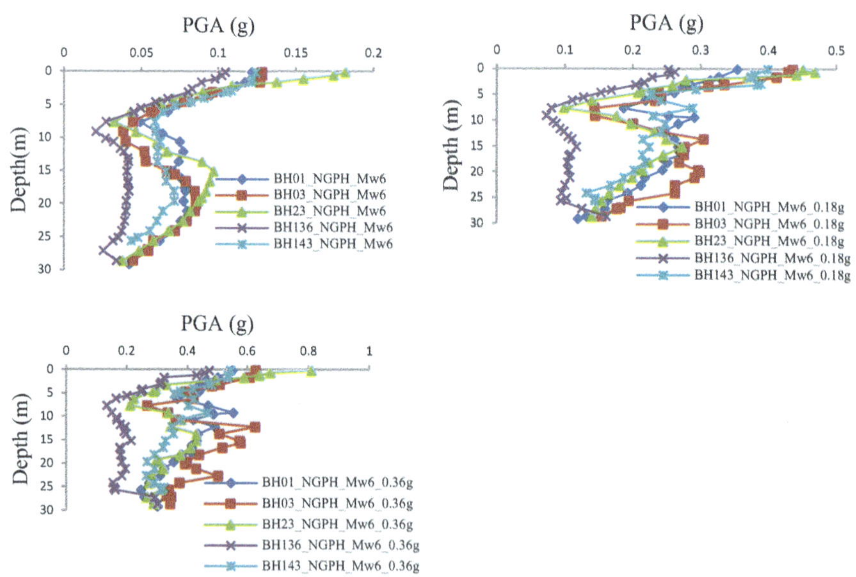

Fig. 1 PGA distribution for input motion 0.04, 0.18 and 0.36 g for AEC area

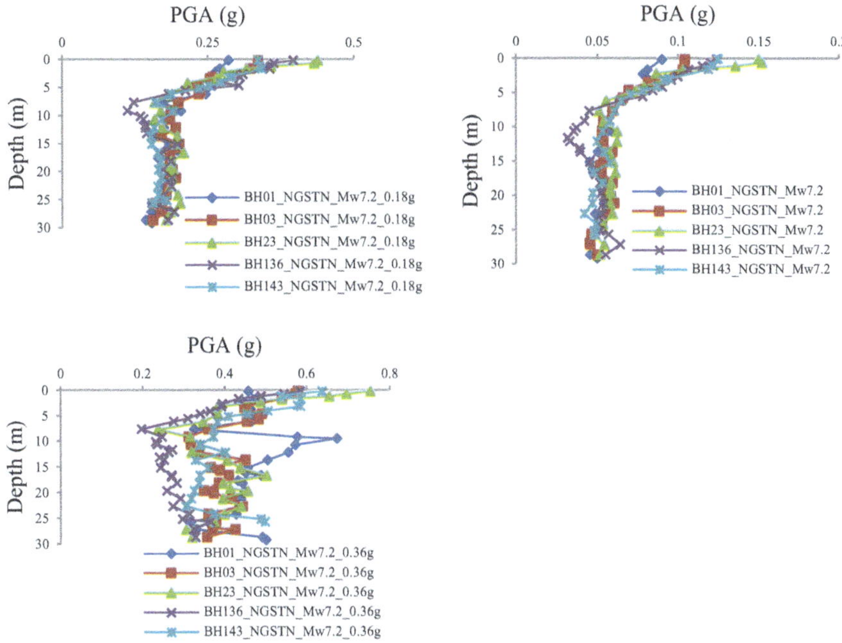

Fig. 2 PGA distribution for input motion 0.05, 0.18 and 0.36 g for AEC area

that though both the motion has almost same bedrock PGA but the motion recorded at Nongstoin station has higher hypocentral distance (400.2 km) than the motion recorded at Nongpoh station (124.1 km).

Similarly, the strain profile of the AEC area is shown in Fig. 3 for the former earthquake and in Fig. 4 for the later earthquake. It is observed that for few locations, the layer having very soft soil shows higher strain value. From both the figures, it can be concluded that when the input motion is scaled up to higher values then the shear strain value also increases correspondingly with respect to depth and follows a similar pattern. Also, it is observed that the depth showing the highest strain value for a particular borehole for a given input motion may vary with the scaled motion. Because of lower hypocentral distance of motion at Nongpoh station, it is observed that the strain range is more for motion recorded at Nongpoh station than Nongstoin station.

The maximum stress ratio distribution is shown in Fig. 5 that represents the effects of local site and strong motion characteristics for input motion 0.04 g and scaled motion 0.18 and 0.36 g recorded at Nongpoh station for Indo-Burma Earthquake, 1997 and Fig. 6 shows the distribution of shear stress ratio with respect to depth for the input motion 0.05 g and scaled up motion 0.18 and 0.36 g recorded at Nongstoin station for Indo-Burma Earthquake, 1988 for AEC locality. From Figs. 5 and 6 we find that shear stress ratio also increases as the input motion gets scaled up to higher. At depths from 3 to 5 m consisting of stiffer soil having higher SPT N value, higher

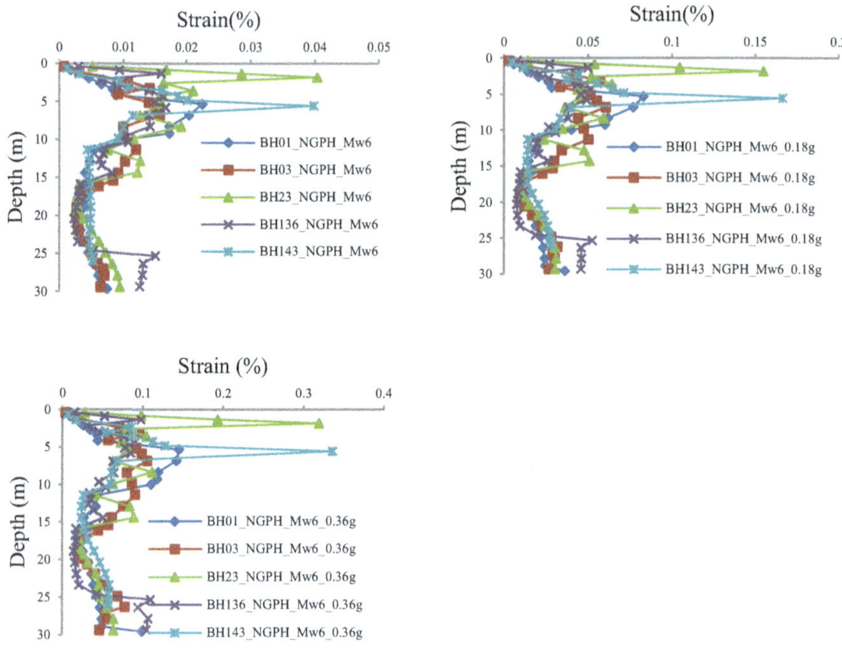

Fig. 3 Shear strain profile of input motion 0.04, 0.18 and 0.36 g for AEC area

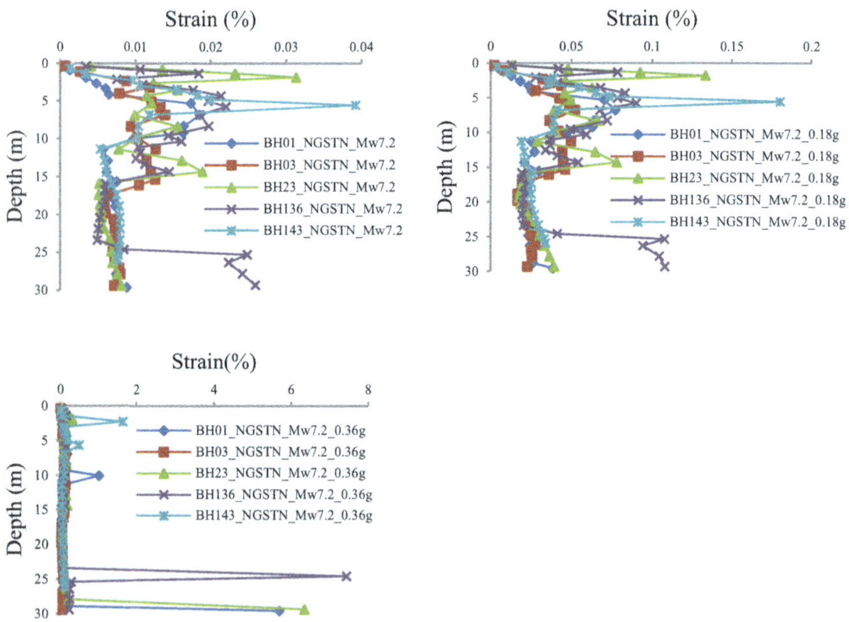

Fig. 4 Shear strain profile of input motion 0.05, 0.18 and 0.36 g for AEC area

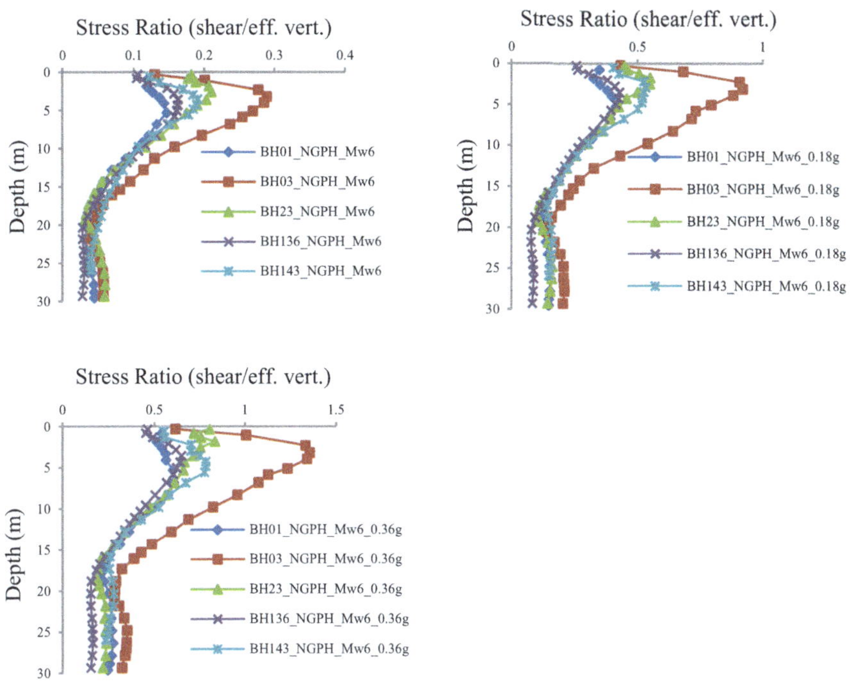

Fig. 5 Distribution of stress ratio with depth for input motion 0.04, 0.18 and 0.36 g

maximum stress ratio is obtained for majority of the sites. A higher stress ratio range is obtained for site due to motion recorded at Nongpoh station than Nongstoin station.

Response spectrum holds a significant role to understand how the structure responds when subjected to strong ground motions for a particular duration. It is important to note that response for a particular location will not be same for a specific duration when the earthquake is different. Therefore, for geotechnical engineers and structural engineers, the reflection of response spectra obtained from the analysis can be utilized to design the structures to be built in specific locations, to assure that the structure can be safe in future when subjected to strong seismic tremor. Response spectrum is described as spectral acceleration with respect to duration of the earthquake for 5% damping. From Table 2, it is seen that in AEC area, for the former earthquake, the maximum spectral acceleration value is 1.17 for 0.04 g motion while for scaled up 0.18 and 0.36 g motion the values are 3.11 and 4.63, respectively, and for the later earthquake the maximum spectral acceleration value is 0.83 for 0.05 g motion while for scaled up 0.18 and 0.36 g motion the values are 2.69 g and 4.47 g, respectively. Similarly, a very high value of spectral acceleration is observed in 0.36 g scaled up input motion for all other locations. It is observed that the maximum spectral acceleration values for 0.36 g in case of Indo-Burma Earthquake 1997, are 3.80, 4.28 and 3.64 g and for Indo-Burma Earthquake 1988, the maximum values are 3.48 g, 3.28 g and 3.32 g for Dharapur and Azara area, VIP and Rani area and

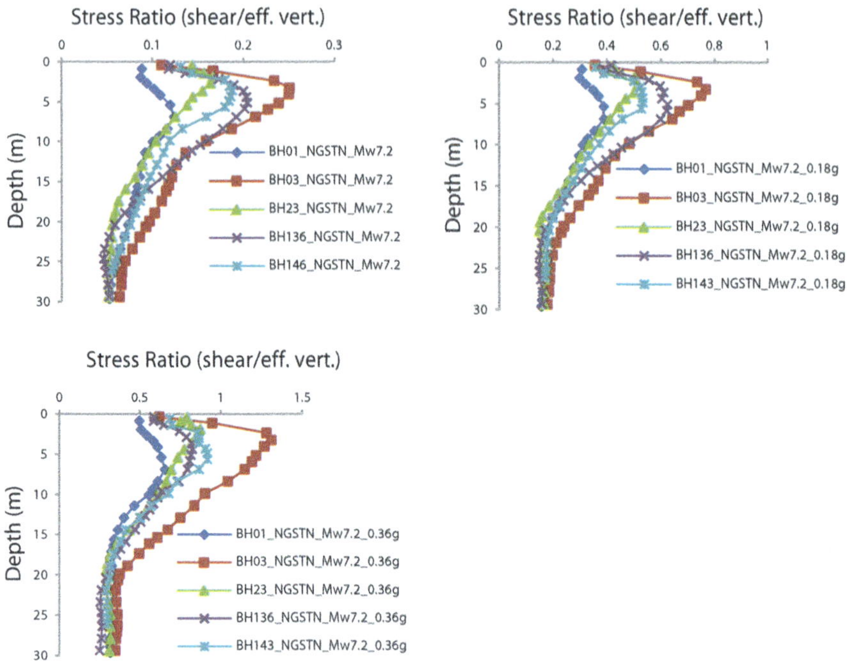

Fig. 6 Distribution of stress ratio with depth for input motion 0.05, 0.18 and 0.36 g

Jalukbari area, respectively. Figures 7 and 8 show the spectral acceleration variation with respect to time period for the respective input motions and compared with the response spectra given by IS 1893 [1] for rock or hard soil in AEC area. It is found that when the input motion is 0.36 g for both the earthquake of different magnitude, the response is very high compared to the other two motions and also it crosses the response spectra given by IS code. Hence, the area will be highly sensitive for that particular time period when the seismic tremor will have PBRA 0.36 g. The motion recorded at Nongpoh station gives a higher spectral acceleration value than motion recorded at Nongstoin station because of lower hypocentral distance of Nongpoh station.

Figure 9 shows the response spectra for 0.36 g motion of Dharapur and Azara area, VIP and Rani area and Jalukbari area for the former earthquake taken into consideration and similar results are obtained for other motions which are not shown in this paper.

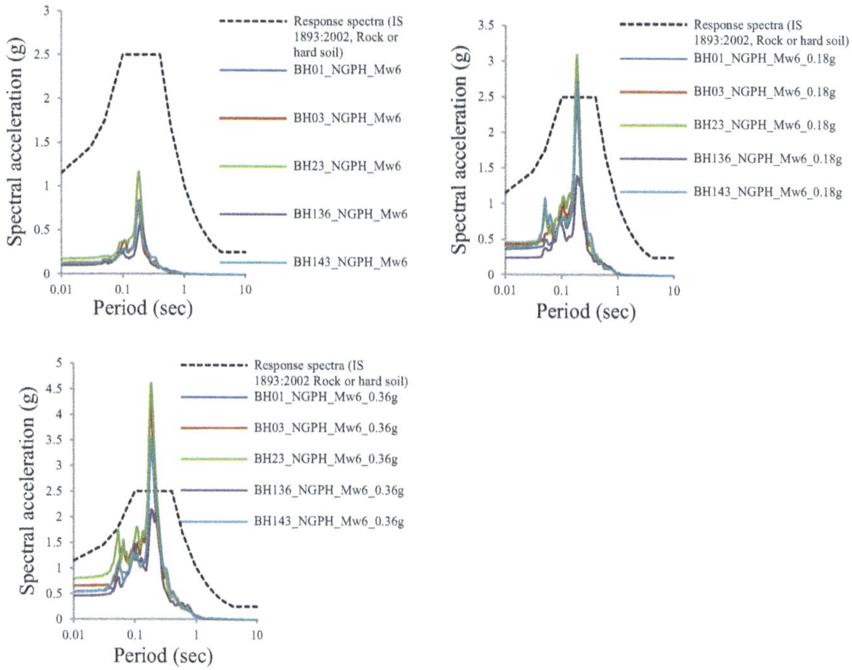

Fig. 7 Spectral acceleration for 0.04, 0.18 and 0.36 g input motion recorded at Nongpoh station for AEC area

7 Conclusions

It is observed that if the soil is soft in nature with low SPT-N value, the PGA value is magnified in the surface. When input motion value is scaled up i.e. by increasing the energy content, keeping other parameters constant, it is observed that the surface PGA value increases for all the sites compared to the original input motion. Response spectra of the soil for all the sites gives an overview of the response shown by soil layer when subjected to various ground motion with respect to time period. The peak spectral acceleration obtained for 0.36 g (scaling) motion for both earthquake shows much higher value than the spectral acceleration of rocky or hard soil sites (IS: 1893 [1]).

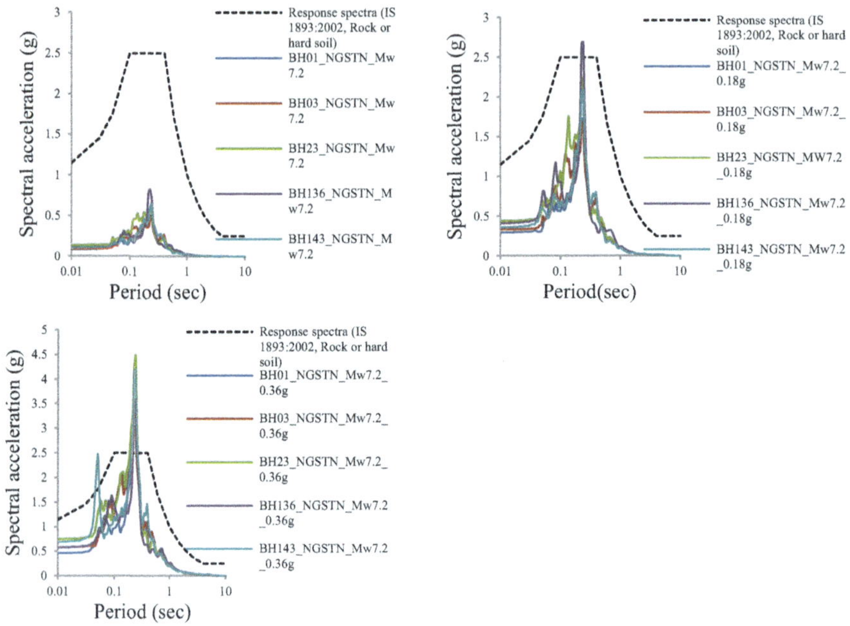

Fig. 8 Spectral acceleration for 0.05, 0.18 and 0.36 g input motion recorded at Nongstoin station for AEC area

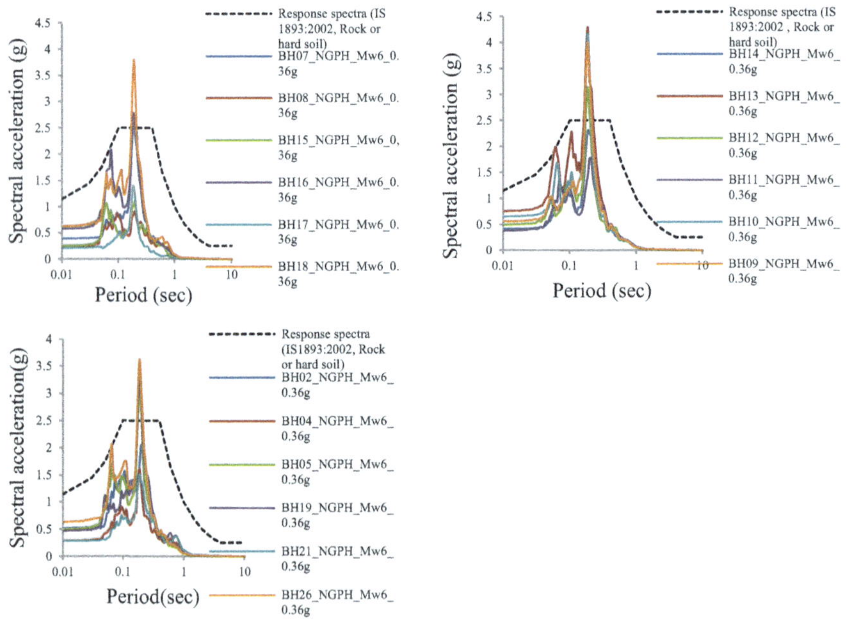

Fig. 9 Spectral acceleration for 0.36 g motion of **a** Dharapur and Azara area, **b** VIP and Rani area and **c** Jalukbari area for Indo-Burma Earthquake, 1997

References

1. IS: 1893 (Part I) (2002) Indian standard criteria for earthquake resistant design of structures, fifth revision, part-1. In: Bureau of Indian Standards, New Delhi
2. Konder RL, Zelasko JS (1963) A hyperbolic stress-strain formulation of sands. In: Proceedings of the 2nd pan American conference on soil mechanics and foundation engineering. Sao Paulo, Brazil, pp 289–324
3. Kramer SL (1996) Geotechnical earthquake engineering. Prentice Hall, New Jersey (NJ), 653
4. Nath SK, Thingbaijam KKS, Raj A (2008) Earthquake hazard in Northeast India: a seismic microzonation approach with typical case studies from Sikkim, Himalaya and Guwahati city. J Earth Syst Sci 117:809–831
5. Sharma B, Rahaman SK (Use of GIS based maps for preliminary assessment of sub- soil of Guwahati City. J Geosci Environ Prot 2016(4):106–116
6. Vucetic M, Dobry R (1991) The effect of soil plasticity on cyclic response. ASCE Geotech J 117(1):89–107

One-Dimensional Ground Response Analysis to Arrive at Surface Peak Ground Acceleration—A Case Study of Golaghat District in Assam

A. F. Siddique, D. Dutta, and A. Deka

Abstract Ground Response Analysis (GRA) is required to be carried out for predicting ground surface motions and to evaluate dynamic properties of soil during an earthquake excitation. In this study one-dimensional equivalent linear (EL) and nonlinear (NL) GRA was carried out for five boreholes located in Golaghat district of Assam, India. This region falls under highly seismic zone 'V'. The soil borehole log obtained from standard penetration test (SPT) confirmed alluvial deposits with layers of both coarse and fine-grained soils. The input motions of 2011 Sikkim earthquake of M_w 6.9 recorded at Gangtok station and 1999 Uttarkashi earthquake of M_w 6.8 recorded at Bhatwari station are considered for the analysis. The results are plotted in terms of peak ground acceleration (PGA), maximum stress ratio and maximum strain with depth, and spectral acceleration over range of periods. On comparison of EL and NL methods, the strain profile follows a similar trend along the depth of borehole from both methods. However on comparison of both the methods, NL methods showed maximum shear strain value. It was depicted that the PGA values for a particular site of interest can be determined directly from the peak horizontal acceleration (PHA) even in the absence of dynamic soil properties of the site. Such PGA values can be directly used for earthquake-induced liquefaction analysis and also for building design purposes.

Keywords Ground response analysis · Equivalent linear method · Nonlinear method · Surface PGA · Spectral acceleration

A. F. Siddique
Department of Civil Engineering, Assam Engineering College, Guwahati 781013, India
e-mail: amaribnesiddiq98@gmail.com

D. Dutta · A. Deka (✉)
Department of Civil Engineering, Central Institute of Technology Kokrajhar, Kokrajhar 783370, India
e-mail: a.deka@cit.ac.in

D. Dutta
e-mail: dikshadutta1401@gmail.com

T. G. Sitharam et al. (eds.), *Local Site Effects and Ground Failures*, Lecture Notes in Civil Engineering 117, https://doi.org/10.1007/978-981-15-9984-2_10

1 Introduction

The acceleration time history of a strong ground motion greatly changes when a motion comes to the ground surface. When the motion travels through different soil layers it gets amplified to the surface. Ground response analysis (GRA) is done to know how much changes will occur in amount of acceleration when a motion travels through different soil layers. In this report 1D equivalent linear (EL) and nonlinear (NL) ground response analysis has been done for five borehole locations of Golaghat district of Assam, India. The region is placed on a very high seismic zone, i.e. zone V [1]. Alluvial deposits with layers of both coarse and fine-grained soils are found in this region. Hence, large modifications in earthquake waves can occur due to variation in soil properties near to the surface of earth. The amplification of clay soil is about 1.5 times higher than sandy soil [2]. Hence, the soil characteristics and response of motion on the surface is very much essential for this region. Based on ground response analysis, the correlation of amplification factor (AF) value as a function of PHA is proposed for building design purposes. The advantage of the correlation is that one can easily determine surface PGA values for the site for any earthquake motions without undergoing ground response analysis.

2 Methodology

In this study both 1D equivalent linear (EL) and nonlinear (NL) methods are used to perform ground response analysis. The input motion that is required for the purpose is obtained from the seismograph station. The acceleration time history of 2011 Sikkim earthquake having bedrock PGA 0.152 g and 1999 Uttarkashi earthquake having bedrock PGA 0.252 g are considered for the purpose of comparative study between two methods. The motions are recorded at Gangtok station and Bhatwari station, India which are rocky sites (site class A; density = 25 kN/m^3 and shear wave velocity = 1500 m/s).

The correlation between PHA and AF is also obtained for the site. For this purpose an additional three earthquakes from the same tectonic regime viz. 1997 Indo-Burma EQ recorded at Nongpoh station (PHA = 0.048 g), Loma Gilroy EQ (PHA = 0.357 g) and Kobe EQ (PHA = 0.82 g) are considered. The input data consist of bulk density, SPT N value, plasticity indexes and depth of water table. The shear wave velocity (V_s) at each 1.5 m interval is calculated from the correlation with SPT N value (N) using the Eq. 1 given by [3].

$$V_s = 97 \times \mathrm{N}^{0.314} \tag{1}$$

As the damping and modulus reduction curves are absent for the site, the curves for sands [4] and for clay [5] are used for EL method [6]. Similarly the curve proposed by Seed and Idriss (mean limit) for sandy soil and [5] for clay are considered for NL method and they are subsequently fitted using MRDF procedure. The stress–strain model [7] is used for performing NL ground response analysis incorporating Masing criteria.

3 Results and Discussion

The results in terms of variation of PGA with depth, maximum stress ratio and maximum strain over depth and spectral acceleration over range of periods are resolved. Good comparative studies between both the methods are also done for the soil profiles of the proposed site. It has been observed that stiffer soil layers usually the deeper soil layers resulted in similar range of PGA by both the methods however it get deviated when reaches to the surface. Figures 1 and 2 show the variation of PGA with depth of five boreholes of Golaghat district of Assam for two different ground motion 2011 Sikkim and 1999 Uttarkashi earthquake. It has been observed that the EL surface PGA lie in the range of 0.27 g–0.37 g and 0.40 g–0.67 g against 2011 Sikkim and 1999 Uttarkashi EQ, respectively. Again NL surface PGA lies in the range of 0.24 g–0.34 g and 0.30 g–0.55 g against 2011 Sikkim and 1999 Uttarkashi EQ, respectively. From Table 1 it has been found that NL method generally gives lower surface PGA values than EL method. This is attributed to changing secant modulus due to dynamic stress–strain characteristics at every time step. Figure 3 shows the comparison of EL and NL methods for BH01 of the site. It has been found that up to 5 m from the depth the variation of PGA from both the methods are almost in similar trend but above 5 m it get deviated and NL method shows slightly lesser value of PGA than EL method.

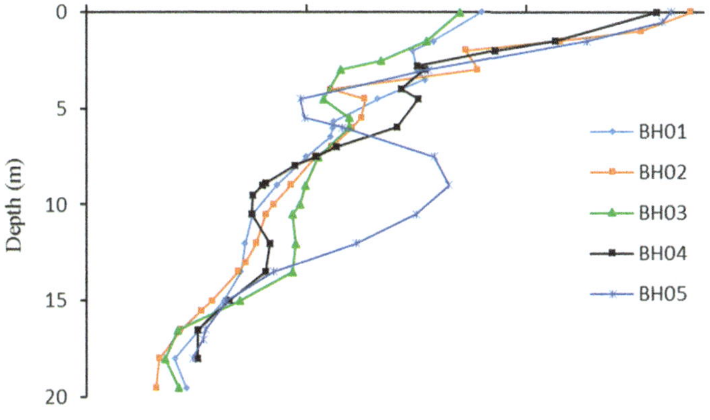

Fig. 1 Variation of PGA with depth against Sikkim earthquake (EL method)

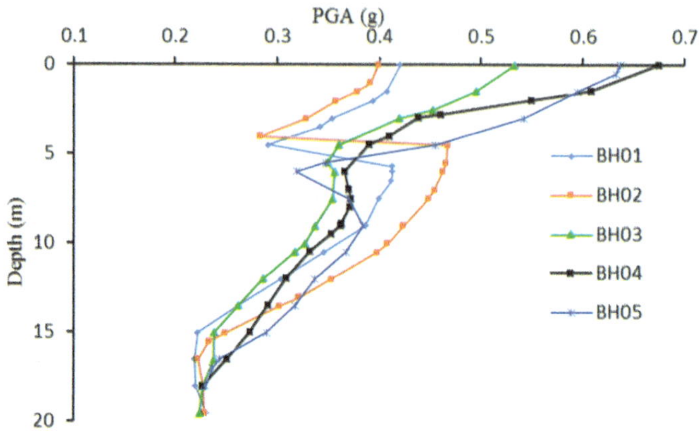

Fig. 2 Variation of PGA with depth against Uttarkashi earthquake (EL method)

Table 1 Surface PGA values of five borehole sites

Borehole	Sikkim 0.152 g		Uttarkashi 0.252 g	
	EL	NL	EL	NL
BH01	0.28	0.24	0.42	0.30
BH02	0.34	0.33	0.40	0.55
BH03	0.27	0.27	0.53	0.42
BH04	0.36	0.34	0.67	0.52
BH05	0.37	0.29	0.64	0.50

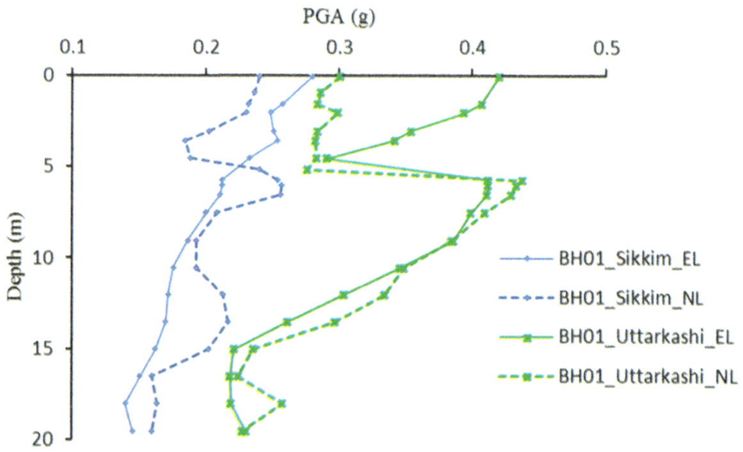

Fig. 3 Comparison of PGA profile at BH01 of Golaghat

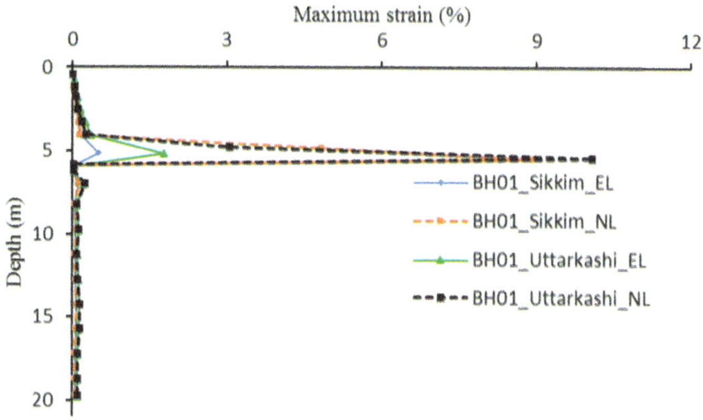

Fig. 4 Variation of strain (%) with depth at BH01 of Golaghat

Figure 4 shows the shear strain profile at Borehole 01 of Golaghat area. On comparison of EL and NL methods, the strain profile follows a similar trend along the depth of borehole from both methods. However on comparison of both the methods, NL methods show maximum shear strain value. Again the borehole experienced maximum strain against Uttarkashi motion compared to Sikkim motion due to high bedrock PHA and high energy content of Uttarkashi motion. The peak of maximum strain value is obtained at about 5 m because of having soft soil (N = 2). Strains in soft soil are generally large due to the low stiffness. Table 2 represents the values of maximum strain at five boreholes site for both EL and NL methods.

A completely opposite picture is seen from the maximum shear stress ratio profile shown in Fig. 5 where the value of shear stress ratio is obtained higher by EL methods than NL methods. On comparing both the motions, Sikkim motion shows a lesser value of shear stress ratio due to lesser energy content and lesser bedrock PGA value of motion.

The response spectra in terms of spectral acceleration (SA) for 5% damping at BH01 of the area is shown in Fig. 6. The SA is obtained against 2011 Sikkim and 1999 Uttarkashi earthquake by using both EL and NL analysis. It is found that EL analysis gives maximum SA value as compared to NL analysis. The SA values for BH01 against two motions are below the SA of [1], Rock or Hard soil. The spectral

Table 2 Maximum strain (%) values at five borehole sites

Borehole	Sikkim 0.152 g		Uttarkashi 0.252 g	
	EL	NL	EL	NL
BH01	0.497	8.904	1.782	10.070
BH02	0.489	0.523	5.072	2.010
BH03	0.101	0.113	0.323	0.388
BH04	0.177	0.168	0.496	3.201
BH05	0.131	0.136	0.634	0.538

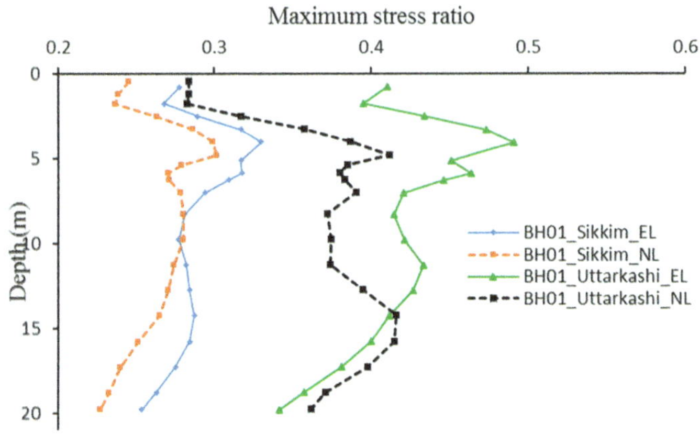

Fig. 5 Variation of maximum stress ratio with depth at BH01 of Golaghat

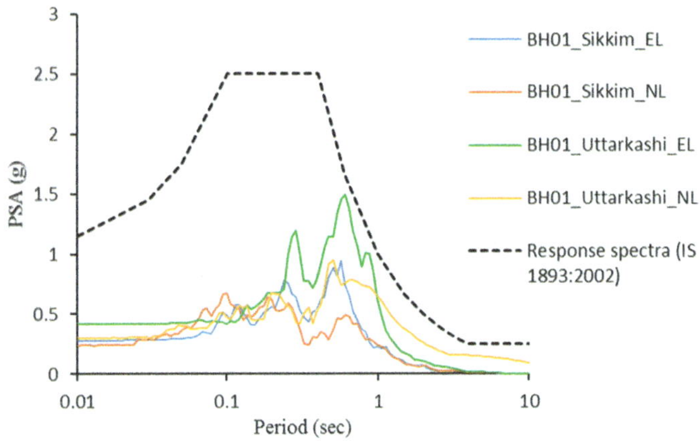

Fig. 6 Spectral acceleration for BH01 using both EL and NL analysis

acceleration values at five boreholes of Golaghat are shown in Table 3. It is obtained that for all the boreholes the SA value against Uttarkashi motion are much higher than Sikkim motion for both EL and NL methods. This is due to high magnitude and high duration of Uttarkashi earthquake motion.

Figure 7 shows the variation of AF with bedrock PHA for five boreholes of Golaghat district during different earthquakes. The Indo-Burma earthquake (1997) having bedrock PHA 0.048 g recorded at Nongpoh station, Sikkim earthquake (2011) having bedrock PHA 0.152 g recorded at Gangtok station, Uttarkashi earthquake (1999) having bedrock PHA 0.252 g recorded at Bhatwari station, Loma Gilroy earthquake having bedrock PHA 0.357 g and Kobe earthquake having bedrock PHA 0.82 g are considered for the analysis. 1D NL GRA was performed for the five boreholes against the above five different earthquakes and amplification factors (AF)

Table 3 Spectral acceleration (g) values at five borehole sites

Borehole	Sikkim 0.152 g		Uttarkashi 0.252 g	
	EL	NL	EL	NL
BH01	0.94	0.67	1.50	0.95
BH02	1.39	1.44	1.58	2.23
BH03	1.28	1.22	2.04	1.67
BH04	1.76	1.63	2.55	1.75
BH05	1.56	1.32	1.81	1.65

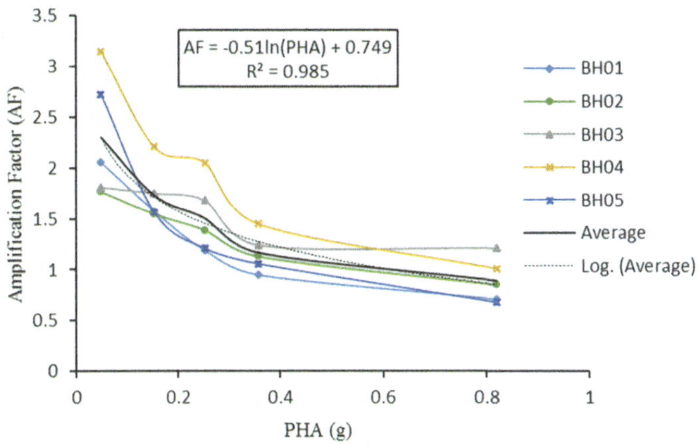

Fig. 7 Plot of AF versus PHA for five boreholes of Golaghat district during different earthquakes

are calculated which is defined as the ratio of surface PGA to bedrock PHA. The AF's are plotted with respect to bedrock PHA for all the boreholes sites. Similar thickness and type of soil deposit might show similar response during an EQ [8]. This might be the reason why the AF variation with respect to PHA values show similar variation pattern for all the selected sites as shown in Fig. 7. To give a general equation between AF and PHA, a correlation between AF and PHA is proposed for Golaghat district as shown in Fig. 7. It has been observed that at PHA < 0.2 g, the AF changes abruptly, however at PHA > 0.2 g the variation of AF is gradual [9]. Therefore, no linear relation is possible between AF and PHA.

A logarithmic function of the form $y = a \ln(x) + b$, correlates dependent variable y with an independent variable x, where a and b are two regression constants. For the present problem a logarithmic function is found to give highest R^2 value. For this a logarithmic function is selected for correlating AF-PHA. The obtained correlation is shown in the Fig. 7. The main intention of this correlation is that by knowing this correlation one can simply determine AF and surface PGA values without doing GRA based on PHA for Golaghat. The surface PGA can be calculated by multiplying AF with PHA values. Based on this proposed correlation the PGA values for a particular site of interest can be determined directly from the PHA even in the absence of dynamic soil properties of the site. Such PGA values can be directly used for earthquake-induced liquefaction analysis and also used for building design purpose.

4 Conclusions

Study of soil characteristics together with the liquefaction potential study is very essential for this region. The following points can be summarised by this study:

- The motion is amplified to the surface when it travels through different soil layers. 1D equivalent linear and nonlinear ground surface analysis has been done for 5 boreholes locations of Golaghat district of Assam, India.
- Amplification of clay soil is 1.5 times higher than sandy soil. A similar range of PGA by both methods was observed, i.e. for stiffer soil layers.
- It was seen that NL method generally gives lower surface PGA values than EL methods.
- Up to a depth of 5 m the variation of PGA from both the methods are almost in similar trend but above 5 m it gets deviated and NL method shows slightly lesser value of PGA than EL methods.
- On comparing both the motion Sikkim motion shows a lesser value of shear stress ratio due to lesser energy content and lesser bedrock PGA value of motion.
- It has been observed that at PHA < 0.2 g, the AF changes abruptly; however at PHA > 0.2 g, the variation of AF is gradual.
  By these conclusions we can easily determine the AF and surface PGA values without doing GRA based on PHA for Golaghat. When AF and PHA values are multiplied we get the surface PGA. Even in the absence of dynamic soil properties the surface PGA values can be calculated by the prescribed correlation.

References

1. IS 1893–Part-1 (2002) Indian standard criteria for earthquake resistance design of structures. BIS, New Delhi
2. Boominathan A, Dodagoudar GR, Suganthi A, Maheswari UR (2008) Seismic hazard assessment of Chennai city considering local site effects. J Earth Syst Sci 117(S2):853–863
3. Imai T, Tonouchi K (1982) Correlation of N value with S-wave velocity and shear modulus. In: Proceedings of 2nd European symposium on penetration testing, Amsterdam, pp 67–72
4. Seed HB, Idriss IM (1970) Soil moduli and damping factors for dynamic response analyses. Technical report EERRC-70-10, University of California, Berkeley
5. Vucetic M, Dobry R (1991) The effect of soil plasticity on cyclic response. ASCE Geotech J 117(1):89–107
6. Hashash YMA, Musgrove MI, Harmon JA. Groholski DR, Phillips CA, Park D (2016) DEEPSOIL version 6.1. User manual, 137 pp

7. Kondner RL, Zelasko JS (1963) A hyperbolic stress-strain formulation of sands. In: Proceedings of the 2nd Pan American conference on soil mechanics and foundation engineering, Sao Paulo, Brasil, pp 289–324
8. Kumar A, Harinarayan NH, Baro O (2017) Nonlinear soil response to ground motions during different earthquakes in Nepal, to arrive at surface response spectra. Nat Hazards 2017(87):13–33. https://doi.org/10.1007/s11069-017-2751-4
9. Kumar A, Baro O, Harinarayan NH (2016) Obtaining the surface PGA from site response analyses based on globally recorded ground motions and matching with the codal values. Nat Hazards 81:543–572. https://doi.org/10.1007/s11069-015-2095-x

Elastic Seismic Design Response Spectra for Deep and Shallow Basin of the Indian Subcontinent

Ketan Bajaj and P. Anbazhagan

Abstract In the present study, new elastic design response spectra (EDRS) for the deep and shallow region of the Indian subcontinent has been proposed for rock and soil site. For determining the EDRS for bedrock, 50 each rock recorded ground motions have been used for Intra and Interplate region. For determining the EDRS for soil sites, 250 and 178 ground motion for deep and shallow sites has been used, respectively. For SI, only few ground motions are available, hence for developing EDRS for shallow sites, ground motions recorded at similar tectonics have been used in the present study. Further, EDRS is derived based on Eurocode, i.e. normalized elastic design response spectra which is based on one parameter, i.e. effective ground acceleration at rock.

Keywords Response spectra · Ground motion · Deep and shallow basin · Seismic site classification

1 Introduction

Local site conditions have great influence on ground surface motion and structural damage caused by an earthquake event. The Indian Subcontinent (IS) has one of the most diverse seismotectonic and seismicity. The high level of seismicity is associated with the Himalaya tectonic province will result in site amplification in the contiguous deep alluvial deposits named Indo-Gangetic Basin (IGB), due to any major earthquake in future. Whereas, low to moderate level of seismicity in the Southern India (SI) causing high amplification due to shallow thin layers. Evidence from past earthquakes suggested that soft soil sites tend to amplify at low frequency, whereas rock sites have significant intensity at high frequencies. Anbazhagan et al. [1] highlighted that local site effect is the major factor that causes the damage due to an earthquake.

K. Bajaj (✉) · P. Anbazhagan
Department of Civil Engineering, Indian Institute of Science, Bengaluru, India
e-mail: ketanbajaj@iisc.ac.in

P. Anbazhagan
e-mail: anabazhagan@iisc.ac.in

T. G. Sitharam et al. (eds.), *Local Site Effects and Ground Failures*, Lecture Notes in Civil Engineering 117, https://doi.org/10.1007/978-981-15-9984-2_11

2001 Bhuj (7.7 M_w), 1999 Chamoli (6.8 M_w), 2011 Sikkim (6.9 M_w), and 2015 Nepal (7.8 M_w) earthquakes are the recent examples that explained the effect of thick deposits on site-specific damage in the IS. Various researchers [e.g. 2–4] have studied the local site effect and estimated amplification factors using the 1D site response study but most of these studies are limited to soil column of 30 m depth. Moreover, in the previous site response studies, the input ground motions were either selected randomly from global database or simulated based on the occurred earthquake scenario. In India very limited attempts have been made to drive the design spectrum with different damping level considering regional recorded data.

In most of the modern seismic design, the estimation of seismic force of a typical structure is based on the 5% damped design response spectrum of recorded data in the region. In India limited attempts have been made to the drive design spectrum with different damping level considering regional recorded data, some attempt has been made by Anbazhagan et al. [5]. Generally, the design spectra of a given site are obtained by modifying the uniform hazard spectrum by considering site factors corresponding to a particular seismic site. Conventionally, the design force is specified via. response spectrum amplitude. However, with the increased complexity of the modern structure and understanding the seismic performance of structure demands, it is now essential to define the amplitude and shape of the design spectra.

This paper aims at the development of new elastic design response spectra for Intra and Interplate region of the Indian subcontinent considering region-specific seismic data. Further, EDRS is derived based on Eurocode, i.e. normalized elastic design response spectra which is based on one parameter, i.e. effective ground acceleration at rock.

2 Study Area

The Indo-Gangetic Basin is the foredeep depression that is situated between the Indian Peninsular shield and the Himalayan region. IGB lies roughly between longitude 74° E, and 88° E and latitude 24° N and 32° N (Fig. 1a). The sediment depth varies from few tens of meter in the south part of the IGB and progressively increasing up to ~5–6 kms in the northernmost part. High neotectonic activity and reactivation of tectonic features and lineaments are acknowledged by various researchers [e.g. 6 etc.]. IGB is contiguous to the most seismically active Himalayan region and experiencing the strong compressional stress conditions. Any large to moderate earthquake in the Himalayan region may result in massive destruction in the IGB due to site amplification and liquefaction.

Southern India is considered as one of the oldest geologically evolved and tectonically stable continental crust of the IS. The seismotectonic of the SI is majorly consist of various faults, ridges, shear zones, and tectonic lineaments. Various researches [e.g. 7, etc.] defined the tectonic feature of the SI and many authors reported the reactivation of fault along the western part of the Peninsular India. Additionally, SI

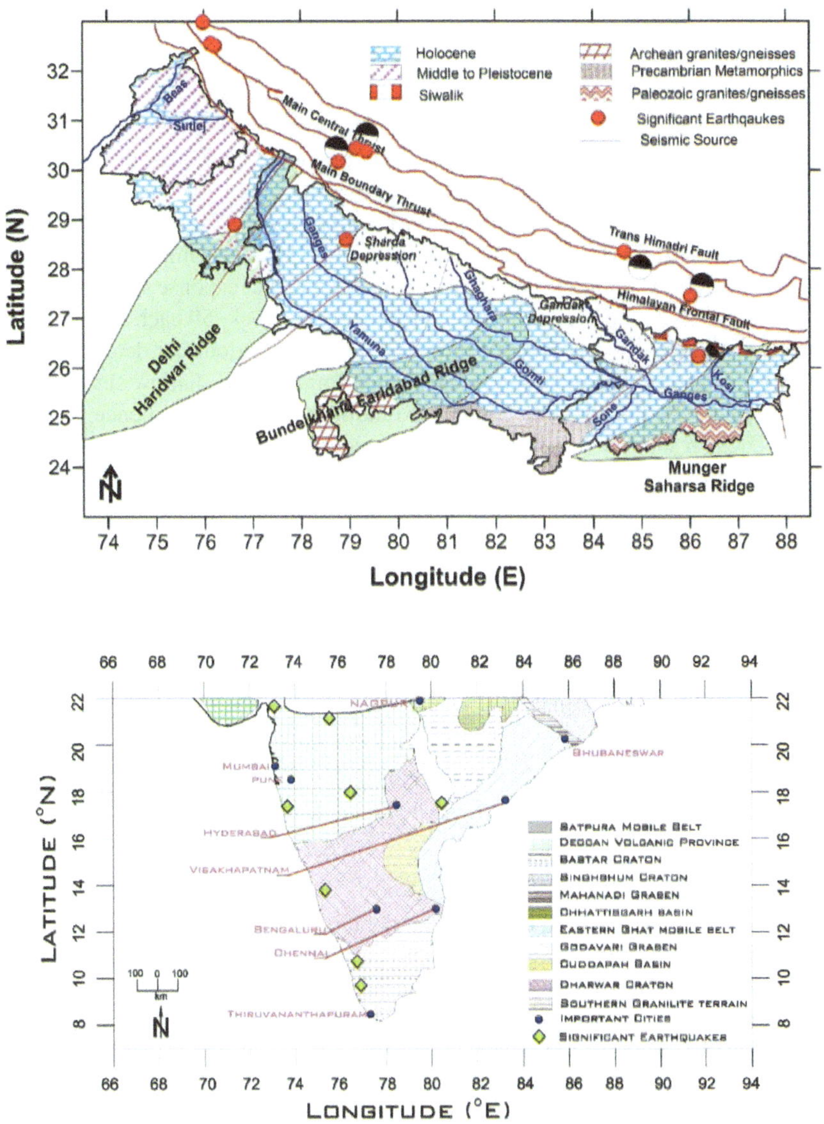

Fig. 1 Study area used in the present study

is having an irregular seismicity. As micro-seismicity is reported in the South Granulite Terrain, Eastern Dharwar craton is surrounded by intermediate seismicity, and Koyna-Warna region and Deccan Volcanic Province has high seismicity. The SI is also marked in Fig. 1b.

3 Methodology

Malhotra [8] concluded that the response of the structure derived using acceleration time history does not correspond to the velocity and displacement time histories. Response of the flexible structures (long period) can be contradictory if computed only using processed acceleration time history [8]. Based on that, Malhotra [8] proposed a methodology to compute elastic response spectra for incompatible acceleration, velocity, and displacement time histories. The following is the procedure recommended by Malhotra [8] for deriving the normalized response spectra and used in the present study. For determining the EDRS for bedrock, 50 each rock recorded ground motions have been used for Intra and Interplate region. For determining the EDRS for soil sites, 250 and 178 ground motion for deep and shallow sites has been used, respectively. For SI, only few ground motions are available, hence for developing EDRS for shallow sites, ground motions recorded at similar tectonics have been used in the present study. After smoothening the recorded ground motion, the elastic design response spectra have been defined considering the following equations

$$0 \leq T \leq T_B : \frac{S_a(T)}{PGA_{rock}} = s.\left[1 + \frac{T}{T_B}.(\beta - 1)\right] \tag{1}$$

$$T_B \leq T \leq T_C : \frac{S_a(T)}{PGA_{rock}} = s.\beta \tag{2}$$

$$T_C \leq T \leq T_D : \frac{S_a(T)}{PGA_{rock}} = s.\beta\frac{T_C}{T} \tag{3}$$

$$T_D \leq T : \frac{S_a(T)}{PGA_{rock}} = s.\beta.T_C\frac{T_D}{T^2} \tag{4}$$

Here, PGA_{rock} is the design ground acceleration at rock-site conditions, S and β are the soil amplification and spectral amplification factors. T_B and T_C are the limits of constant acceleration branch and T_D is the beginning of the constant displacement range of the spectrum. Distributing the spectra into two regions helps in better representation of the shape of the response spectra. β, T_B, T_C, and T_D are determined based on the shape of the normalized response spectra. S and β have been determined considering the methodology proposed by Pitilakis et al. [9]. Typical EDRS is given as Fig. 2. The amplification factor corresponding to acceleration, velocity, and displacement is denoted as α_A, α_V, and α_D. T_1, T_2, T_3, T_4, T_5, and T_6 in Fig. 7.2 is denoted as control periods (where the straight-line segments meet).

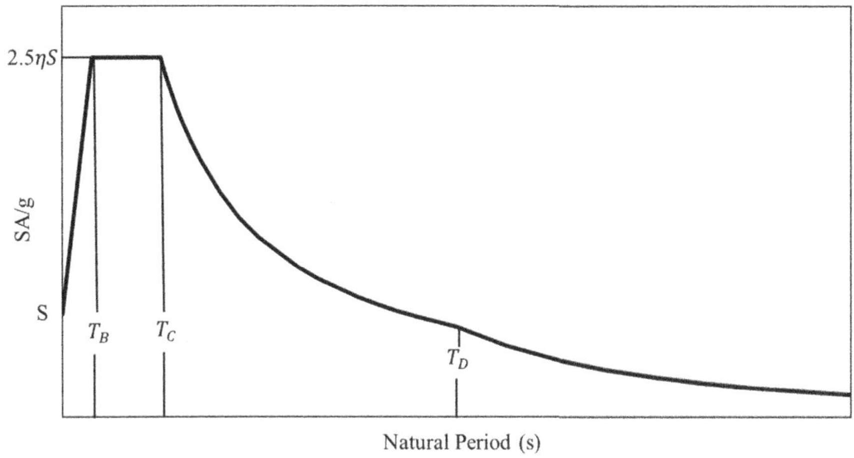

Fig. 2 Typical EDRS derived in this study

4 Result and Discussion

Pitilakis et al. [9] defined different approaches in determining the soil amplification (S). In this study, it is calculated considering the period-independent amplification factor with respect to rock site. The period-independent amplification factor for soil site and different magnitude at an interval of 0.5 with respect to rock is calculated as

$$S = (I_{soil}/I_{rock}).(1/SR) \tag{5}$$

here, SR is the spectral ratio, I_{soil} and I_{rock} are the spectral intensities for soil and rock, respectively. Housner [10] defined the response spectrum intensity as

$$I = \int_{0.05}^{2.5} PSV(\xi, T)dT \tag{6}$$

i.e. the area under the pseudo velocity response spectrum between periods 0.05 to 2.5 s. To compute the I_{rock} for IGB and SI, the area under the PSV curve has been determined. Similarly, the area under the PSV curve for soil site is computed and I_{soil}/I_{rock} has been determined. SR reflects the difference in shape of soil spectra with respect to rock spectra, when all the spectra are normalized and have the same ordinate at starting period. The spectral amplification factor (β) is analogues to α_A (see Fig. 3) is also determined for both the regions and given as Table 1. From Table 1, it has been observed that S is 1.22 in case of IGB and 1.13 in case of SI for soil site. Similarly, β is 2.91 in case of IGB and 2.21 in case of SI, for soil.

The parameters β, T_B, T_C, and T_D have been determined using the normalized spectrum. These factors are the result of fitting the smooth spectrum to the median

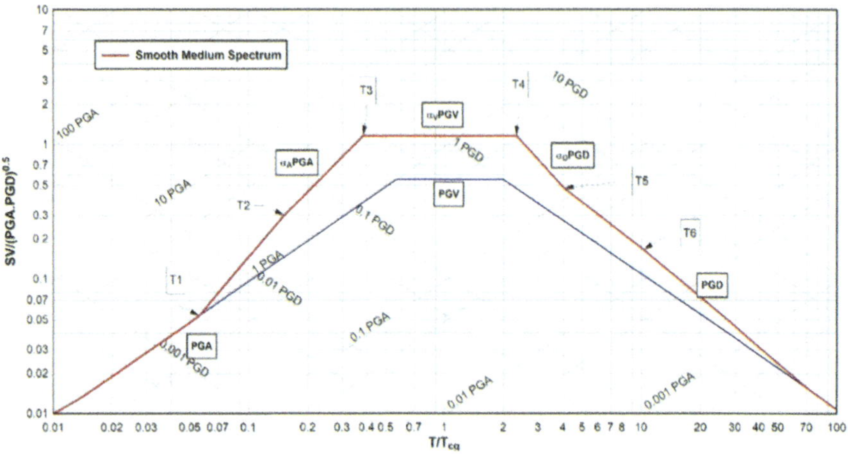

Fig. 3 Typical smooth medium response spectrum considering

Table 1 Parameters of the proposed design response spectrum for 5% damping

Parameters	Soil		Rock	
	IGB	SI	IGB	SI
T_B	0.16	0.12	0.15	0.08
T_C	0.52	0.29	0.38	0.28
T_D	1.90	1.41	2.33	1.52
S	1.22	1.13	1.00	1.00
β	2.91	2.21	2.29	2.81

normalized spectrum (see Fig. 3). The values are given in Table 1. The derived spectrum for the IGB and SI is given as Fig. 4a, b, respectively. The proposed spectra in this case depend on the effective ground acceleration at rock sites (i.e. PGA) and soil amplification factors (S, β), whereas control periods remain constant for the corresponding site class. The derived EDRS is also compared with design spectrum given in Indian standard code (IS:1893) and given in Fig. 4.

5 Conclusion

In the present study, new elastic design response spectra for the deep and shallow soil sites in Inter and Intraplate regions of the Indian subcontinent have been proposed. EDRS for bedrock is determined by considering 50 each rock recorded ground motions for Intra and Interplate region. For determining the EDRS for soil sites, 250 and 178 ground motion for deep and shallow sites has been used, respectively.

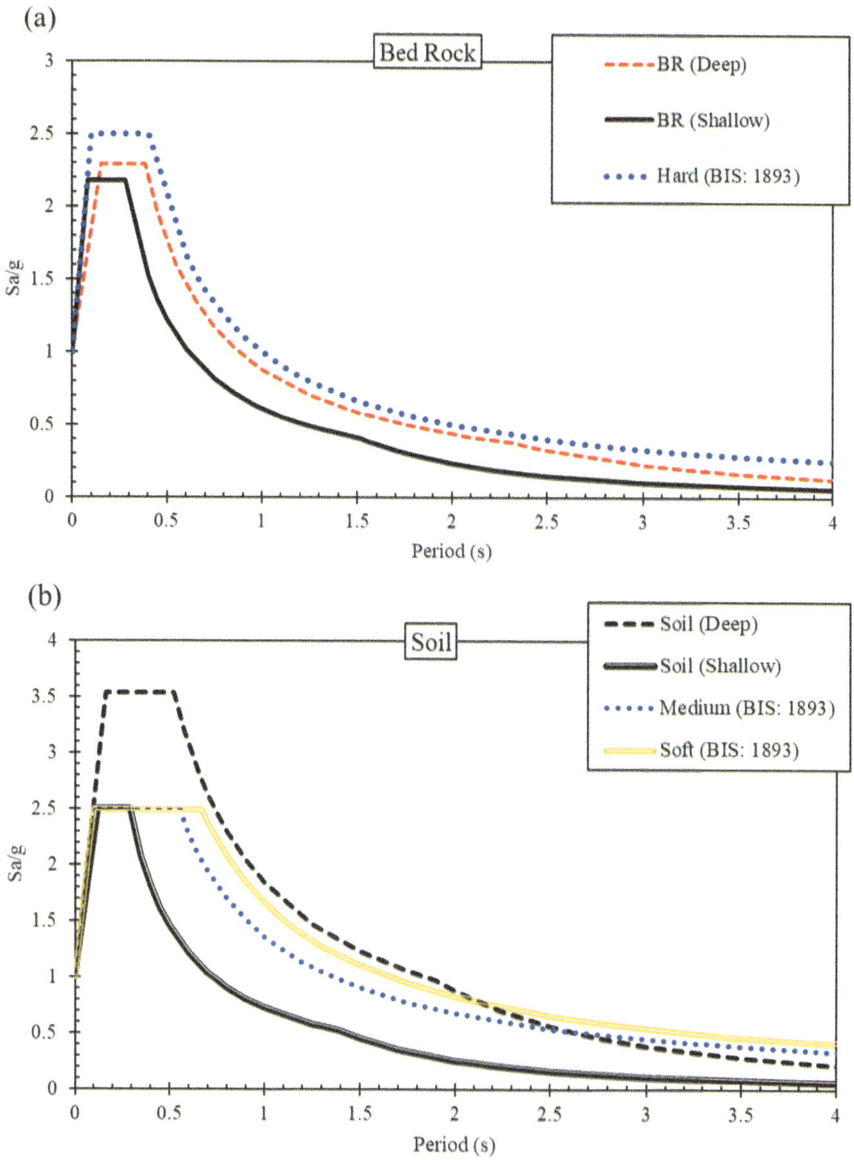

Fig. 4 Design response spectra for deep and shallow basin of India for bedrock and soil sites

EDRS is derived based on Eurocode, i.e. normalized elastic design response spectra which is based on one parameter, i.e. effective ground acceleration at rock. Response spectra for IS:1893 are underestimating the spectral values at long and short periods.

References

1. Anbazhagan P, Kumar A, Sitharam TG (2010) Site response of Deep soil sites in Indo Gangetic plain for different historic earthquakes. In: Proceedings of the 5th international conference on recent advances in geotechnical earthquake engineering and soil dynamics, vol 3.21b. San Diego, California, p 12
2. Boominathan A, Dodagoudar GR, Suganthi A, Maheshwari RU (2008) Seismic hazard assessment of Chennai city considering local site effects. J Earth Syst Sci. 117(S2):853–863
3. Anbazhagan P, Sitharam TG (2008) Site characterization and site response studies using shear wave velocity. J Seism Earthq Eng. 10(2):53–67
4. Naik N, Choudhury D (2013) Site specific ground response analysis for typical sites in Panjim city, Goa. In: Proceedings of Indian geotechnical conference, Roorkee, India
5. Anbazhagan P, Bajaj K, Reddy GR, Phanikanth VS, Yadav DN (2016) Quantitative assessment of shear wave velocity correlations in the shallow bedrock sites. Indian Geotech J 46(4):381–397
6. Singh IB (1996) Geological evolution of Ganga plain – an overview. J Paleont Soc India 41:99–137
7. Ramasamy SM (2006) Remote sensing and active tectonics of south India. Int J Rem Sens 27(20):4397–4431
8. Malhotra PK (2001) Response spectrum of incompatible acceleration, velocity and displacement histories. Earthquake Eng Struct Dyn 30(2):279–286
9. Pitilakis K, Riga E, Anastasiadis A et al (2013) New code site classification, amplification factors and normalized response spectra based on a worldwide ground motion database. Bull Earthq Eng 11(4):925–966
10. Housner GW (1952) Spectrum intensities of strong-motion earthquakes. In: Proceedings of symposium on earthquakes and blast effects on structures, earthquake engineering research institute, pp 20–36

Case Studies on Preshaking and Reliquefaction Potential for Different Earthquakes in Japan

Gowtham Padmanabhan and B. K. Maheshwari

Abstract The paper examines the effect of liquefaction potential of the site due to normal earthquake and sequential earthquakes occurred in Japan. Two earthquakes have been considered (a) 2016 Kumamoto sequential earthquakes with mainshock ($M_w = 7.3$) and a couple of foreshock in the previous days ($M_w = 6.2$ and $M_w = 6.5$); and (b) 2005 Fukuoka-ken Seiho-oki earthquake ($M_w = 7.0$), the earthquake has only mainshock. With the help of available literatures, damages associated with the liquefaction was examined with reference to effect of preshaking, liquefaction history and earthquake pattern. It is evident that, liquefaction effects of the 2005 earthquake are larger than those associated with Kumamoto sequential earthquakes. It is inferred that liquefaction resistance and reliquefaction potential of the site was highly influenced by the seismic preshaking, liquefaction history and earthquake pattern. This is the most probable explanation for the higher liquefaction resistance exhibited by the Kumamoto earthquake. Preshaking effect is highly associated with earlier earthquake history or significant foreshocks as in the 2016 Kumamoto earthquakes. The occurrence of sequential earthquakes induces reliquefaction phenomenon in the site. It is concluded that, liquefaction history and preshaking effect of the site contributed to the enhanced liquefaction resistance of the site.

Keywords Sequential earthquakes · Liquefaction history · Preshaking · Reliquefaction

1 Introduction

Earthquakes are known for its complex nature, as its return period and frequency cannot be predicted accurately. It has been known to occur successively at the same site within a short period of time, as shown by the examples from Christchurch

G. Padmanabhan (✉) · B. K. Maheshwari
Department of Earthquake Engineering, IIT Roorkee, Roorkee, India
e-mail: gowtham@eq.iitr.ac.in

B. K. Maheshwari
e-mail: bkmahfeq@iitr.ac.in

T. G. Sitharam et al. (eds.), *Local Site Effects and Ground Failures*, Lecture Notes in Civil Engineering 117, https://doi.org/10.1007/978-981-15-9984-2_12

earthquake 2010–2011, Japan earthquake 2011, Nepal earthquake 2015, Kumamoto earthquake 2016, Indonesia earthquake 2018 and recent Philippines earthquake 2019 [1, 2, 8, 10, 16, 20].

Soil liquefaction is an intricate phenomenon occurs due to the generation of excess pore water pressure due to seismic shaking under undrained soil conditions. Liquefaction of saturated loose sands during earthquakes has been the cause of severe damage to various buildings, embankments and retaining structures. The liquefaction was first reported in 1964 Niigata earthquake, Japan and in Good Friday earthquake, Alaska in the same year and has been extensively studied since [12, 15, 23]. Some recent earthquakes with the major liquefaction damages are the 2010 Christchurch, 2011 Japan, 2015 Nepal and the recent 2018 Indonesia earthquakes.

Sequential earthquakes are in the form: (1) couple of strong earthquakes in short interval (2) significant foreshocks followed by mainshock (3) mainshock followed by couple of aftershock (4) combination of both foreshock and aftershock. Sequential earthquakes resulted in the preshaking of the site, that alters the geological properties of the soil. From the experimental studies, it is concluded that preshaking was responsible for the densification of loose sand deposits resulted in enhanced liquefaction resistance [3–7]. Seismic preshaking of a zone results in a decreasing risk of future liquefaction [3, 4]. Each previous instance of liquefaction allows soil to build up resistance to future liquefaction in terms of densification or change in orientation of soil deposits. Reliquefaction phenomenon was clearly observed during the 2011 Japan sequential earthquake, in which many man-made and natural sites reliquefied. Based on above literature, it can be concluded that, very limited work has been reported for the effect of sequential earthquakes on the liquefaction potential. This has been examined in the present study as discussed in the scope.

2 Scope of the Paper

The paper is an attempt toward understanding the influence of preshaking and liquefaction history in developing liquefaction resistance. The potential of sequential earthquakes in inducing the reliquefaction and the resistance associated with this phenomenon is presented. Case studies considering the seismic preshaking and earthquake and liquefaction history was rarely reported.

The existing reports focused on earthquake parameters and characteristics, damages associated with earthquakes. This paper is a first of its kind in comparing the two earthquakes based on the earthquake pattern, preshaking history, liquefaction history and the liquefaction damages associated with the site.

3 Case Study I—Kumamoto Sequential Earthquakes

3.1 Sequential Earthquakes, Location and Characteristics

A sequential earthquake is a sequence of seismic events occurring in a local area within a relatively short period of time. The successive earthquakes were recorded for a period of two months. From USGS data it is evident that, several foreshocks and aftershocks hit the same region along with mainshock. Black dots represent microearthquakes with magnitude <5 and yellow circles represent medium earthquakes with magnitude greater than or equal to 5 and lesser than 6, Pink circles represent the strong events which are accounted in the study with magnitude ≥6 [17].

Table 1 shows the detailed list of the significant events which contributed for the preshaking and reliquefaction. Earthquakes with magnitude ≥6 are considered as major events that have potential to induce preshaking and reliquefaction in the site.

The Kumamoto earthquake was observed in the year 2016 with a sequence of two foreshocks of moment magnitude 6.5 on 14th April at 21.26 local time, the hypocentral parameters include 32.74 N 130.81 E at a depth of 11 km and 6.2 on 15th April at 00:03 local time and the location of the focus was 32.70 N 130.78 E at a hypocenter depth of seven kilometers (USGS 2016). A mainshock of moment magnitude 7.0 at 01:25 local time, at a location of 32.75 N 130.76E and a depth of 12 km was occurred on 16th April [17, 18, 19]. The three major earthquake events each had high-intensity ground motions with recorded PGAs ranging from 0.2 to 1.2 g. The 2016 Kumamoto earthquake ground motions were sufficient enough to induce required Cyclic Stress Ratio (CSR) intensity and duration for liquefaction in the Kumamoto plain areas.

Though major events alone are considered in the present study, the events with magnitude ≤6 are also acknowledged for the contribution in preshaking effect and resulted in the development of liquefaction resistance and reliquefaction. Table 2 provides the details of earthquake events based on magnitude scale, which was recorded in the Kumamoto region for a period of two months from April 01 to May 31, 2016. A total of 1157 events were observed in the specified duration within the Kumamoto region.

Table 1 Details of Kumamoto earthquakes, 2016 [17]. Location: Kumamoto Chiho of Kumamoto Prefecture

Date and time	Moment magnitude
14 April 2016 21:26 JST (12:26 UTC)	6.5
15 April 2016 00:03 JST (14 April 2016 15:03 UTC)	6.2
16 April 2016 01:25 JST (15 April 2016 16:25 UTC)	7.3

Table 2 Number of seismic observations in Kumamoto region [18, 19]

Moment magnitude	Number of earthquakes
7	1
6	2
5	12
4	80
3	224
2	461
1	377

3.2 Seismic Preshaking History of Site

For better understanding of seismic preshaking history of the site, sequential occurrence of earthquakes or foreshock/aftershock events alone are not sufficient. The earthquake history of the corresponding site should also be considered for accurate prediction of preshaking effect in minimizing damages associated with earthquake-induced liquefaction [1].

Table 3 shows the earthquake history of the Kumamoto region. Kumamoto region is identified as seismically active zone in Japan. 15 major events with moment magnitude $\geq$6 were recorded in the region since sixteenth century witness the seismic potential of the site [13, 14].

Table 3 Earthquake history of the Kumamoto region

S. No	Date dd/mm/yyyy	Region and name	Moment magnitude
1	01/05/1619	Higo and Yatsushiro	6.0
2	21/06/1625	Kumamoto	6.1
3	28/10/1707	Houei earthquake	8.4
4	19/12/1723	Higo, Bungo and Chikugo	6.5
5	29/08/1769	Higo, Bungo and Chikugo	7.8
6	21/05/1792	Unzendake	6.4
7	24/12/1854	Ansei Nankai earthquake	8.4
8	28/07/1889	Kumamoto	6.3
9	19/11/1941	Hyuga-nada sea	7.2
10	21/12/1946	Nankai earthquake	8.0
11	23/01/1975	Northern edge of Mount Aso	6.1
12	17/01/1995	Southern Hyogo Prefecture earthquake	7.3
13	14/04/2016	Kumamoto	6.5
14	15/04/2016	Kumamoto	6.2
15	16/04/2016	Kumamoto	7.0

3.3 Liquefaction History of Site

Earlier to the 2016 earthquake, Kumamoto city was indicated as being a moderate to significant liquefaction vulnerable zone [10].

The liquefaction data of the Kumamoto region was shown in Table 3. From the table, it is clear that, Kumamoto city and the surrounding regions are highly susceptible to liquefaction. Around 11 cities are considered in present study to evaluate the liquefaction damages associated with the sequential earthquakes. The earlier earthquakes recorded in the corresponding region shown in Table 3 induced higher liquefaction susceptibility.

3.4 Liquefaction Damages

Liquefaction seems to have been a concern in the 2016 sequential Kumamoto earthquake; given the high-intensity ground motions.

Despite higher liquefaction susceptibility and high-intensity ground motions, the damages associated with the liquefaction are limited when compared to earlier earthquakes. The factors hindered the liquefaction damages are (a) Occurrence of reliquefaction, which develops the liquefaction resistance of the sand deposits by the densification effect (b) Effect of seismic preshaking has a significant role in increasing the liquefaction resistance due to the reorientation of sand particles.

4 Case Study II—Fukuoka-ken Seiho-oki Earthquake, 2005

4.1 Earthquake Location and Characteristics

The strong earthquake occurred at 10:53 am on March 20, 2005 in the Kyushu region, about 70 km west of shimonoseki city in Yamaguchi prefecture. The hypocentral parameters include 33.90 N 130.20 E at a depth of 9 km [14]. The depth of the earthquake was extremely shallow, with a presumed Richter scale of 7.0. Fukuoka is not as seismically active as many other parts of Japan, and was known prior to the earthquake as one of Japan's safest locations in terms of natural disasters; the previous earthquake, a magnitude 5, had occurred over a hundred years ago and it had been centuries since the city had experienced a serious earthquake [21, 22]. A quake measuring about seven on the Richter scale struck three hundred years ago, but no major quake has struck after that event [14].

4.2 *Liquefaction Damages*

Liquefaction was observed mainly in the reclaimed lands of the Hakata bay area. Severe damages were identified in the structures and pavements due to liquefaction. The earthquake-induced liquefaction damages were found to be very severe and resulted in loss of human life and economic resources [14].

The considered region is free from seismic preshaking, reduced liquefaction susceptibility, absence of foreshocks/aftershocks. Still, liquefaction damages associated with the Fukuoka-ken Seiho-oki Earthquake found to be significant. This shows that, liquefaction resistance of the sand deposits depends on several factors.

5 Discussions

1. In the present study, two sites with different geological compositions, seismic history, liquefaction history and earthquake pattern were considered. The extent and severity of the liquefaction damages were found to be higher in 2005 Fukuoka-ken Seiho-oki earthquake, when compared to the 2016 Kumamoto earthquakes. It is due to the reorientation of the sand particles and effect of densification observed in the site.
2. The liquefaction resistance of the sand deposits increases with increase in number of earthquakes recorded in the site. This pattern was observed in the 2016 Kumamoto sequential earthquakes. Three earthquakes with magnitude >6 hit the same region, seismic preshaking takes place and resulted in the enhanced liquefaction resistance. Whereas in case of 2005 Fukuoka-ken Seiho-oki Earthquake, foreshocks/aftershocks are not observed.
3. The reliquefaction phenomenon is observed in the 2016 Kumamoto earthquake as a result of successive foreshocks and mainshock. Densification mechanism is associated with the occurrence of reliquefaction. As a result, liquefaction resistance of sand deposits increased to a greater extent.
4. Earthquake shaking was so strong in 2016 Kumamoto region comparatively to 2005 Fukuoka-ken Seiho-oki earthquake, still the damages associated with liquefaction are lesser. It is strongly suggested that, there must be other factors that hindered liquefaction potential in Kumamoto region.
5. All of these discussions are valid only for the specific site and magnitude of earthquakes considered in the study. Further research is needed to investigate the combined influence of seismic preshaking and occurrence of reliquefaction in reduced liquefaction damages in the sand deposits.
6. The work presented in the paper may be helpful for analyzing and comparing liquefaction history and preshaking effect in liquefaction resistance and reliquefaction potential of sandy soils.

6 Conclusions

From this study, following conclusions can be drawn.

1. The liquefaction resistance of the sand deposits was highly influenced by the intense seismic preshaking history of the site.
2. The reliquefaction potential of the sites strongly depend on the characteristics of the earthquake motions.
3. Successive earthquakes or combination of foreshocks/aftershocks events are responsible for occurrence of reliquefaction phenomenon.
4. A site liquefied by earthquake loading can also be reliquefied by a subsequent earthquake even when the magnitude caused by a later was less than that caused by previous earthquake.
5. The liquefaction history of the site also contributed to liquefaction resistance, higher the liquefaction history; higher will be the liquefaction resistance.

References

1. Anderson DJ (2019) Understanding soil liquefaction of the 2016 Kumamoto earthquake
2. Cubrinovski M et al (2011) Soil liquefaction effects in the central business district during the February 2011 Christchurch earthquake. Seismol Res Lett 82(6):893–904
3. El-Sekelly WE (2014) The effect of seismic preshaking history on the liquefaction resistance of granular soil deposits. Rensselaer Polytechnic Institute
4. El-Sekelly W, Abdoun T, Dobry R (2015) Liquefaction resistance of a silty sand deposit subjected to preshaking followed by extensive liquefaction. J Geotech Geoenviron Eng 142(4):04015101
5. El-Sekelly W et al (2018) Experimental simulation of the effect of preshaking on liquefaction of sandy soils. In: Physical modelling in geotechnics, vol 2. CRC Press, pp 949–953
6. Goda K et al (2014) Sensitivity of tsunami wave profiles and inundation simulations to earthquake slip and fault geometry for the 2011 Tohoku earthquake. Earth Planets Space 66(1):105
7. Goda K et al (2015) The 2015 Gorkha Nepal earthquake: insights from earthquake damage survey. Front Built Environ 1:8
8. JMA. https://www.jma.go.jp/jma/indexe.html. Accessed 12 Dec 2019
9. Kayen RE et al (2017) Geotechnical aspects of the 2016 MW 6.2, MW 6.0, and MW 7.0 Kumamoto earthquakes. Geotechnical Extreme Events Reconnaissance Association
10. Kiyota T et al (2017) Geotechnical damage caused by the 2016 Kumamoto earthquake, Japan. ISSMGE Int J Geoeng Case Histor 4(2):78–95
11. Kokusho T et al (2015) Energy-based liquefaction potential evaluation and its application to a case history. In: Proceedings of 6th international conference on earthquake geotechnical engineering, Christchurch, NZ, Paper
12. Kramer S (1996) Geotechnical earthquake engineering. Prentice-Hall, Inc., New Jersey, pp 348–422
13. Mukunoki T et al (2016) Reconnaissance report on geotechnical damage caused by an earthquake with JMA seismic intensity 7 twice in 28 h, Kumamoto, Japan. Soils Found 56(6):947–964

14. Nagase H et al (2005) Zoning for liquefaction and actual damage during the 2005 Fukuokaken-Seiho-Oki earthquake. In: Geotechnical engineering for disaster mitigation and rehabilitation: (With CD-ROM), pp 293–298
15. Prakash S (1981) Soil dynamics. McGraw-Hill Companies
16. Towhata I, Goto H, Kazama M, Kiyota T, Nakamura S, Wakamatsu K, Wakai A, Yoshida N (2011) On gigantic Tohoku Pacific earthquake in Japan. Earthquake News, Bull Int Soc Soil Mech Geotech Eng 5(2)
17. USGS. https://earthquake.usgs.gov/earthquakes/eventpage/usp000djku/executive. Accessed 5 Dec 2019
18. USGS. https://earthquake.usgs.gov/earthquakes/eventpage/official20110311054624120_30/executive. Accessed 1 Dec 2019
19. USGS. https://earthquake.usgs.gov/earthquakes/eventpage/us20005hzn/executive. Accessed 12 Dec 2019
20. Wakamatsu K (2000) Liquefaction history, 416–1997, Japan. In: Proceedings of 12th world conference on earthquake engineering, Paper no 2270
21. Yasuda S, Matsumura S (1991) Microzonation for liquefaction, slope failure and ground response during earthquakes in Fukuoka City. Proc 4th Int Conf Seis Zonat 3
22. Youd TL (1984) Recurrence of liquefaction at the same site. Proc 8th World Conf Earthq Eng 3. Prentice-Hall Inc
23. Youd TL, Izzat MI (2001) Liquefaction resistance of soils: summary report from the 1996 NCEER and 1998 NCEER/NSF workshops on evaluation of liquefaction resistance of soils. J Geotech Geo-environ Eng 127(4):297–313

Seismic Response of Basal Geogrid Reinforced Embankments Supported on a Group of Vertical and Batter Piles

Radhika M. Patel, B. R. Jayalekshmi, R. Shivashankar, and N. R. Surya

Abstract Basal geogrid reinforced embankments supported on vertical piles are proven to be a feasible and effective solution for constructing embankments over thick soft clay deposits and bridge approaching embankments. These solutions minimize the lateral displacements, total and differential settlements of embankment crest and toe by transmitting embankment loads into the deeper stratum through pile foundations and arching action of geogrid. Basal geogrid reinforcements provide good restraint against lateral spreading of the toe. Providing batter piles near the toe will further enhance this restraint against lateral spreading. Not many studies are available in literature on performance of batter piles below embankment toe, especially under seismic excitations. The present study aims to find the advantages of providing batter piles below embankment toe under seismic excitations. A 6 m high basal geogrid reinforced embankment having 1 V:1.5H side slope constructed over 28 m thick soft clay is considered for the 3-Dimensional finite element analysis. The soft clay is stabilized with 22 m long 300 mm diameter vertical and batter piles spaced at three times the pile diameter. Embankment crest vertical displacements, toe horizontal displacements, maximum differential settlements at the crest and crest lateral accelerations are analysed for different batter angles of 0°, 5°, 10°, 15°. Analysis of results reveals that larger the batter angle more is the reduction of toe horizontal displacements.

Keywords Pile foundations · Embankments · Seismic analysis · Finite element method

1 Introduction

Inclined (or batter) piles are often used to resist lateral thrust in offshore constructions. In the pile-supported embankments in relatively very soft grounds, these batter piles are used below embankment toe to prevent lateral spreading of toe. The addition

R. M. Patel (✉) · B. R. Jayalekshmi · R. Shivashankar · N. R. Surya
Department of Civil Engineering, NITK, Surathkal, Mangalore 575025, Karnataka, India
e-mail: radikagmpatel@gmail.com

T. G. Sitharam et al. (eds.), *Local Site Effects and Ground Failures*, Lecture Notes in Civil Engineering 117, https://doi.org/10.1007/978-981-15-9984-2_13

of basal geogrid in the embankment eliminates the necessity of batter piles below embankment toe (BS 8006-2010) [3] under static loading conditions. Most of the studies from literature on pile-supported embankments under static loading conditions considered only vertical piles [4, 16, 17, 19, 20, 23]. Numerical analysis on pile-supported embankments considered unit cell models instead of full 3-Dimensional models [2, 5, 6, 10], thereby the presence or absence of batter pile is ignored. Thach et al. [21, 22] and Han et al. [11] analysed geogrid reinforced pile-supported embankments subjected to cyclic loading. To investigate the pile-pinning effects for the embankment supported over piles, dynamic centrifuge model tests were performed by Armstrong et al. [1]. Wang and Mei [24] studied the seismic performance of micropile-supported embankment. Patel et al. [18] analysed the seismic behaviour of end bearing and floating pile-supported embankment. Many studies reported the beneficial and detrimental effects of these batter piles subjected to dynamic loading conditions [8, 9, 25]. The presence of batter piles below embankment toe in a pile-supported embankment under dynamic loading conditions is beneficial or detrimental need to be studied yet.

Hence the present study aims to find the seismic behaviour of geogrid reinforced embankment supported over vertical and batter piles. The results are analysed based on the toe horizontal displacements, crest vertical displacements, differential settlements and lateral accelerations at crest.

2 Methodology

2.1 Numerical Analysis

6 m high embankment made of pulverized fuel ash (PFA) having 38 m base width constructed over 28 m thick soft marine clay is considered for the time-history analysis. Hard stratum exists below the clay layer. The embankment geometry is shown in Fig. 1.

2.2 Idealization of Soil

The properties of PFA, surface fill [16], soft marine clay which exists in Cochin region [12, 14] and hard soil considered for the analysis are listed in Table 1. PFA and surface fill were modelled using mohr–coulomb material model with drained condition. The soft marine clay and hard soil were modelled using mohr–coulomb material model including consolidation.

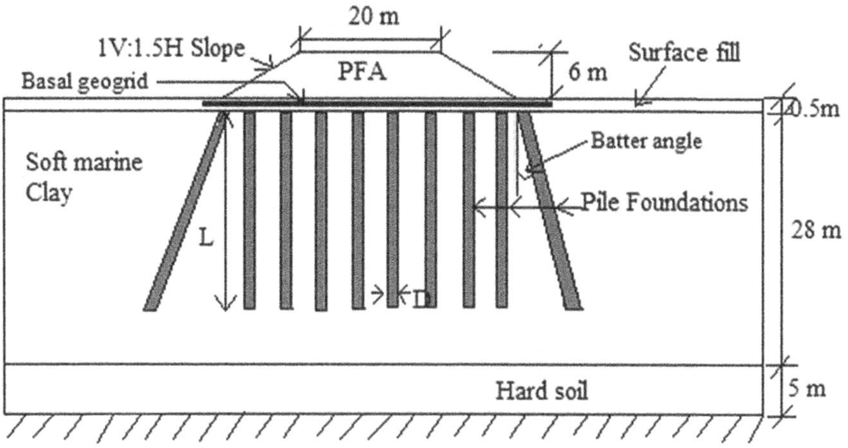

Fig. 1 Pile-supported embankment geometry

Table 1 Soil properties considered

Soil type	Unit weight (kN/m^3)	Young's Modulus (MN/m^2)	Poisson's ratio	Cohesion (kN/m^2)	Friction angle	Permeability (m/s)	v_p (m/s)	v_s (m/s)
PFA	18.5	20	0.3	10	30°	–	119.5	63.9
Surface fill	18.5	7	0.3	15	28°	–	70.7	37.8
Soft Marine clay	14	4	0.45	12.5	2°	5×10^{-10}	103.7	31.3
Hard soil	21	250	0.3	50	40°	5×10^{-8}	400.3	213.9

2.3 *Idealization of Pile Foundations*

Piles of length (L) 22 m having 300 mm diameter (D) and arranged in a 3D spaced square grid pattern were considered. Below embankment toe, a batter pile with a batter angle of 0°, 5°, 10° and 15° are considered. Piles were modelled as linear elastic isotropic material with a modulus of elasticity corresponding to M20 grade concrete, unit weight of 25 kN/m^3 and Poisson's ratio of 0.15.

2.4 Idealization of Geogrid Reinforcement

Geogrid with tensile modulus of 4000 kN/m and Poisson's ratio of 0.3 with linear elastic isotropic material property was used as basal geogrid.

2.5 Modelling

3-dimensional finite element modelling was performed using general-purpose finite element software ANSYS. PFA, surface fill and piles were modelled using SOLID185 element. SOLID185 is an eight noded element with three degrees of freedom at each node: translations in the nodal x, y and z directions. Soft marine clay and hard soil were modelled using CPT215 element. CPT215 is a coupled pore-pressure mechanical solid element. It is defined by eight nodes having four degrees of freedom at each node: translations in the nodal x, y and z directions, and one pore-pressure degree of freedom at each corner node. Geogrid was modelled using SHELL181 element with membrane effect. It is a four noded element with 3 translational degrees of freedom at each node. A slice of 3-Dimensional finite element model of embankment, pile, geogrid and viscous boundaries are shown in Fig. 2.

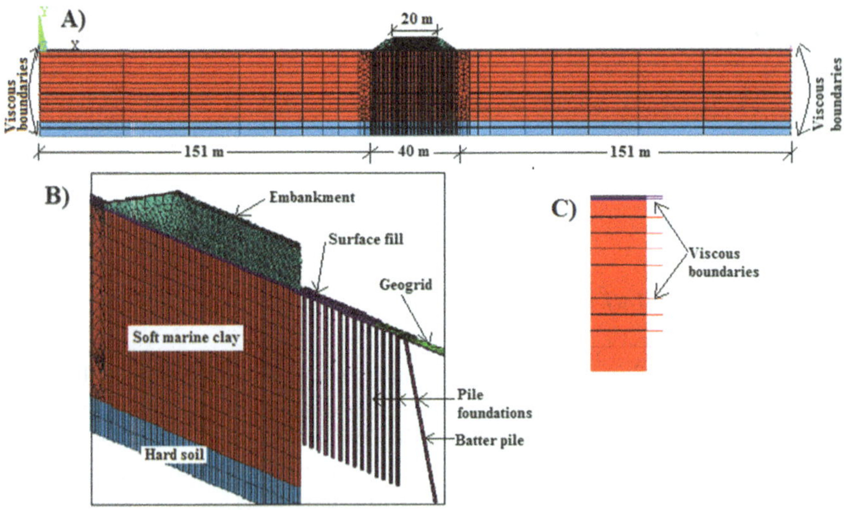

Fig. 2 **a** 3-Dimensional finite element model of embankment. **b** Enlarged view of geogrid reinforced pile-supported embankment. **c** Enlarged view of viscous boundaries

2.6 Boundary Conditions

For time-history analysis, Ghosh and Wilson [7] suggested that the lateral boundary should be at a distance of four times the base width of the embankment so that the waves propagated from the soil cannot reflect back. To simulate the infinite soil medium, viscous boundaries were applied for the lateral boundaries using spring-damper element given by Kianoush and Ghaemmaghami [15]. The equation of motion with additional damping matrix C^* can be written as follows when the viscous boundaries are taken into account.

$$[M]\{\ddot{u}(t)\} + [C]\{\dot{u}(t)\} + \left[C^*\right]\{\dot{u}(t)\} + [K]\{u(t)\} = -[M]\left\{\ddot{u}_g(t)\right\} \quad (1)$$

where, [M] is the structural mass matrix, [C] is the structural damping matrix, [K] is the structural stiffness matrix, $\{\ddot{u}_g(t)\}$ is the ground acceleration vector, $\{\ddot{u}(t)\}$ is the nodal acceleration vector, $\{\dot{u}(t)\}$ is the nodal velocity vector, $\{u(t)\}$ is the nodal displacement vector and [C*] is the special damping matrix that is considered as follows,

$$\left[C^*\right] = \begin{bmatrix} A_n \rho v_p & 0 & 0 \\ 0 & A_{t1} \rho v_s & 0 \\ 0 & 0 & A_{t2} \rho v_s \end{bmatrix}$$

where, v_p and v_s are the dilatational and shear wave velocity of the considered medium (Table 1), ρ is the density of soil medium, A_n, A_{t1} and A_{t2} are the fields controlling the viscous dampers and the subscripts n and t represent normal and tangential directions in the boundary.

2.7 Loading

The present study considers embankments situated at sites classified as earthquake zone III under seismic zoning of India. Hence the embankment is subjected to time-history of accelerations corresponding to Indian Standard (IS-1893) [13] code spectrum for Zone III scaled to 0.35 g. The time duration corresponding to bracketed duration was used for the analysis. The acceleration time-history record of IS code spectrum for Zone III scaled to 0.35 g is shown in Fig. 3. For time-history loading the macro developed by Sahin, 2010; was used in ANSYS software.

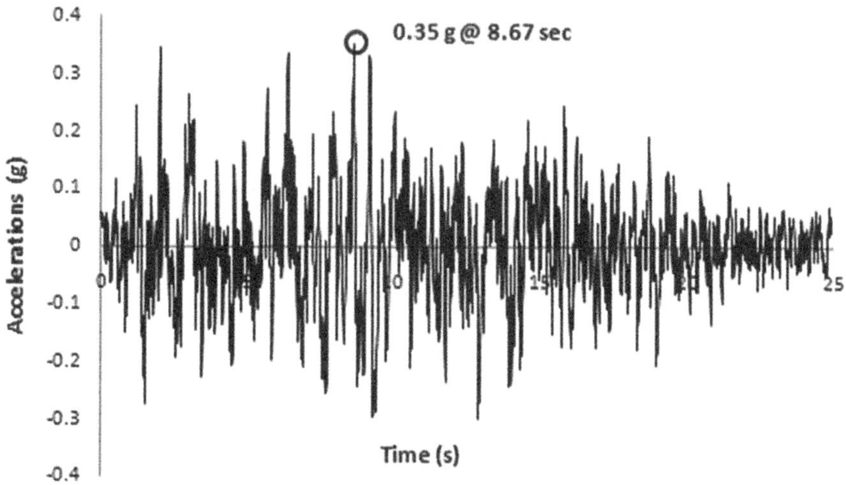

Fig. 3 Acceleration time-history record of IS Code Zone III ground motion scaled to 0.35 g

3 Results and Discussions

3.1 *Crest Vertical Displacements*

Figure 4 shows the time history of embankment vertical displacements at crest centre and crest edge for embankments supported over vertical and batter piles. Table 2 shows the maximum vertical displacements at embankment crest due to IS code zone III ground motion. For embankment toe supported on 5° batter piles, the crest centre and crest edge vertical displacements are about 2% lesser when compared with the toe supported over other batter piles. It is also observed from Fig. 4 that, for embankment toe supported over 0° and 5° batter piles crest centre vertical displacements are less than the crest edge vertical displacements. But in 10° and 15° batter pile-supported embankment, crest centre vertical displacements are more than the crest edge vertical displacements.

3.2 *Differential Settlements at Crest*

From Fig. 4 and Table 2 it is noticed that embankment supported over 5° batter piles experiences less differential settlements. It is also observed from Table 2 that increase in batter angle beyond 5° increases the differential settlements at crest.

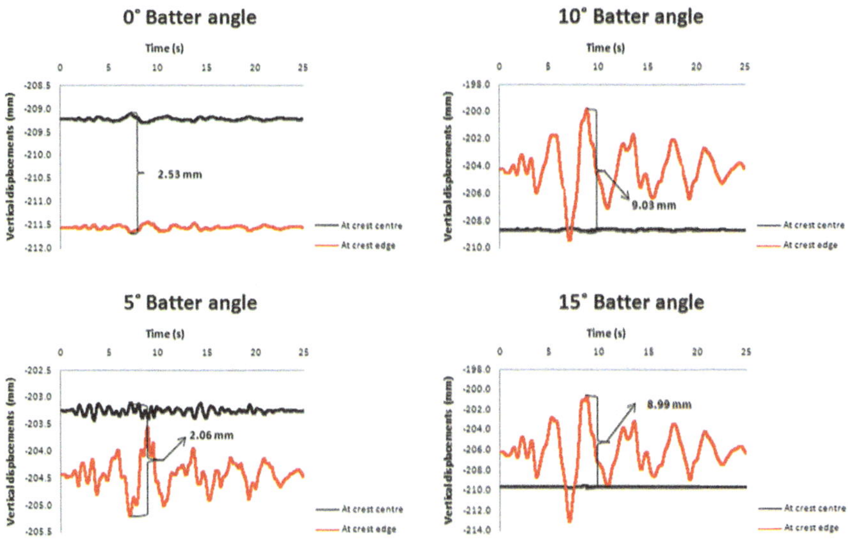

Fig. 4 Time history of embankment vertical displacements at crest centre and edge for embankments supported over vertical and batter piles

Table 2 Maximum vertical displacement and differential settlement at embankment crest

Batter angle	Maximum vertical displacement (mm)		Maximum differential settlement (mm)
	At crest centre	At crest edge	
0°	209.12	211.65	2.53
5°	203.4	205.18	2.06
10°	208.8	209.4	9.03
15°	209.73	212.98	8.99

3.3 Toe Lateral Displacements

Figure 5 shows the time history of toe lateral displacements for embankments supported over vertical and batter piles. From Fig. 5 it is observed that increase in batter angle reduces the toe lateral displacements. About 4.2% reduction in toe lateral displacements is seen by increasing the batter angle from 0° to 15°. Toe lateral displacements for 15° batter pile-supported embankment without basal geogrid are 2.5% higher than the embankment supported only with vertical piles and basal geogrid.

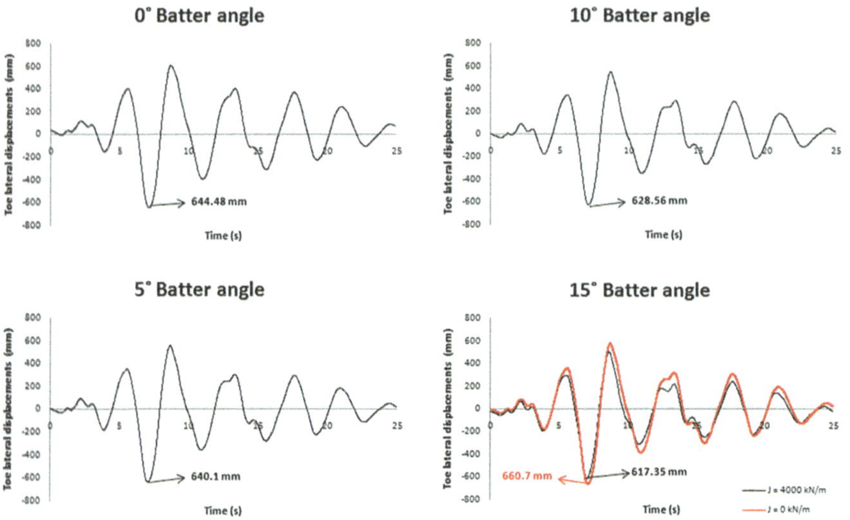

Fig. 5 Time history of toe lateral displacements for embankments supported over vertical and batter piles

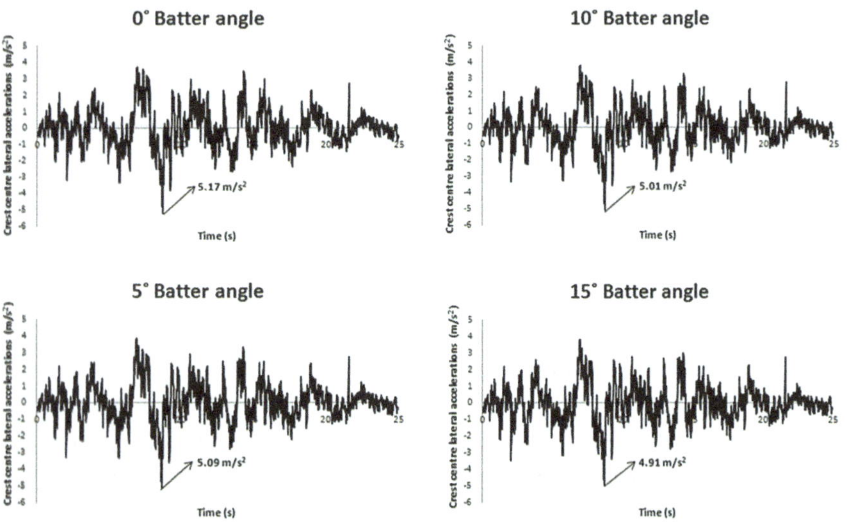

Fig. 6 Time history of crest lateral accelerations for embankments supported over vertical and batter piles

3.4 Lateral Accelerations at Crest

Figure 6 shows the time history of crest lateral accelerations for embankments supported over vertical and batter piles. From Fig. 6 it is noticed that increase in batter angle reduces the crest centre lateral accelerations. About 3% reduction in crest lateral accelerations were observed by increasing the batter angle from 0° to 5° and about 5% reduction in crest lateral accelerations were observed by increasing the batter angle from 0° to 15°.

4 Conclusions

3-Dimensional finite element analysis was conducted for the embankment supported over soft marine clay improved with vertical and batter piles along with basal geogrid subjected to IS code zone III ground motions.

From the analysis it is revealed that, higher batter angles (10°, 15°) lead to higher vertical displacements at crest centre than at crest edge. This also increases the differential settlements at crest.

Toe lateral displacements reduce with increase in batter angle. About 4.2% reduction in toe lateral displacements is seen by increasing the batter angle from 0° to 15°.

Toe lateral displacements for embankment supported over vertical piles and basal geogrid are less when compared with the embankment supported over vertical and 15° batter piles without basal geogrid.

Crest centre lateral accelerations reduce with increase in batter angle. More reduction in crest lateral accelerations occurs for 5° batter pile-supported embankment.

References

1. Armstrong RJ, Boulanger RW, Beaty MH (2013) Liquefaction effects on piled bridge abutments: centrifuge tests and numerical analyses. J Geotech Geoenviron Eng ASCE 139(3):433–443
2. Ariyarathne P, Liyanapathirana DS (2015) Review of existing design methods for geosynthetic-reinforced pile-supported embankments. J Soils Found 55:17–34
3. BS 8006: code of practice for strengthened/reinforced soils and other fills (2010) British Standard Institution, UK
4. Briançon L, Simon B (2011) Performance of pile-supported embankment over soft soil: full-scale experiment. J Geotech Geoenviron Eng 138(4):551–561
5. Bhasi A, Rajagopal K (2014) Geosynthetic-reinforced piled embankments: comparison of numerical and analytical methods. Int J Geomech ASCE 15(5):04014074
6. Bhasi A, Rajagopal K (2015) Numerical study of basal reinforced embankments supported on floating/end bearing piles considering pile-soil interaction. Geotext Geomembr 43:524–536
7. Ghosh S, Wilson EL (1969) Analysis of axi-symmetric structures under arbitrary loading (No. 69–10) EERC report

8. Gerolymos N, Giannakou A, Anastasopoulos I (2008) Evidence of beneficial role of inclined piles: Observations and summary of numerical analyses. Bull Earthq Eng 6(4):705–722
9. Giannakou A, Gerolymos N, Gazetas G, Tazoh T, Anastasopoulos I (2010) Seismic behavior of batter piles: elastic response. J Geotech Geoenviron Eng 136(9):1187–1199
10. Han J Gabr MA (2002) A numerical study of load transfer mechanisms in geosynthetic reinforced and pile supported embankments over soft soil. J Geotech Geoenviron Eng ASCE 128(1):44–53
11. Han GX, Gong QM, Zhou SH (2014) Soil arching in a piled embankment under dynamic load. Int J Geomech 15(6):04014094
12. IRC:113-guidelines for the design and construction of geosynthetic reinforced embankments on soft subsoils (2013)
13. IS 1893(Part 1): criteria for earthquake resistant design of structures (2016)
14. Jose BT, Sridharan A, Abraham BM (1988) A study of geotechnical properties of Cochin marine clays. Mar Georesour Geotechnol 7(3):189–209
15. Kianoush MR, Ghaemmaghami AR (2011) The effect of earthquake frequency content on the seismic behavior of concrete rectangular liquid tanks using the finite element method incorporating soil–structure interaction. Eng Struct 33(7):2186–2200
16. Liu HL, Charles W, Ng W, Fei K (2007) Performance of a geogrid-reinforced and pile-supported highway embankment over soft clay: case study. J Geotech Geoenviron Eng ASCE 133(12):1483–1493
17. Liu KW, Rowe RK, Su Q, Liu B, Yang Z (2017) Long-term reinforcement strains for column supported embankments with viscous reinforcement by FEM. Geotext Geomembr 45(4):307–319
18. Patel RM, Jayalekshmi BR, Shivashankar R (2019) Seismic response of Basal geogrid reinforced embankments supported over floating and end bearing piles. In: Silvestri, Moraci (eds) Earthquake geotechnical engineering for protection and development of environment and constructions, pp 4620–4628. Rome, Italy. ISBN 978-0-367-14328-2
19. Smith M, Filz G (2007) Axisymmetric numerical modeling of a unit cell in geosynthetic-reinforced, column-supported embankments. Geosyn Int 14(1):13–22
20. Shen P, Xu C, Han J (2017) Model tests investigating spatial tensile behavior of simulated geosynthetic reinforcement material over rigid supports. J Mater Civ Eng 30(2):04017288
21. Thach PN, Liu HL, Kong GQ (2013) Evaluation of PCC pile method in mitigating embankment vibrations from a high-speed train. J Geotech Geoenviron Eng 139(12):2225–2228
22. Thach PN, Liu HL, Kong GQ (2013) Vibration analysis of pile-supported embankments under high-speed train passage. Soil Dyn Earthquake Eng 55:92–99
23. Wachman GS, Biolzi L, Labuz JF (2009) Structural behavior of a pile-supported embankment. J Geotech Geoenviron Eng 136(1):26–34
24. Wang Z, Mei G (2012) Numerical analysis of seismic performance of embankment supported by micropiles. Mar Georesour Geotechnol 30(1):52–62
25. Wang Y, Orense RP (2019) Numerical analysis of inclined pile group performance in liquefiable sands. Silvestri, Moraci (eds) Earthquake geotechnical engineering for protection and development of environment and constructions. Rome, Italy, pp 5630–5637. ISBN 978-0-367-14328-2

Seismic Response of Low Height Embankment over Soft Foundation Clay

Debabrata Ghosh, Narayan Roy, and R. B. Sahu

Abstract Low height embankments are widely used for construction of highways and rural roads in India. Sometimes, it becomes necessary to construct such types of embankment on soft foundation soil under unavoidable circumstances. In addition to the stability under static loading, seismic excitation very often plays a major role in the stability of such an embankment. So, to study the behaviour of such embankment under seismic loading becomes very important as the failures may lead to huge loss of property and life. The present study analyses the response of a low height embankment overlying soft foundation soil under seismic loading with the help of finite difference analysis using FLAC. The Mohr-Coulomb failure criterion is used to define the soil constitutive model for both embankment and foundation soil. A stiffer half-space has been considered below the soft foundation soil and the analysis has been performed varying the shear wave velocity of the stiffer half-space. Low to moderate level of shaking has been applied at the base of the stiffer half-space by scaling the original input motion of 1987 Loma Prieta earthquake.

Keywords PGA · Soft foundation soil · Embankment · Seismic excitation · Excess pore pressure · Mohr-Coulomb

1 Introduction

Earthen embankment is a common civil engineering structure that serves a wide range of purpose, like, roadways, railways, earthen dams, etc. The sudden failure of an embankment as in the case of seismic event causes huge loss of property and life. Sometimes low height embankments are built on soft foundation clayey soil for construction of highways and rural roads without having an alternate choice. So, even if these kinds of embankments are safe against static failure but under seismic excitation they may undergo substantial deformations due to strength degradation. Ansal and Erken [1] reported that degradation of underlying soil does not occur if

D. Ghosh · N. Roy (✉) · R. B. Sahu
Jadavpur University, Kolkata, India
e-mail: narayan.roy04@gmail.com

T. G. Sitharam et al. (eds.), *Local Site Effects and Ground Failures*, Lecture Notes in Civil Engineering 117, https://doi.org/10.1007/978-981-15-9984-2_14

the number of equivalent load cycles is less even though the stress level exceeds the critical level. They indicated that performance of an embankment depends upon the frequency of the dynamic loading. Egawa et al. [2] conducted an extensive experimental study on the behaviour of embankment on soft peaty ground under seismic excitation generated by centrifuge and found that the deformation increases as the loading frequency approaches the natural frequency. Okamura and Tamamura [3] investigated the seismic stability of embankment on soft clay and peat deposits. The deformation of the embankment was shown to be concave type with maximum settlement at the crest due to the higher excess pore pressure during shaking. Rahardjo [4] discussed some case histories on embankment failures in Indonesia and showed that the main cause of bearing capacity failure of the foundation soil was excess pore pressure generation during the cyclic loading. Gordan et al. [5] investigated the dynamic behaviour of short embankment on soft soil using FE method simulating the earthquake loading for specified PGA and duration and the response was shown to be a function of modular ratio of embankment soil and foundation clay. Most of above studies are for embankment overlying a deep soft clay deposit. However, in general, in alluvial deposit, the top thick soft clay layer is generally underlain by a deep deposit of stiff/very stiff/hard silty clay/clayey silt/dense/very dense sand layer. When earthquake waves propagate upward through the stiff soil layer to the soft foundation soil, the motion that will finally reach to embankment may substantially get modified and affect the response of the embankment. With this background the present study highlights the seismic response of a 5 m high embankment built on 10 m thick soft foundation clay overlying a stiffer half-space. To study the influence of different bed-rock conditions, three different half-space shear wave velocities of stiffer layer have been considered. Also in order to study the effect of level of shaking, input motions with three different level of shaking have been considered by scaling the original input motion for each case of half-space velocity.

2 Methodology

Numerical modelling and analysis for a 5 m high embankment underlain by a soft foundation soil of 10 m thick have been performed using the finite difference software FLAC (version: 7.0). The soft foundation soil is again underlain by a half-space of stiff clay layer considered down to 40 m depth and the input motion has been applied at the base of this stiff half-space. Figure 1 depicts the embankment with foundation soil considered in the study. The engineering properties considered for the embankment and foundation soil layers are given in the Table 1.

For the stiffer half-space, three different shear wave velocities, 300, 500 and 800 m/s, have been considered for the analysis so as to find out their influence on the response of the embankment. Again for each half-space case, three different acceleration time histories with varying PGA values have been considered. For this purpose, the original time history has been scaled to three different PGA values. So,

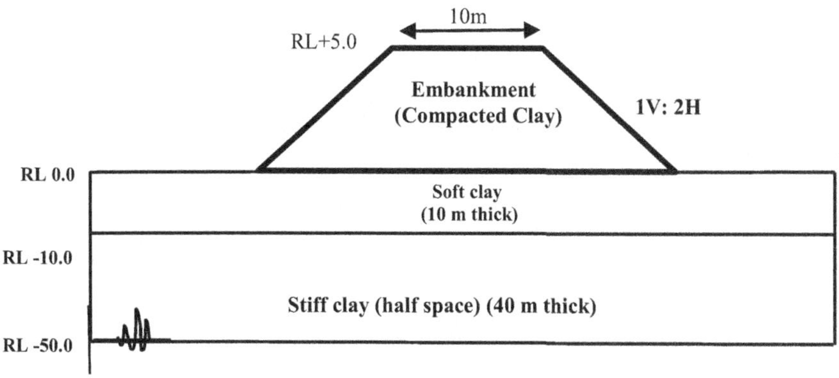

Fig. 1 Schematic of the considered embankment with foundation soil

Table 1 Properties of embankment and foundation soil

Properties	Unit	Embankment	Foundation soil	Half-space
Unit weight	kg/cum	1900	1700	2000
Cohesion (c)	kPa	40	32	70/80/100
Friction angle	Degree	20	0	0
Poison's ratio		0.35	0.45	0.35
Shear modulus (G)	MPa	42.75	17	180/500/1280
Porosity		0.35	0.4	0.3
Permeability (k)	m/s	10^{-7}	10^{-6}	10^{-7}
Shear velocity (Vs)	m/s	150	100	300/500/800

Table 2 Combinations of study considered in the analysis

Sl. No.	Model name	Embankment Vs (m/s)	Foundation Vs (m/s)	Half-space Vs (m/s)	Input motion PGA
1	Model 1	150	100	300	0.07g
2	Model 2	150	100	300	0.15g
3	Model 3	150	100	300	0.25g
4	Model 4	150	100	500	0.07g
5	Model 5	150	100	500	0.15g
6	Model 6	150	100	500	0.25g
7	Model 7	150	100	800	0.07g
8	Model 8	150	100	800	0.15g
9	Model 9	150	100	800	0.25g

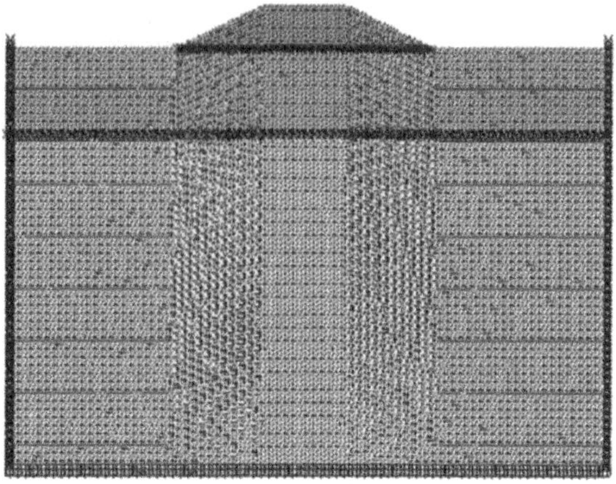

Fig. 2 A sample embankment model with generated mesh considered in the study

total nine combinations of models have been analysed in the current study. Table 2 presents the combinations of models analysed in the study.

At first, the static equilibrium of the embankment is checked. Figure 2 presents a sample discretized model using FLAC. The Mohr-Coulomb soil model is adopted for the analysis of the problem. The entire model is divided into the small rectangular/triangular elements for analysis with finite difference. The size of the element is decided on the basis of maximum considered frequency of the input motion and the minimum shear velocity of foundation soil. The element size has been decided considering the combination that causes the worst effect from the stability point of view. Now from acceleration spectrum analysis, the maximum considered frequency is obtained approximately as 20 Hz. Kuhlemeyer and Lysmer [6] showed that for an accurate representation of the wave transmission through the soil model, the spatial element size, Δl, must be smaller than approximately one-tenth to one-eighth of the wavelength associated with the highest significant frequency.

$$\text{wavelength } (\lambda) = \frac{\text{shear velocity}}{\text{significant maximum frequency}} = \frac{100}{20}\,\text{m} = 5\,\text{m}$$

Maximum linear dimension of mesh element $\Delta l \leq (\lambda/10) = (5/10)\,\text{m} = 0.5\,\text{m}$.

Hence, the rectangular mesh with element size of 0.5 m × 0.5 m has been considered in the modelling. For static analysis, the usual boundary condition with fixity at a distance of 20 m from the toe and 50 m from the base of embankment has been considered. For nonlinear dynamic analysis, the vertical boundaries are modelled in such a way that the reflection of the wave does not happen and the waves can radiate out to minimize the geometry effect. The water table is considered at the bottom of embankment, i.e. at the ground level (at RL 0.0).

Now the static analysis has been performed and the stability of the embankment is evaluated in terms of factor of safety, and a pattern of potential base failure is found for all cases as the base foundation soil is soft (Fig. 3). The displacement contours

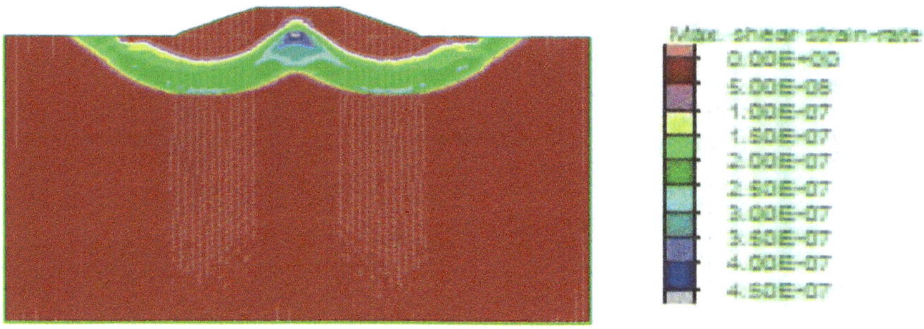

Fig. 3 Failure pattern after the static analysis of Model 1

show the maximum static displacements in between 20 and 23 mm at the crest of the embankment for all the models. The static factor of safety for all the considered models has been found greater than 1.5.

Now, once the static equilibrium is checked, the dynamic analysis is performed using three different acceleration time histories with PGA values 0.07g, 0.15g and 0.25g. The acceleration time history is of 40 s duration and the predominant frequency of the motion is 1.0 Hz. The input motion can be given mainly in three different ways: as acceleration time history, velocity time history and stress wave history. The input motion is given at level RL-50.0 m, i.e. throughout the bottom edge of half-space in form of a stress wave which corresponds to varying particle velocity with respect to the time period of 40 s. The expression for stress wave is given below.

$$\sigma = 2\rho C v_s \tag{1}$$

σ = stress generated, ρ = mass density, C = shear wave velocity, v_s = particle velocity.

Initially, the model is analysed without considering any kind of damping and then considering hysteresis damping. A small amount of Rayleigh damping (0.2%) is considered combined with hysteresis damping. At first, all the models are analysed without the embankment portion. Then the embankment loading is applied after which the static analysis is done. It is assumed that the seismic loading acts on the model after the completion of consolidation of the soft foundation clay due to embankment.

After static analysis, the stress wave history corresponding to the PGA magnitudes of 0.07g, 0.15g and 0.25g is applied at the base of the models. The acceleration time history is shown in Fig. 4.

After the completion of analysis for dynamic loading, different parameters like displacement vectors, displacement contours, stresses, etc., are evaluated for determining the response of the models under dynamic loading. A vertical displacement contour of Model 9 is shown in Fig. 5 that shows the maximum vertical displacement occurs near the top edge of the embankment. This model is subjected to input motion with PGA 0.25g and Half-space velocity 300 m/s.

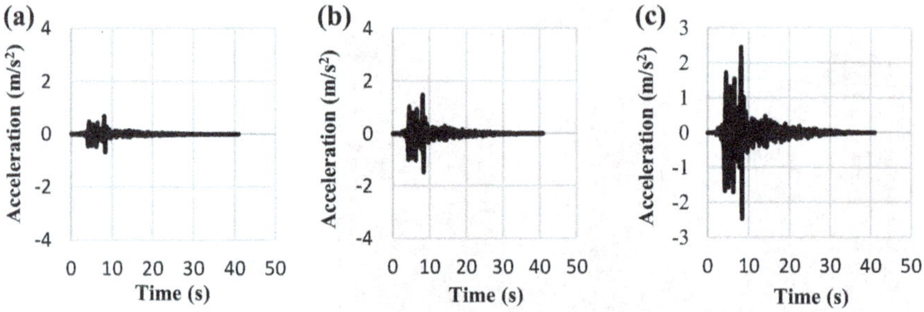

Fig. 4 Input motions considered in the study with PGA **a** 0.07g **b** 0.15g **c** 0.25g

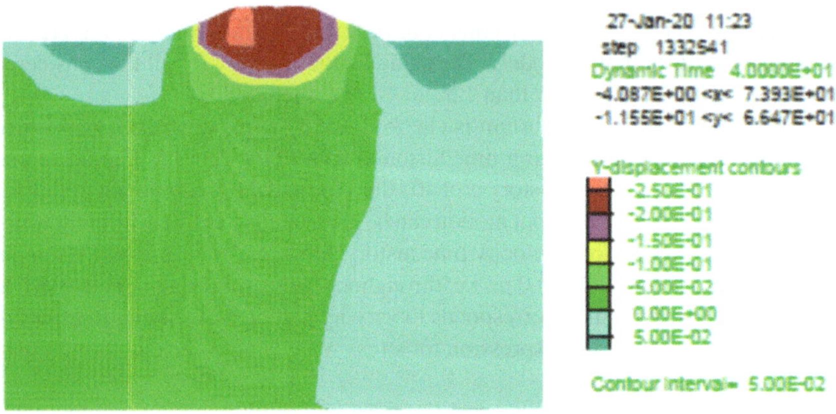

Fig. 5 Vertical displacement contour of embankment with Model 9

3 Results and Discussion

In the static analysis, the factor of safety is calculated for all the models and it ranges between 1.55 and 1.56. The failure pattern shows a probable base failure or the foundation soil would fail first and it appears to be upward movement of the base soil near the toe of the embankment. The base failure envelope is tangential to the interface of the foundation soil and stiff half-space. In Fig. 3 the shear strain vector of the Model 1 is shown after static analysis and factor of safety is found to be 1.556. The base failure pattern is clear from the Fig. 3. After ensuring the static stability the dynamic analysis is performed.

3.1 *Variation of Amplification*

After the dynamic analysis, the variations of PGA at different depth have been obtained and presented as amplification ratio with depth. Amplification ratio presents the ratio between PGA at a particular depth to the bed-rock PGA. Figures 6 and 7 present the variation of horizontal acceleration and PGA amplification ratio with depth for half-space velocity 800 m/s and 500 m/s, respectively, along the mid-width from bottom of model to crest of the embankment considering hysteresis damping condition. The figures clearly depict the increase in amplification ratio towards the surface. On the other hand, weaker motion exhibits higher amplification in comparison with strong motion, as in the case of strong motion the effect of

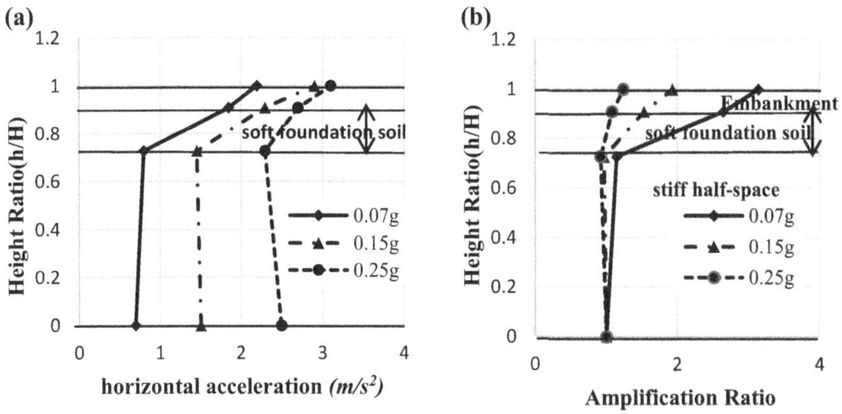

Fig. 6 **a** Horizontal acceleration versus height ratio **b** Amplification ratio versus height ratio from RL-50 for half-space Vs = 800 m/s) (with hysteresis damping) (Horizontal acceleration is in m/s^2)

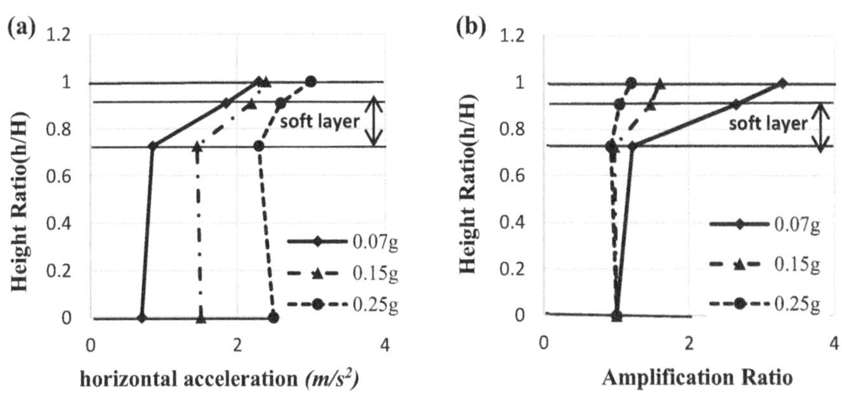

Fig. 7 **a** Horizontal acceleration versus height ratio **b** Amplification ratio versus height ratio from RL-50 with half-space Vs = 500 m/s with hysteresis damping (horizontal acceleration is in m/s^2)

Table 3 Vertical displacement at the middle of the crest after loading period

Sl. No.	Model name	Disp. static (mm)	Disp. (undamped) dynamic (mm)	Disp. (hysteresis) dynamic (mm)
1	Model 1	23	118	130
2	Model 2	23	156	176
3	Model 3	23	185	225
4	Model 4	21	88	83
5	Model 5	21	170	190
6	Model 6	21	230	260
7	Model 7	20	78	68
8	Model 8	20	189	215
9	Model 9	20	258	270

induced soil non-linearity in the soil layer is more prominent. The results exhibit that the soft foundation layer modifies the motion significantly in comparison with stiffer half-space for all the considered cases. So, the seismic response of an embankment constructed on soft foundation soil might be quite critical as the motion will get amplified substantially by the soft soil layer.

3.2 Variation of Vertical Crest Displacement

The variations in vertical crest displacements, at the middle of the embankment, have been presented in Table 3. Displacements have been tabulated from three different types of analysis, static, undamped and hysteresis damping analysis. The results show that displacements increase from static to undamped and it further increases in hysteresis analysis. Further, the analysis points that as the level of shaking increases the crest displacement of the embankment increases considerably; like for Model 7, 8 and 9 in case of hysteresis damping the vertical crest displacement for PGA 0.07g, 0.15g and 0.25g the displacement are 68 mm, 215 mm and 270 mm, respectively. The results also show that for low level of shaking, i.e. PGA 0.07g, when the half-space velocity increases the crest displacement decreases. For stronger shaking (PGA 0.15 and 0.25g), with the increase in half-space velocity the crest displacement is found to increase for a particular input motion.

3.3 Variation of Excess Pore Pressure

The comparison of excess pore pressure is shown in Fig. 8a, b and c for Model 7, 8 and 9, respectively, with input ground motion of PGAs 0.07g, 0.15g and 0.25g, respectively, for half-space velocity of 800 m/s. From the figures it is evident that

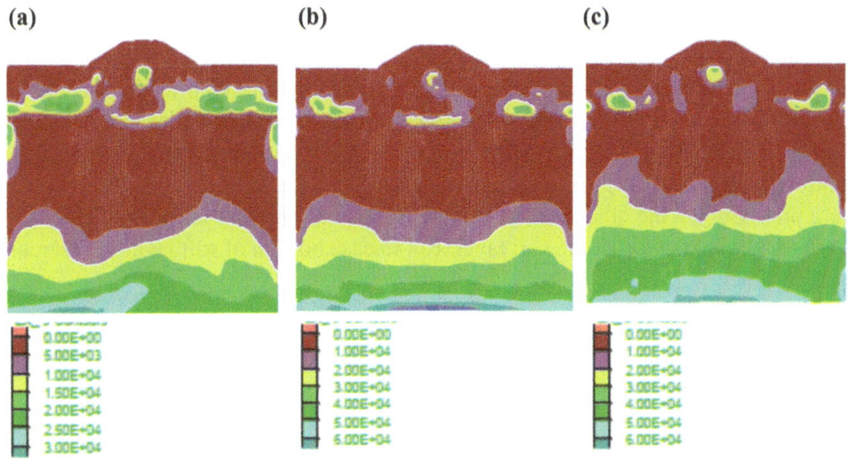

Fig. 8 Excess pore pressure contour after the dynamic analysis considering hysteresis damping for: **a** Model 7 **b** Model 8 and **c** Model 9

the excess pore pressure is maximum near the point of application of input ground motion and it gradually decreases while travelling upwards. Again, in the soft soil layer, a little built up of excess pore pressure is observed and this built up of excess pore water pressure is concentrated mainly in both the sides, away from the middle of the embankment, which tends to reduce the effective stress of the portion and could lead to instability.

Figure 9 depicts the variation of pore water pressure with time at the bottom, middle and top of the soft foundation soil layer. From the figure it is observed that

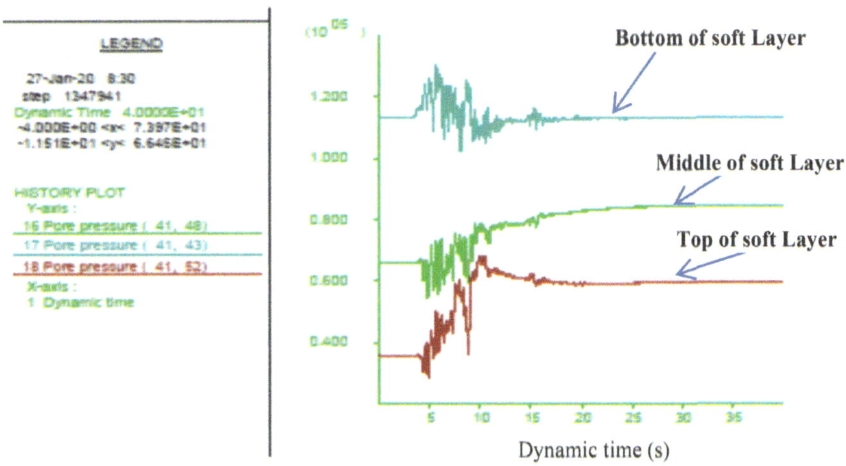

Fig. 9 Pore pressure variation at bottom, mid and top of soft foundation soil layer of Model 8 with time under hysteresis damping condition; X-axis: time (s); Y-axis: Pore pressure (Pa)

built up of excess pore water pressure is maximum at the top of the soft layer; whereas minimum excess pore water pressure is observed at the bottom of the soft layer. The reason probably is the substantial amplification of ground motion while it reaches at the top of the soft soil layer. From the Fig. 9, it can be found that at middle of soft layer, the static pore water pressure is around 67 kPa and after application of input motion (with PGA 0.15g) it rises to around 85 kPa, which is almost 25% increase in pore pressure. On the other hand, the increase in pore pressure is around 75% (from 35 to 61 kPa) at the top of the soft layer. But in case bottom of soft layer the increase is negligible.

4 Concluding Remarks

From the above analysis, following conclusions can be drawn:

(i) The analysis clearly shows a probable mode of failure is base failure due to the soft foundation soil and the crest settlement is in between 20 and 23 mm under static condition for all the models.
(ii) Analysis clearly depicts the increase in amplification ratio towards the surface. Weaker motion exhibits higher amplification (3–3.5) in comparison with strong motion (1–2). This decrease in amplification in case of strong motion is due to the effect of soil non-linearity which is appearing more prominent for higher level of shaking.
(iii) The vertical displacement of crest is found to increases for same property model with the increase in level of shaking as the input motion with higher shaking induces higher magnitude of stress wave with higher PGA. With the introduction of hysteresis damping the crest settlement is found to increase.
(iv) The excess pore pressure is found to decreases from bottom of stiff soil layer towards top of the model except in the soft foundation soil layer where little built up of excess pore water pressure is observed. In the soft soil layer, at the top and middle the pore pressure increment are around 75% and 25%, respectively, whereas, the pressure increment at the bottom is negligible.

References

1. Ansal AM, Erken A (1989) Undrained behaviour of clay under cyclic stresses. J Geotech Eng 115
2. Egawa T, Nishimoto S, Tomisawa K (2004) An experimental study on the seismic behaviour of embankments on peaty, soft ground through centrifuge model test. In: 13th world conference on earthquake engineering, Vancouver, B.C, Canada, Paper 36
3. Okamura M, Tamamura S (2011) Seismic stability of embankment on soft soil deposit. Int J Phys Model Geotech 11(2):50–57
4. Rahardjo PP (2014) Geotechnical failures case histories of construction on soft soils forensic investigations and counter measures in Indonesia. Int J Integr Eng 6(2):11–23

5. Gordan B, Adnan A, Aida MAK. Soil saturated simulation in embankment during strong earthquake by effect of elasticity modulus, vol 2014, Article ID 191460, 7 pp. Hindawi Publishing Corporation Modelling and Simulation in Engineering. http://dx.doi.org/10.1155/2014/191460
6. Kuhlemeyer RL, Lysmer J (1973) Finite element method accuracy for wave propagation problems. J Soil Mech Found Div 99(5):421–427

Influence of Toe Cutting on Seismic Response of a Typical Hill Slope in North-East India

Rubi Chakrabory and Arindam Dey

Abstract In hilly regions of India, excavation of slope is a common practice for construction of roadways. Toe cutting of slopes is repeatedly carried out for construction of new roads or widening of existing roads with rapid growth in population needing urbanization. In earthquake prone hilly regions such as North Eastern part of India, seismic activity can play the role of a major triggering agent for catastrophic slope failure due to non-engineered and impromptu toe excavation. In majority of the cases, effect of toe cutting on hill slope stability under earthquake condition is studied based on Limit Equilibrium (LE) pseudo-static method, wherein the safety is measured in terms of a time independent single factor of safety (FoS) value. However, pseudo-static approach, in which earthquake is represented by a constant inertia force acting on the slope, cannot simulate the earthquake condition accurately and may overestimate the earthquake force. Hence, in this paper rigorous dynamic analysis is performed for safe and economical excavation of toe regions of the slope before road construction. The study reveals that the conventional pseudo-static analyses provide conservative results, which when used as a basis of mitigation measures, will lead to uneconomical stabilization technique.

Keywords Toe cutting · Limit equilibrium · Pseudo-static method · Factor of safety · Dynamic analysis

1 Introduction

The primary means of transportation and communication in hilly regions of North-East India are the highways constructed by toe excavation of the hill slopes. Therefore, road cuts are often carried out for widening or construction activities along the hill slopes. But due to inaccessibility in hilly terrain, lack of detailed site investigation

R. Chakrabory
Department of Civil Engineering, NIT Meghalaya, Shillong 793003, India

A. Dey (✉)
Department of Civil Engineering, IIT Guwahati, Guwahati 781039, Assam, India
e-mail: arindam.dey@iitg.ac.in

T. G. Sitharam et al. (eds.), *Local Site Effects and Ground Failures*, Lecture Notes in Civil Engineering 117, https://doi.org/10.1007/978-981-15-9984-2_15

may lead to improper mitigation measures. Therefore, failure of non-engineered cut slopes is a common phenomenon in this part of the country, causing a high monetary loss and safety issues very often. The primary reason of slope failure is the reduction in confining stress in natural slope after excavation. Geometry of slope structure, soil shear strength and position of ground water table are the prime design parameters for slope excavation. For excavation of slopes with cohesionless soil, slope stability does not depend on height of cut and therefore slope angle is the most crucial parameter of design. For slopes with cohesive soils, the height of the excavation is the primary design parameter. For c-φ and saturated soils, both height and angle of slope are important in slope excavation design.

Development of roadway along hill slopes is subjected to very high risk taking into consideration the various past hazards reported around the world [7, 8]. Stark et al., in 2005 [9] investigated the cause of distress to a single-family residence located adjacent to a major highway cut slope. It was reported that the distress was caused due to a landslide induced by excavation made to widen an existing highway. Umrao et al., in 2011 [11] investigated the stability of road cut in rock slopes at five different locations along NH-109 in Himalayan territory of India, by the means of continuous slope mass rating (CSMR) technique. Kainthola et al., in 2015 [3] studied a high basaltic and lateritic road cut hill slope in their natural state at Mahabaleshwar, India. They investigated several critical parameters from the study i.e., mode of failure, factor of safety, shear strain rate, displacement magnitudes, etc. Mahanta et al., in 2016 [5] investigated stability of slopes in areas susceptible to failure determined from hazard zonation studies in the vicinity of Luhri village (NH-305), India. Finite element method (FEM) was used to investigate the deformation mechanism associated with such slope failures. Chakraborty and Dey, in 2018 [1] investigated the effect of vertical toe excavation on hill slope stability by means of deterministic LEM for a wide range of geotechnical, hydraulic, and seismic parameters. Based on the study recommendations were given regarding the critical horizontal extent of vertical toe cutting in hill slopes depending on the type of slope and different geotechnical parameters. In all the studies so far, effect of toe cutting on hill slope stability under earthquake condition is studied based on Limit Equilibrium (LE) pseudo-static method [1, 2, 4], wherein the safety is measured in terms of time independent single factor of safety (FoS) parameter. In pseudo-static method earthquake is represented by a constant time independent inertia force acting on the slope. Pseudo-static method overestimates the earthquake force and cannot simulate the earthquake condition accurately. Hence, rigorous dynamic analyses need to be performed for more accurate assessment of seismic response of slope in different stages of toe excavation and corresponding safety parameters, leading to risk-free and economical excavation of toe regions of the slope before road construction.

In this paper, a dynamic analysis is performed and compared with the outcomes from pseudo-static analysis, to assess the slope stability more precisely for various horizontal extent of vertical toe cutting under earthquake conditions. Firstly, the static and pseudo-static LE analyses of the slope, for different horizontal extent of vertical toe cutting (b_t) are performed using GeoStudio Slope/W v2007. Dynamic analysis is conducted using GeoStudio Quake/W v2007 and the stresses generated are incorporated in Slope/W module for evaluating the finite element FoS and its variation for the entire duration of the assigned earthquake time history. It has been observed that the pseudo-static analyses provide conservative results, which when used as a basis of mitigation measures, will lead to large structural dimensions of the chosen stabilization technique.

2 Static and Pseudo-static LEM Analysis for Slope Toe Excavation

In this section of the paper, a typical hill slope (Slope height = 40 m, Slope inclination = 30°, Cohesion = 50 kPa, Angle of internal friction = 20°, Unit weight = 18 kN/m^3) is analyzed under static and pseudo-static condition using LEM by Slope/W v2007. Defining the location of critical slip surface is of high importance in LEM analysis. There are various methods to specify the locations of trial slip surfaces in SLOPE/W 2007, namely Entry and Exit method, Grid and Radius method, Fully specified slip surfaces. In this study, Entry and Exit method is used. In this technique, the location of the trial slip surfaces can be specified with their likely entry and exit ranges, thus facilitating a large range of slip surfaces to be analyzed for assessing the critical stability of the slope. Among the different limit equilibrium methods available in SLOPE/W 2007, Morgenstern-Price method [6] is used and shear strength of soil materials is simplistically represented by Mohr-Coulomb's failure criterion.

In geotechnical earthquake engineering pseudo-static condition refers analyzing the seismic response of any geotechnical structure by simply adding a permanent time independent inertia force representing the earthquake force to a static LEM analysis. For present pseudo-static analysis, horizontal acceleration considered is 0.12g, which is equal to the peak horizontal acceleration (PHA) of the earthquake time history assigned for dynamic analysis in Sect. 3.

The results obtained for static and pseudo-static analysis are given in Table 1. From Table 1 it is noticed that in static condition, the FoS value is above unity (i.e., slope is safe) for a high range of horizontal extent of toe excavation (approximately

Table 1 Results of static and pseudo-static analyses

Horizontal extent of vertical toe cutting (b_t) (m)	Static FoS	Pseudo-static FoS
0	1.43	1.14
5	1.39	1.11
10	1.33	1.07
15	1.25	1
20	1.14	0.93
25	1.02	0.85
30	0.92	0.77

b_t= 25 m). Whereas, pseudo-static analysis reports a safe toe cutting up to 15 m. Therefore, it is understood that earthquake force can be a crucial agent to trigger failure in unplanned excavated slope for road construction or widening to such an extent (in this case beyond 15 m) without proper mitigation measures.

Among different techniques available in literature for slope stability study under dynamic loading, LEM pseudo-static is the simplest approach as it is based on simple mathematical calculations leading to less computational time. However, there are some inherent limitations associated with LEM based pseudo-static method. In LEM the failure mechanism is predetermined, it does not allow the failure surface to develop through the actual weakest path. Moreover, pseudo-static approach highly overestimates the earthquake loading acting on a geotechnical structure. As high risk and economy is associated with highway construction planning and design, relying completely on LEM based pseudo-static analysis for safety is not justifiable. To overcome the above-mentioned limitations, in next section of the paper a dynamic FEM based slope stability study is carried out where the test slope structure is assigned an earthquake time history for more accurate assessment of seismic response of the slope structure subjected to toe excavation in various stages.

3 Dynamic Analysis for Slope Toe Excavation

In this section of the paper, the same test slope (Elastic Modulus = 5000 GPa, Poisson's ratio = 0.3, Damping ratio = 0.1) as described in Sect. 2 is analyzed under equivalent linear dynamic condition by Quake/W v2007. For finite element based FoS calculation for the entire duration of the assigned earthquake time history (Fig. 1), Quake/W v2007 is coupled with Slope/W v2007. El-Centro seismic motion is used in the present study, which is scaled to a PGA similar to that experienced in the Northeastern region of India. Further, although the actual recorded motion was of longer durations, however, only 10 s motion is used in the present study that is obtained as the bracketed duration of the original recorded motion. The bracketed duration is obtained by applying the proposition by Trifunac and Brady [10]. The

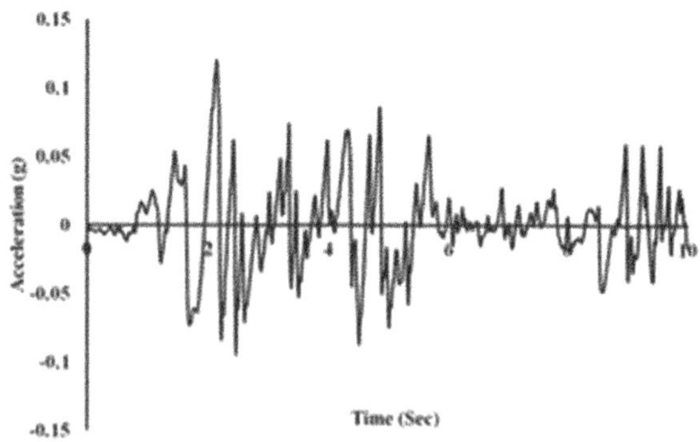

Fig. 1 Earthquake time history

stresses generated due to the earthquake motion assigned at the bottom of the slope structure in Quake/W analysis are incorporated in Slope/W for evaluation of time-dependent variation of FoS at different stages of toe cutting operation. Determining the location of critical failure surface having the lowest FoS is one of the key issues in a slope stability analysis. In LEM the location of the slip surfaces is predetermined, whereas FEM is capable to seek out the actual weakest path through the soil domain. Hence, FEM represents the failure mechanism more precisely.

The equivalent linear dynamic model is an approximation of the actual non-linear behavior of soil. With the Equivalent Linear model, a dynamic analysis is conducted in Quake/W with the specified *Gmax* value. Quake/W steps through the entire earthquake record and find the peak shear strain value for each iteration. Then the shear modulus value is modified according to the specified *G* reduction function as shown in Fig. 2a, and the process is repeated. This iterative process is continued until the required *G* modifications are within a specified tolerance. Figure 2b shows that *G* is constant during one iteration through the earthquake record and the change in slope shows the reduction in *G* value in next iterations. The cyclic shear strain is estimated from the FE analysis. The estimated shear strain together with the *G* reduction function and the specified *Gmax* value are used to evaluate new *G* values for each iteration. The damping ratio in QUAKE/W also can be specified as a constant or as a function. Similar to the *G* reduction function, the damping ratio is a function of the cyclic shear strain. Damping ratio is considered as constant in present analysis.

The FoS variation with time estimated from FEM based dynamic analysis is shown in Fig. 3. Dynamic analysis (Fig. 3) shows that up to 20 m of toe cutting FoS does not fall below unity for the assigned earthquake motion having a peak horizontal acceleration of 0.12g. Beyond 20 m of excavation, FoS falls below unity several times during the earthquake. It is understood that even if the FoS reaches the stability limit for certain instances of time, it does not lead to the failure of the slope, owing to the stress reversals in the following time instances. Hence, dynamic analysis reports that a toe excavation of 20 m may be carried out and beyond 20 m proper safety measures are required. Whereas, the same slope structure shows a safe

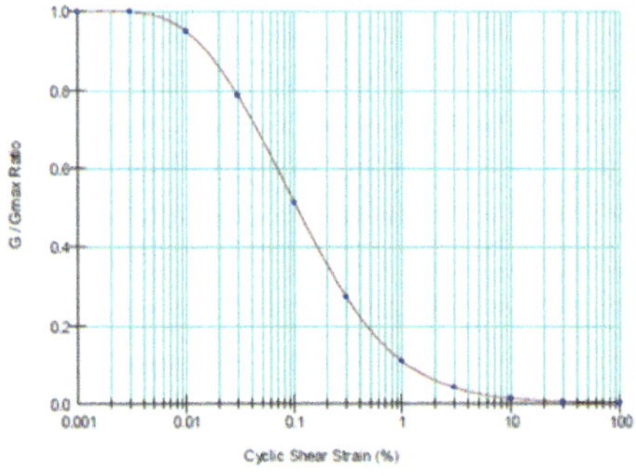

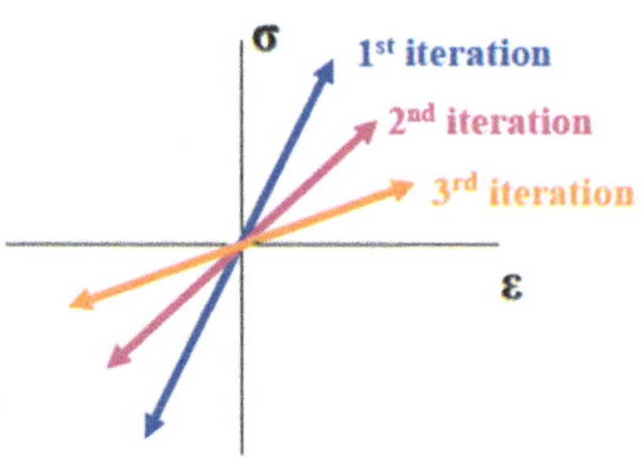

Fig. 2 **a** *G* reduction function, **b** change of *G* value for each iteration through the earthquake (adopted from Quake/W Engineering Book, 2007)

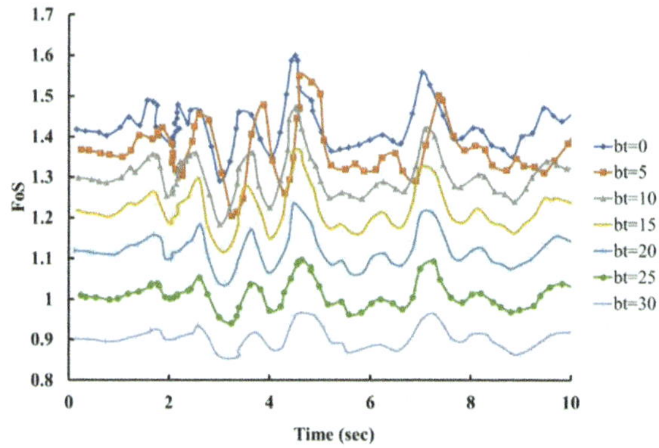

Fig. 3 FoS variation with time for dynamic analysis

horizontal excavation extent of 15 m under pseudo-static earthquake condition having horizontal acceleration equal to the peak horizontal acceleration of the assigned earthquake time history in dynamic analysis. Hence, pseudo-static analysis provides conservative results leading to uneconomical safety measures. It is worth mentioning that different other seismic motions could have been chosen for better understanding of the effect of seismicity on the stated problem. However, as the present study focuses primarily on the difference in the outcomes as obtained by LEM and FEM methods, the application of several other motions is not included in the present study.

4 Conclusions

In this paper, a report is provided against the safe excavation of the toe of a typical hill slope that is carried out based on pseudo-static LEM and FEM based dynamic earthquake analyses. The results from both the analyses are compared. As limit equilibrium methods do not exhibit the evolution of failure mechanism due to toe cutting, finite element (FE) studies have been conducted to delineate the development of failure mechanism with progressive toe cutting of the hill slope. A rigorous dynamic FE analysis illustrated the time-dependent factor of safety (FoS). It is understood that even if the FoS reaches the stability limit for certain instances of time, it does not lead to the failure of the slope, owing to the stress reversals in the following time instances. Under such conditions, it has been observed that the pseudo-static analyses provide conservative results, which when used as a basis of mitigation measures, will lead to large structural dimensions of the chosen stabilization technique.

References

1. Chakraborty R, Dey A (2018) Effect of toe cutting on hillslope stability. In: Anirudhan IV, Maji V (eds) Geotechnical applications. Lecture notes in civil engineering, vol 13. Springer, Singapore
2. Choudhury D, Basu S, Bray JD (2007) Behavior of slopes under static and seismic conditions by limit equilibrium method. Geo-Denver, Denver, CO
3. Kainthola A, Singh PK, Singh TN (2015) Stability investigation of road cut slope in basaltic rockmass, Mahabaleshwar, India. Geosci Front 6(6):837–845
4. Ling HI, Leshchinsky D, Mohri Y (1997) Soil slopes under combined horizontal and vertical seismic accelerations. Earthq Eng Struct Dyn 26:1231–1241
5. Mahanta B, Singh HO, Singh PK, Kainthola A, Singh TN (2016) Stability analysis of potential failure zones along NH-305, India. Nat Hazards 83(3):1341–1357
6. Morgenstern NR, Price VE (1965) The analysis of the stability of general slip surfaces. Geotechnique 15(1):79–93
7. Singh PK, Kainthola A, Singh TN (2014) Influence of rock mass parameters on the stability of high hill slopes. In: Indorock 2014, New Delhi, pp 577–587
8. Singh R, Umrao RK, Singh TN (2014) Stability evaluation of road-cut slopes in the Lesser Himalaya of Uttarakhand, India: conventional and numerical approaches. Bull Eng Geol Environ 73:845–857
9. Stark TD, Arellano WD, Hillman RP, Hughes RM, Joyal N, Hillebrandt D (2005) Effect of toe excavation on a deep bedrock landslide. Managing 244–255
10. Trifunac MD, Brady AG (1975) A study on the duration of strong earthquake ground motion. Bull Seismol Soc Am 65(3):581–626
11. Umrao RK, Singh R, Ahmad M, Singh TN (2011) Stability analysis of cut slopes using continuous slope mass rating and kinematic analysis in Rudraprayag district, Uttarakhand. Geomaterials 1(3):79–87

August, 2019 Landslide Events in Kinnaur, H.P.—An Assessment of Earthquake and Landslide Consequences Using Satellite Data

Madan A. Mohan, Vidya Sagar Khanduri, and Amit Srivastava

Abstract On August 18, 2019, the Kinnaur region in the Indian state of Himachal Pradesh experienced many landslides due to heavy rains. In H.P.'s Kinnaur district, National Highway 5 was blocked, heavy rainfall triggered many landslides in isolated regions of Himachal Pradesh, blocking large number of roads, it was blocked near Ribba village of Kinnaur district causing heavy traffic jam. According to reports, this event resulted in the death of 22 people and a significant loss of property. This study presents the results of the Kinnaur landslide damage assessment derived from the analysis of very high-resolution images (VHR) received from different satellites. These datasets were obtained through a coordinated effort by the government organisation major disasters. The damage is mainly attributed to rock slides from the area of the slope that later became debris flows by sanding the material along the exit zone.

Keywords Landslides · Heavy rainfall · VHR · Earthquake · Damage assessment

1 Introduction

Topographical, meteorological, physical and anthropogenic factors are the major causes of slope failures. Topographical factors include terrain, curvature, drainage, slope, elevation, landslide spatial distribution and frequency control. Meteorological factors consist of rainfall intensity, rainfall magnitude and climatic conditions. Physical factors consist of type of soil, structure, texture, roughness and its underlying geology. Anthropogenic factors include human activities like over grazing of animals, cutting of trees, forest fires, excavation of slopes for construction of

M. A. Mohan (✉) · V. S. Khanduri · A. Srivastava
School of Civil Engineering, Lovely Professional University, Jalandhar, Punjab, India
e-mail: chinnu2madan@gmail.com

V. S. Khanduri
e-mail: vidya.18579@lpu.co.in

A. Srivastava
e-mail: amit.22986@lpu.co.in

T. G. Sitharam et al. (eds.), *Local Site Effects and Ground Failures*, Lecture Notes in Civil Engineering 117, https://doi.org/10.1007/978-981-15-9984-2_16

roads, mineral mining, laying of pipelines, etc. [11]. There are many records kept regarding occurrence of earthquake and rainfall data causing heavy destruction in Kinnaur district. Due to the development in urbanisation, construction of roads and dams, environmental changes and global warming, the terrain condition is changing leading to landslides. So, to take preventive measures and reduce risk Landslide Hazard Zonation is prepared with the help of very high-resolution images (VHR) taken from different satellites. Landslide Hazard Zonation is the process in which an area is selected and it is divided into different subclasses based on the landslide susceptibility which consists of different parameters and doing the assessment of earthquake induced landslides. Since it consists of lot of factors to be considered, the analysis of the Landslide Hazard Zonation becomes a complex work and requires skilled and experienced person. Landslide Hazard Zonation can be done in both qualitative and quantitative approach [14]. The quantitative approach includes logistic regression, linear regression and discriminant regression and qualitative approach includes parameter weighted method, site visitation and recording, landslide hazard mapping, soil testing.

The objective of the study is

- To investigate and understand the impact of landslides in Ribba region which lead to economic and physical loss by using landslide susceptible mapping.
- To evaluate the hazard associated with Ribba landslide by using very high-resolution images, rainfall and earthquake data.
- To determine the geotechnical properties of the soil sample collected from Ribba landslide region.
- To suggest slope stability analysis and its frequency by using kinematic and finite element analysis.

2 Area and Location

Kinnaur is a Trans-Himalayan region which is located in the bank of Sutlej and tributary of Baspa. It is the east frontier district of Himachal Pradesh. The total area of Kinnaur is about 6401 km^2. Ribba is a village panchayat located in Kinnaur district, Himachal Pradesh, India. The coordinates of Ribba are longitude 78.364 and latitude 31.584. In eastern part of Kinnaur lies Ngari region located in Tibet. In southern region, Douladhar ranges which separates Shimla district of Himachal Pradesh and Uttarkashi district of Uttar Pradesh. Srikhand Dhar makes boundary between Kuku and Kinnaur in the western region. In norther region Spiti is located which shares Indo-Tibetan border (2017) (Fig. 1).

The elevation of Kinnaur varies between 1220 and 3050 m. But in mountainous region the average height of the peaks is 5500 m which consist of ice and snow throughout the year. Kinnaur district consists of rivers, glaciers, deep valleys gorges and very high mountains.

The mountains of Kinnaur are full of craggy rocks of massive heights perilous gradients in spurs. The mountains are covered with huge boulders and cliffs which

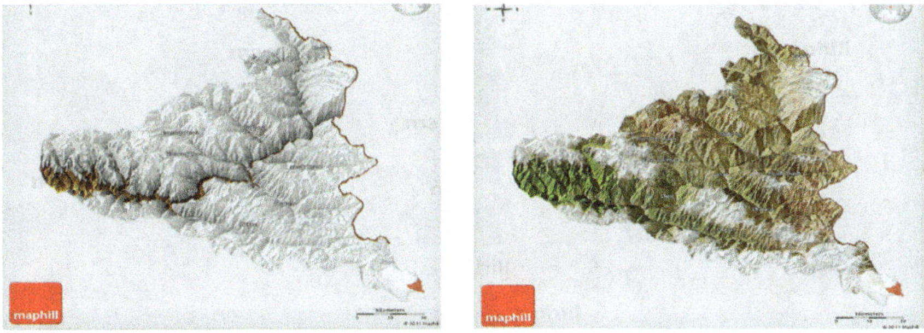

Fig. 1 Physical and satellite map of Kinnaur district [8]

makes a dangerous location during landslides. Rainfall occurs throughout the year in this region. As the district is located in the higher region, it has temperate climate with longer winters from October to May compared to remaining parts of India (2017).

From 2011 census records of India, Kinnaur district population is 84,121, with population growth rate of 7.61%. The district consists of 65 panchayats with 12,503 households. Due to tourism and verity of culture tourist from different parts of the world are attracted to this region (2017).

2.1 Zone of Landslide in Ribba Village

The landslide of Ribba village is located at 31°59′20″N and 78°36′49″E, along which another landslide zone is located called Pawari landslide zone where the landslides are common. These two zones are located near the bank of Sutlej River. Recently on August 18, 2019 due to heavy rainfall a massive landslide was occurred in Ribba landslide zone due to which National Highway 5 was blocked. The width and length of landslide zone are 2.58 km and 859 m, respectively, and has an area of about 2.21 km^2. [9] (Fig. 2).

The landslide occurred at MSL of 1993 m and at an altitude of 2850 m. The surrounding villages are Rarang, Akpa, Panah and Purbani. In this landslide zone the northern part is a cliff with debris flow and rock boulders. In the southern zone there is a valley of Sutlej river. By geomorphic characteristics, this zone has steep angle of slope with a difference of 50° and 65° in the northern region and 40° and 55° in the southern region [1].

3 Frequency of Earthquake and Rainfall

The rainfall data can be obtained from meteorological centre, Shimla which can be used to observe the pattern of rainfall occurring in Kinnaur district. With this data

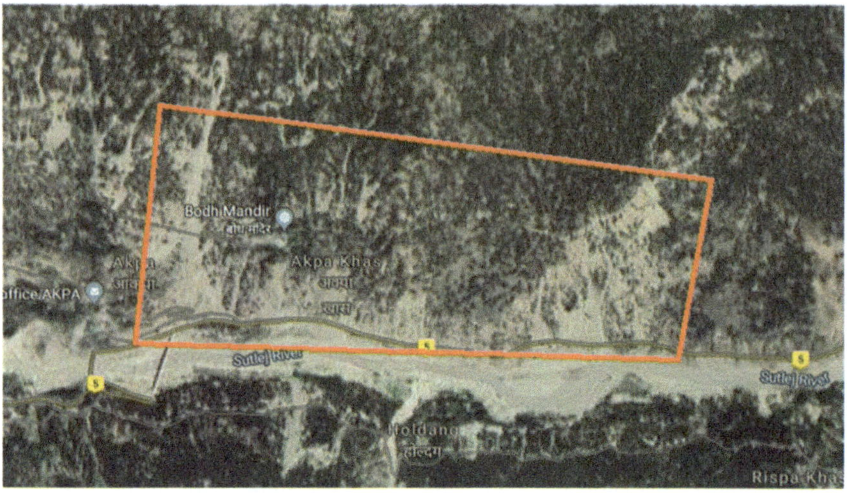

Fig. 2 Ribba village landslide zone [9]

the affected area of landslide is accessed. The earthquake data is compiled from 'The tribune' [13].

The largest earthquake occurred in Kinnaur district is '1975 Kinnaur Earthquake' with a magnitude of 6.8 and causing extensive damage in whole Himachal Pradesh. Its epicentre was in Kinnaur district in the southern east of Himachal Pradesh. Earthquakes are common in Himachal Pradesh which can be varied from low intensity to high intensity. Since the state of Himachal Pradesh lies toward the southern boundary of Himalayan region belt, the collision between Eurasian plate and Indian plate causes ground movements. There are a lot of incidents of earthquake recorded in this region by Indian meteorological department.

The district receives rainfall from the month of June till September. The marginal shift in the monsoon pattern has been noticed over the period of years. The average annual rainfall in the district is 816 mm. However much of the rainfall is received in parts of lower Kinnaur. There is a progressive decrease in rainfall as one goes from west to east. The parts of upper Kinnaur receive more snowfall than the rainfall. The district receives heavy snowfall from November end or early December till March or sometimes April. During this period, the area remains totally cut off from rest of the area. The higher peaks of Kinnaur district are completely covered with snow throughout the year (Tables 1 and 2).

4 Data Preparation for Landslide Hazard Zonation

The required data of landslide can be obtained from Geological Survey of India, National Remote Sensing Centre which gives the satellite images. By using GPS and Google Earth the present situation of landslide can be obtained for mapping. With

Table 1 Earthquake data of past 10 years [7]

S.no	Date	Intensity	Epicenter	Casuality	Remarks
1	28, May 2010	4.8	Shimla dist	No	
2	02, Oct 2012	4.5	Border of districts of Chamba, Lahaul and Spiti	No	
3	12, Nov 2012	4.1	Baijnath in Douladhar ranges	No	
4	05, Jun 2013	4.8	Border of the districts of Chamba and Lahaul-Spiti	No	The region falls in seismic zone-IV
5	26, Apr 2015	7.9	Nepal	Killed-36 Injuries-200	
6	27, Aug 2016	4.6	Kullu Region	No	
7	27, Oct 2017	4.4	Mandi Region	No	
8	25, May 2018	3.6	Kinnaur Dist	No	
9	03, Feb 2019	3.5	Kangra Dist	No	
10	03, May 2019	4.2	North east of Mandi	No	

Table 2 Rainfall data of Kinnaur district of past six years [4]

Year	2014	2015	2016	2017	2018	2019
Jan	78.1	67.2	16.1	135	1.8	81.8
Feb	140.7	166.7	44.3	38.6	36.8	154.6
Mar	69.5	228.2	79.2	69.5	74.6	102.2
Apr	52.9	32.7	105.2	121.1	40.1	9.4
May	25.6	44.1	15.2	17.4	36.0	69.4
Jun	7.3	44.1	10.1	39.4	24.9	38.2
July	25.9	35	29.8	41.9	51.9	19
Aug	8.6	23.3	49.2	27.5	24.5	53.7
Sep	10.2	37.8	27.2	51.6	78.8	9.5
Oct	1.4	4.5	1.7	0.1	0.0	0
Nov	0.2	8.9	0.0	21.5	35.7	74
Dec	39.3	18.0	5.0	29.9	13.3	0
Total	426.5	710.5	383	593.5	645.1	611.8

the collected information, by using a software called ARC GIS 10 a thematic layer of the landslide can be developed by considering the factors of land cover, slope angle, type of soil, curvature, density of drainage, geological features and relief. By using previous records of landslides occurred in the Kinnaur district with different properties like landslide pattern, geological characteristics, runoff distribution and earthquake pattern an inventory is prepared. In Kinnaur District a total of 6322 landslides have been occurred in the entire district. Using the image processing techniques and field investigations in detail gives the type of landslide, its mechanism, failure pattern and its causes. There are two main aspects of hazard analysis, Aspect and Slope [6]. As the distance of change in slope increases the slope value varies, if the slope value is more then it represents the steeper terrain and vice versa. The slope value can be calculated by

$$\text{Slope (In degrees)} = 57.3 \tan^{-1} \sqrt{(\mathrm{d}z/\mathrm{d}x)^2 + (\mathrm{d}z/\mathrm{d}y)^2}$$

The slopes which have been modified is classified into five zones with different land use pattern. Now the steepest slope direction gives the aspect and can be calculated by

$$\text{Aspect} = 57.3 \tan^{-1} 2[(\mathrm{d}x/\mathrm{d}x) + (\mathrm{d}z/\mathrm{d}y)]$$

x and y are the direction of slope which are represented by dz/dx and dz/dy [6].

5 Stability Analysis

Since the landslide zone of Ribba village consist of rock mass and boulders the stability analysis is done to determine the slope deformation and rock failure mode. Now the rock failure mode can be determined by using kinematic analysis based on stereonet and slope deformation is obtained by using shear strength reduction technique and plain strain (FEM).

5.1 Rock Failure Mode by Kinematic Analysis

With the help of stereographical projection method, the rock exposure, joint orientation data and stereo plot can be obtained in the northern part. The kinematic analysis gives the possible failure mode and direction in a rock mass by using angular relationship data between slope surface and discontinuities [14].

5.2 Finite Element Analysis

In the finite element analysis, a two slope model is used to determine the failure criteria of slope. Generalised Hooke brown model gives the failure criteria for slope 1 by using the following formula

$$\sigma_1 = \sigma_3 + \sigma_{ci}[m_b(\sigma_3/\sigma_{ci}) + s]\,\Delta(a)$$

σ_1	Maximum principal stress.
σ_3	Minimum principal stress.
σ_{ci}	Intact compressive strength (mpa).
m_b, s and a	Hoek–Brown parameters.

Now Generalised Hooke brown model is dependent on a parameter called Geological Strength Index (GSI) of rock mass. The GSI can be obtained by correlating discontinuity condition and rock mass blocky nature which has a standard chart proposed by Cai [3].

After the finding out the failure criteria of slope 1, Mohr coulomb criteria is used to determine debris flow in slope 2 with the formula

$$\tau = c + \sigma \tan\theta$$

τ	shear stress or shear strength of material.
c	cohesion of material.
θ	angle of internal friction.

By using these two failure criteria, a model is prepared in CartoSat-1 Digital Elevation Model which is used for execution of the slope parameters for finite element analysis.

6 Soil Properties

By performing soil experiments on the soil sample collected from the Ribba landslide it is found that the soil is poorly graded with coarse and very fine grade category.

- The average specific gravity of the soil is 2.47.
- By performing sieve analysis, the following graph is plotted in which the value of coefficient of uniformity is 16.84 and coefficient of curvature is 0.31.
- In Atterberg limits, Liquid Limit (LL) is 23.339%, Plastic Limit (PL) is 19.67% and shrinkage limit (SL) is 17.05%.
- From standard proctor test, a compaction curve is plotted to obtained optimum moisture content (OMC) of 25.1% and maximum dry density (MDD) of 1.744 gm/cc (Figs. 3, 4 and 5).

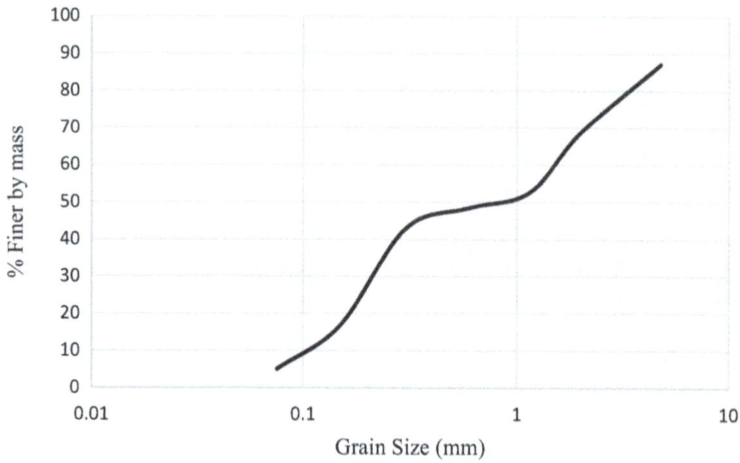

Fig. 3 Particle size distribution

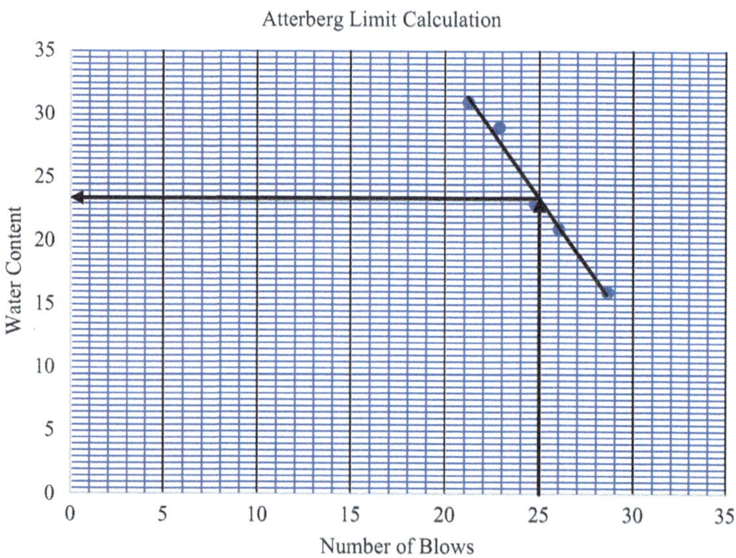

Fig. 4 Atterberg limit calculation

7 Landslide Susceptibility Mapping

There are different factors of mapping which are responsible for landslide susceptibility. The previous landslides data and information are accessed so that these can be used for future analysis of landslides to be occurred. A landslide inventory is used to prepare a whole map showing the old and currently active landslides and

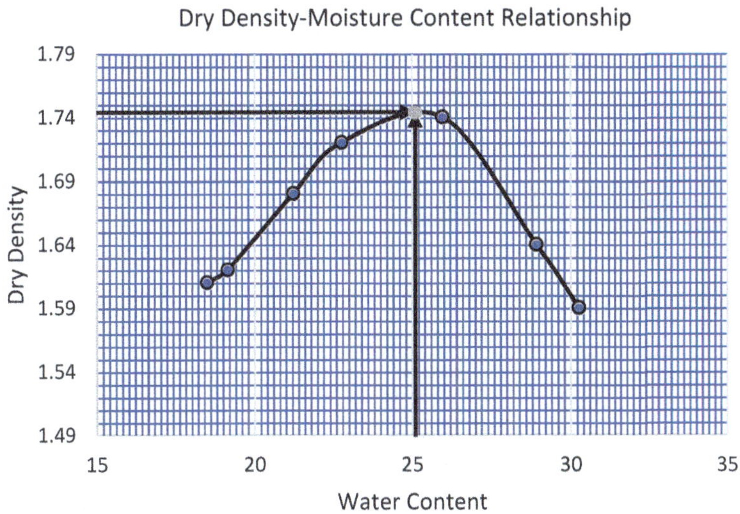

Fig. 5 Determination of optimum moisture content and maximum dry density

to perform further investigation in those regions. Thematic layers of these map are prepared by overlaying the different factors showing geological, morphological and land use conditions [5]. Finally, a model is prepared on the landslide susceptibility for assessment which can be qualitative or quantitative approach. Landslide vulnerability includes different aspects based on physical, land use and land cover, people perception, comprehensive vulnerability (Fig. 6).

7.1 Morphometric Properties

Due to heavy rainfall in this region, a high amount of volume of water is discharged with greater velocity which leads to erosion of soil. As the slope has steeper gradient the flow of water along with debris is high and this flow washes away the pavement constructed and finally into the Sutlej river. The bifurcation ratio is defined as the ratio of number of streams of given order to number of streams to next order. The bifurcation ratio gives us the physical loss of the soil mass (Fig. 7).

Here the Sutlej River has a low bifurcation value which says that it has a controlled bedrock. Many streams are formed from higher ridge, and due to continuous flow, the boulders and soil gets washed away into Sutlej basin.

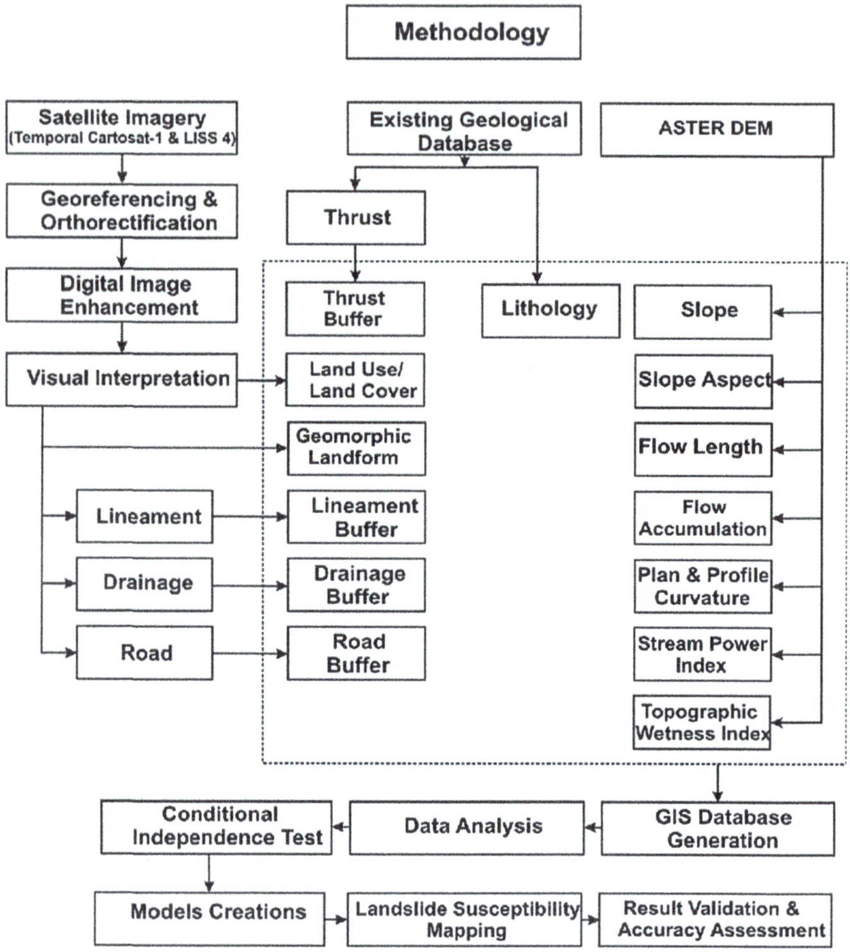

Fig. 6 Landslide susceptibility mapping

Fig. 7 Flow of rainwater along with debris and soil blocking NH 5

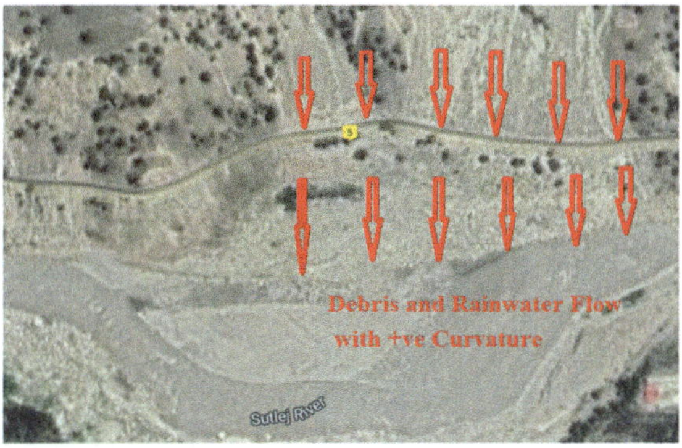

Fig. 8 Positive curvature of landslide

7.2 *Curvature of Landslide*

The deceleration and acceleration of the flow due to profile curvature affect the nature of the landslide. If the curvature has a negative value then it shows that the surface is upward convex and the flow will be decelerated. If positive then the surface is upward concave with accelerated flow (Fig. 8).

Here the landslide occurred at Ribba has a positive curvature due to which high amount of movement of soil mass, boulders and debris took place.

7.3 *Lithology of the Location*

The lithology of the location is obtained from the geological survey of India and google maps. It is found that the area of Kinnaur district is mostly covered with type of peat with the mixture of limestone, phosphate, quartzite, greenish grey sandstone with area of 29%. The affected area was about 1.2% which consists of boulders, peat, sediments of sandstone and clay. The area of forest covers with limestone, shale which are highly vulnerable to floods, soil erosion and landslides [1].

7.4 *Hydroelectric Projects*

Construction of hydroelectric project is a major issue which finally leads to slope failures and landslides. If surrounding of the Sutlej is considered it is observed that slope instability is common causing sliding of soil mass and boulders. There are four hydroelectric projects constructed in Sutlej basin and are different in type which were under investigation, clearance, one was commissioned and another was under construction. It was found that the maximum number of landslides were occurred due to under construction type and were causing risk. As the continuity of construction

Fig. 9 Hydroelectric project [10]

activities, the chances of landslides in the surrounding region are increasing. The wildlife sanctuary, forestland and topsoil and many villages were threatened under hydroelectric projects [2] (Fig. 9).

When these hydroelectric projects are constructed blasting technique and tunnelling is done which will cause ground disturbances and finally causing landslides.

7.5 Occurrences of Earthquake

Land movements due to earthquake in Kinnaur region have an intensity between 4.0 and 6.1. The average magnitude recorded is 4.5 and low intensity earthquakes commonly occur in this region. By comparing with the Mercalli intensity scale Kinnaur district falls under IV to V. Every time when a low intensity earthquake is occurred, the tremors can be felt in the surrounding region, with cracks developed in buildings, pavement failure, spalling, settlement, utility disturbances. People living in the surrounding also explained about how they experienced those tremors (Fig. 10).

Even due to blasting operations and tunnelling construction they felt those vibrations. Due to these conditions every village in Kinnaur district comes under earthquake vulnerability zone.

8 Questionnaire Survey

A questionnaire survey was constructed in Ribba landslide zone to gather the information about the landslide occurred based on different parameters. Out of 70 people 50 people told that the vibrations which they observe are due to blasting operations

Fig. 10 Small range seismic waves due to blasting and tunnelling operations

and tunnelling activities. When heavy rainfall occurs and a huge amount of water is released through dam causing heavy flow of river finally leads to slope failure told by 43 people. All 70 people told that the loss of property and life were common in this region. 27 people told that most of the landslides are both overburden and boulders sliding. Some of them also stated that there are no dumping site and the waste materials were dumped near river or beside roads. Some told that due to deforestation activities, the soil in this region is getting eroded.

9 Conclusion and Discussions

From the study it was found that the Ribba village zone is highly vulnerable to landslide because of its clumsy geophysical nature. Most of the landslides in this region are occurring due to blasting operations as the slopes are steeper so failure occurs easily. By considering morphological factors it can be stated that this region comes under highly vulnerable region. The use of hydropower plant was causing disturbances in soil and also ecological imbalance in Sutlej valley. Further construction activities near the steeper region should be avoided. The recent landslide occurred in Ribba village lead to 22 persons death, and many households got destroyed. The Ribba landslide zone comes under high vulnerability with an impact of 70% failures if any disaster occurs. For further studies it is necessary to consider detailed study of slope stabilisation and rock mechanics and these research works should be undertaken by experienced institutes and departments. The following pie chart shows the research work done or are to be carried out in different areas by considering previous research papers or review papers. Total of eight papers have been found related to Kinnaur district landslides in which these have been divided into six categories in which further investigation can be carried out (Fig. 11).

It can be seen that most of the work has been done in assessment of impact of landslides causing economic and physical loss and investigation of slope failure, ground movements due to blasting operations. Even a little work was also carried in vulnerability assessment in Kinnaur district but still the proper data has been not

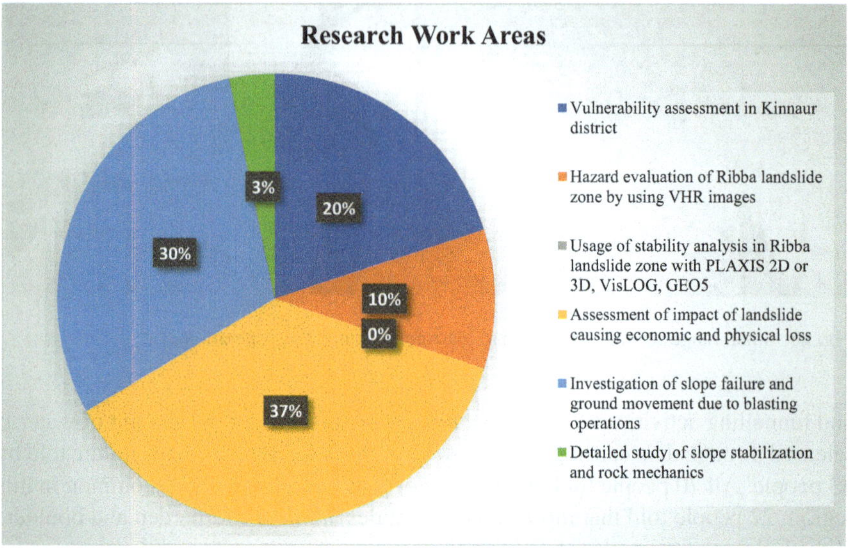

Fig. 11 Pie chart showing research work done or to be carried out in ribba landslide zone [1–3, 5, 6, 10, 12, 14]

provided in any of their research work regarding Ribba landslide zone. So, the main area of work can be carried out in slope stabilisation and rock mechanics analysis. Moreover, no stability analysis has been done in Ribba landslide zone by using PLAXIS 2D or 3D, VisLOG and GEO5 software.

References

1. Amit Jamwal NK (2019) Use of geographic information system for the vulnerability assessment of landscape in upper Satluj basin of district Kinnaur, Himachal Pradesh, India. Geol Ecol Landscapes
2. Asher M (2019) Hydropower projects ruining the lives in the himalayas
3. Cai M (2004) Estimation of rock mass deformation modulus. J Rock Mech Min Sci
4. centre M (2019) https://www.weathershimla.nic.in
5. Kundan Digambar Rangari RS (2017) Macro scale (1:50,000) landslide susceptibility mapping in parts of Kullu, Mandi, Chamba, Kangra and Lahaul & Spiti districts of Himachal Pradesh. Spec Publ J Eng Geol
6. Laxmi Devi VRS (2019) Comparative evaluation of GIS based landslide hazard. J Geol Soc India
7. List of earthquakes in India (2019) Wikipedia. Retrieved from https://en.wikipedia.org/wiki/List_of_earthquakes_in_India
8. maphill (n.d.) Maps of kinnaur
9. Maps G (2019) Google Maps
10. Negi BB (2014) Environmental impect of making hydropower projects in kinnaur district of himachal pradesh

11. Prakash S (2013) Earthquake related landslides in the Indian Himalaya. Springer, pp 327–334
12. Suneet Naithani AS (2018) Mapping of natural hazards and expected incidences in Great Himalayan National Park Conservation Area, Himachal Pradesh. Indian J Ecol
13. The Tribune (2019). Retrieved from The tribune
14. Vipin Kumar VG (2018) Hazard evaluation of progressive Pawari landslide zone, Satluj valley, Himachal Pradesh, India. Springer

Geotechnical Investigation of Landslide of Atharamura and Baramura Hill, Tripura

Kousik Adak and Sujit Kumar Pal

Abstract Landslide is a natural disaster that occurs when rock, earth or debris flows to downward slope due to gravity after being detached from slope underneath. It is essential to know the geotechnical properties to analyse the causes of landslide. For the present study geotechnical investigation has been carried out on the slopes of Atharamura and Baramura Hill, Tripura. Results obtained from laboratory test and field investigation revealed that the underlying causes of the landslide could be (a) the geological formation of those hills (b) permeability through the different layers of soil (c) shear strength of the soil (d) cutting of hills slope for reconstruction and widening of the road. Landslide triggered due to heavy precipitation during the monsoon season.

Keywords Landslide · Debris flow · Geotechnical property · Shear strength · Precipitation

1 Introduction

In India about 10% of total population was affected by natural disaster in 2019. Among this, landslides also play an important role in affecting the people of India. Geological Survey of India says that 15% of the land in our country is hazardous to landslide. North-east portion of the country is vulnerable to the landslide hazard. It is a very common phenomenon in the hilly region of Tripura especially in North and Dhalai districts. Almost every monsoon due to heavy rainfall total hilly region is mostly affected by the landslide. Due to its geological instability, non-tectonic activity and extreme rainfall infiltration during monsoon season [1, 2, 3]. As the North-east portion is the developing region there is new road and railways construction along the unstable slope of the hilly region. This road construction and various infrastructural

K. Adak · S. K. Pal (✉)
Department of Civil Engineering, NIT Agartala, Tripura, Jirania 799046, India
e-mail: skpal1963@gmail.com

K. Adak
e-mail: kousik7258@gmail.com

T. G. Sitharam et al. (eds.), *Local Site Effects and Ground Failures*, Lecture Notes in Civil Engineering 117, https://doi.org/10.1007/978-981-15-9984-2_17

works repeatedly increase the number of landslides in those areas [4]. In the time of rigorous field survey most of the landslide occurred alongside of the NH-44 which connecting Triupra state with Assam. Due to landslide mainly in monsoon time when rainfall is moderate to heavy (where moderate is defined as 2.6–7.6 mm per hour $\geq$ 7.6 mm) blocking and damaging of NH-44 is a common occurrence.

As landslide is a complex phenomenon worldwide it is very difficult to say the exact causes of it. Most of the publications are based on landslide hazard zonation mapping deals with landslide susceptibility mapping [5]. But field investigations also helpful to know the type and analysis of landslide as we have seen that there is very heterogeneity in geotechnical properties of soil in a very small area.

It also helps us to find the further remedial of landslide. Based on field investigation and laboratory similar case studies have been carried out [6–8, 9]. Our present study is the outcome of field investigation and the laboratory testing of soil collected from the landslide areas which help us to analyse the causes of landslide.

2 Experimental Test Programme and Set-Up

Experimental test programme of present study is based on the laboratory experiment. Our study is mainly to know about the geotechnical property of the landslide area and to find out the basic causes of landslide. Mainly we have visited those landslide prone areas in the monsoon season. The soil sample is collected from four different sites from Atharamura hills and Barramura hills. Using cylindrical sampler of size 10 cm diameter and 30 cm long we have taken undisturbed soil and also collected some disturbed sample using plastic bag. On July 9, locations 1 and 2 201,923.81^0 N,91.58^{0}E and 23.88^0 N,91.78^{0}E near Kulai Rain Forest besides NH 44 Shillong-Agartala-Sabrum road; and some soil samples are collected in November 2019, locations 3 & 4 from Baramura Hill, of North Tripura district. After collecting the samples we have done laboratory test programme given below (Figs. 1 and 2):

(a) Specific Gravity Grain Size Analysis (b) Atterberg limit determination (c) Direct shear test (d) Compaction test (e) Permeability test.

3 Test Procedure

The tests were carried out systematically and the following steps were followed:

Specific Gravity

The specific gravity test is conducted as per IS: 2720 (Part 3/Sec-1) 1980. The specific gravity of soil shall be calculated from the following formula:-

$$G = [\{W_2 - W_1/(W_2 - W_1)\} - (W_3 - W_4)\}]$$

Fig. 1 Soil sample collection

Fig. 2 Landslide affected area

W_1 = Weight of the density bottle (g)

W_2 = Weight of the density bottle + soil (g)

W_3 = Weight of density bottle + soil + water (g)

W_4 = Weight of the density bottle full with water (g)

Grain Size Analysis

Grain size analysis has been done to find out the classifications and the group of the soil on the basis of ASTM D6913. The particle size distribution can be explained as a plot of percentage finer to the particle size plot. This is determined to know the exact type of the soil. This is followed the following three steps. i. Dry sieve analysis; ii. Wet sieve analysis; iii. Hydrometer analysis. However for sandy soil dry sieve analysis is enough to have the properties. For dry sieve analysis the dry soil has been taken and sieving is carried out by the standard IS sieves. And the plot is generated. Then the wet sieving is done with the soil solution. And the retained part from 0.075 mm sieve is taken for the hydrometer analysis. In case of hydrometer analysis the soil particles are allowed to settle with time and using stokes law the particle size distribution is generated to have the full profile of grain size distribution of soil.

Atterberg Limits

The Atterberg limits are a basic measure of the critical water contents of a fine grained soil, such as its liquid limit, plastic limit and shrinkage limit. Depending on the water content of the soil, it may appear in four states: solid, semi-solid, plastic and liquid.

- Liquid limit (LL): The boundary between the liquid and plastic states;
- Plastic limit (PL): The boundary between the plastic and semi-solid states;
- Shrinkage limit (SL): The boundary between the semi-solid and solid states;

Direct Shear Test

This test is also conducted to find out the 'c' and 'ϕ' of the soil having more angle soil is first subjected to some amount of normal stress. Then the shearing stress is given until the soil sample fails or deforms too much. This one set of experiments will give a point in the plot of normal stress (i.e. 'σ') to the failure shearing stress (i.e. 'τ').

$$\tau = \phi +$$

In the laboratory the direct shear test is conducted with the soil sample of skin friction (i.e. silt, sand, etc.) (Fig. 3).

Permeability Test

The permeability has been determined on the basis of ASTM D 2434. The property of the soil which permits water (fluid) to percolate through its interconnecting voids is called permeability. The coefficient of permeability may be determined both in the laboratory and field by direct test. I did falling head permeability test in laboratory.

In laboratory we determine the permeability of soil by using falling head permeability test.

Compaction Test

The compaction test is conducted as per ASTM D698 & D1557. The test was conducted on soil sample. Water was added to the mixtures for determining the maximum dry density (MDD) and optimum moisture content (OMC) of the sample on the basis of light compaction (Fig. 4).

Fig. 3 Direct shear test apparatus

Fig. 4 Standard proctor test apparatus

4 Test Results and Discussion

After performing all the above tests on the collected sample from four different sites we are able to find out the geotechnical properties of those samples. We present the properties of the soil in tabular format.

Physical Properties of the Soil Samples (Table 1)

Engineering Properties of the Soil Samples (Table 2)

5 Test Results and Discussion

The results of this present study fully depend on the geotechnical and geological properties of soil which effect the causes of landslide. So the causes of landslides are discussed below:

Table 1 Physical properties of the soil sample

Physical prop. of soil	Location 1	Location 2	Location 3	Location 3
Specific gravity	2.65	2.62	2.62	2.65
Sand (%)	88.75	88.64	85.29	86.28
Still (%)	6.68	6.65	6.6	6.5
Clay (%)	4.57	4.71	8.11	7.22
Cu & Cc	6.2 & 1.68	6.5 & 1.65	6.2 & 1.5	7.13 & 2.5
Classification	Silty sand	Silty sand	Silty sand	Silty sand
L.L (%) & P.L (%)	25.25 & 22.08	28.28 & 21.09	30.12 & 25.03	29.23 & 26.65

Table 2 Engineering properties of the samples

Engineering properties	Sample 1	Sample 2	Sample 3	Sample 4
C & ϕ	c-23.58 KN/m^2	c-20.62 KN/m^2	c-19.52 KN/m^2	c-20.58 KN/m^2
	ϕ -30.52^0	ϕ -28.23^0	ϕ -25.69^0	ϕ -31.53^0
Maximum dry density (gm/cc)	1.54 gm/cc	1.68 gm/cc	1.59 gm/cc	1.62 gm/cc
Optimum moisture content (%)	18	20.61	19.23	23.25

Effect of Soil Characteristics on Landslide

The rock information we get from Bhuban teratory information. The soil information we get from the laboratory test the soil is unconsolidated sandy materials with occasional intercalation of moderate to poorly sorted silt or clay layers. After site investigation we saw that the whole slope is composed of highly weathered and fragile shale and mudstone. In some parts of the slope, coarse grained sand with mud balls was also found. The whole landslide scar is mainly composed of huge thickness of laminated siltstone, shale with narrow bands of sandstone; occasionally lenticular structure of medium to coarse sandstone with mudstone. Basically the whole area is composed of mainly silty sand soil that's why the area is prone to landslide (Fig. 5).

Effects of Cohesion and Shear Strength

From the results obtained from the direct shear strength test we have seen that cohesion value is in the range of 19–24 kN/m^2. So the value of cohesion is very less. It is one of the causes of instability of slope. As the cohesion is the main component for resistance of soil to sleep then the angle of internal friction also triggers the landslide (Fig. 6).

Effect of Permeability on Landslide

Collected samples from the study area we have done falling head permeability test for different layer of soil. We have seen that the permeability of upper layer soil is less than the permeability of lower layer soil. For this after rainfall the water gets logged

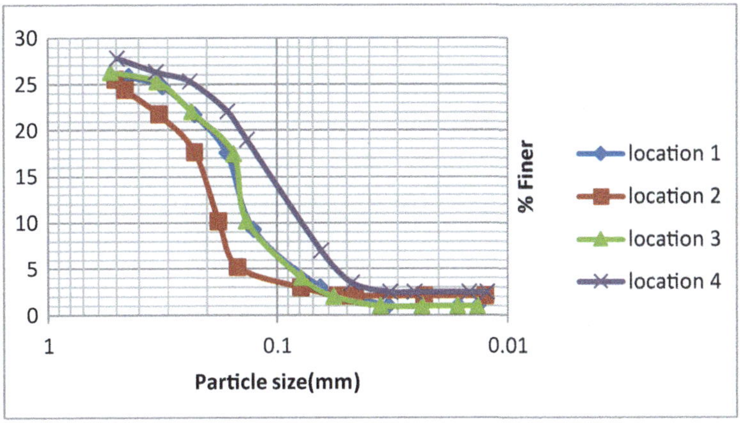

Fig. 5 Particle size distribution curve of silty sand

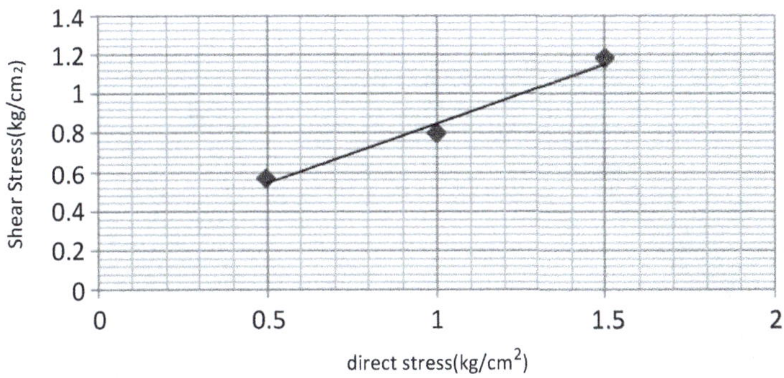

Fig. 6 Shear stress versus normal stress curve

in the upper layer soil naturally pore water pressure is developed in this region. The upper layer of the soil gets sheared off due to the pore water pressure in this region. We also have seen that permeability is decreasing day by day in the soil. Because of this after few more days of rainfall there will be a chance of shear failure to be started (Fig. 7).

Effects of Vegetation Cover on Landslide

Vegetation cover has an effective impact on resistance of slope failure. Due to increasing population infrastructural works going on the slope cutting trees. During the time of site investigation we have seen that due to less vegetation landslides are more on those areas (Fig. 8).

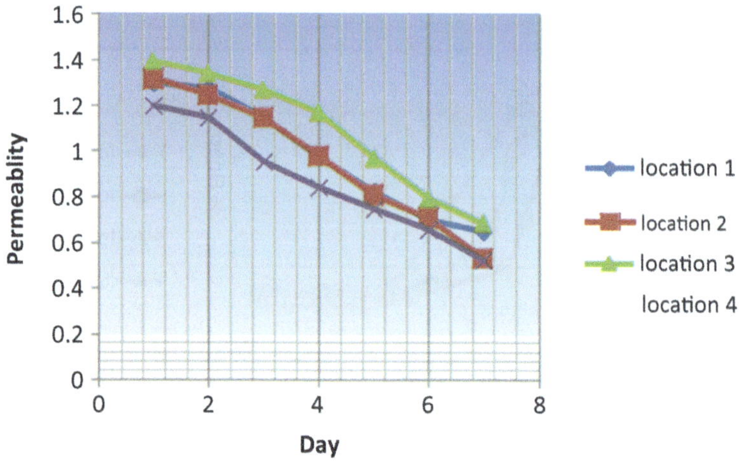

Fig. 7 Permeability in 10^{-4} mm versus no of days curve

Fig. 8 Landslide area with less vegetation cover

Effects of Vertical Slope Cutting on Landslide

Basically in hilly areas, roads and other infrastructural works are done by cutting the slope without precautions. Due to less skill sometimes we have seen during our site investigation they have cut the slope vertically for the widening of pre-existing road.

Table 3 Stability analysis of different locations

	Location 1	Location 2	Location 3	Location 4
C & ϕ	c-23.58 KN/m^2	c-20.62 KN/m^2	c-11.52 KN/m^2	c-10.58 KN/m^2
	ϕ −30.52	ϕ −28.23	ϕ −25.69	ϕ −31.53
SLOPE ANGLE (β)	70	65	85	85
Height in m	6.5	7.5	6	7.2
Unit weight ()	15.56	17.52	18.32	15.53
$\overline{\tan \phi}$	0.395	0.29	0.21	0.15
tan Ø	0.38	0.45	0.59	0.7
FOS	1.55	1.19	0.81	0.876

6 Slope Stability Analysis

We used Taylor's stability method to analyse the stability of slopes by the data obtained from field survey and experiment conducted on laboratory. From this we have seen that locations 3 & 4 have the factor of safety less than one that means those slopes are unstable and prone to landslides (Table 3).

7 Conclusion

Landslide is a complex phenomenon and very deep knowledge is required to analyse the causes of landslide. This has mainly happened when there is a loosely deposited siltstone and sandstone and occasionally sandstone with limestone. Rainfall is also a triggering factor for occurrence of this hazard. As it is a complex phenomenon a deep knowledge is required to take some mitigation steps. It is very difficult to predict the causes of landslide only by checking the geotechnical properties. Grain size and shear strength have the main effects on the occurrence of landslides. The result obtained from the present study is very helpful for slope management thus its helps to do infrastructural works on those landslide prone areas.

References

1. Anbarasu K, Sengupta A, Gupta S & Sharma S P (2010) Mechanism of activation of the Lanta Khola landslide in Sikkim Himalayas. Landslides, 7(2), 135–147.
2. Dahal R K, Hasegawa S, Nonomura A, Yamanaka M, Masuda T & Nishino K (2008) GIS-based weights-of-evidence modelling of rainfall-induced landslides in small catchments for landslide susceptibility mapping. Environmental Geology, 54(2), 311–324.
3. Wei Y, Wu X, Cai C, Wang J, Xia J, Wang J & Yuan Z (2019) Impact of erosion-induced land degradation on rainfall infiltration in different types of soils under field simulation. Land Degradation Develop 30(14):1751–1764

4. Ghosh S, van Westen CJ, Carranza EJM, Jetten VG (2012) Integrating spatial, temporal, and magnitude probabilities for medium-scale landslide risk analysis in Darjeeling Himalayas, India. Landslide 9(3):371–384
5. Mandal B, Mandal S (2018) Analytical hierarchy process (AHP) based landslide susceptibility mapping of Lish river basin of eastern Darjeeling Himalaya, India. Adv Space Res 62(11):3114–3132
6. De SK, Jamatia M, Bandyopadhay S (2008) Geo-technical investigation of Mirik Landslide, Darjiling Himalayas. Geomorphology in India. Prayag Pustak Bhawan, Allahabad, pp 207–216
7. Dutta KK (1966) Landslips in Darjeeling and neighbouring hill slopes in June 1950. Bull GSI, Ser B 15(1):7–30
8. Paul DK (1973) Report on geo-technical investigation of hill slope stability and alignment of border road under project Swastik in Darjiling district, West Bengal and Sikkim. Unpublished Geological Survey of India Report
9. Basu SR (1985) Some considerations on recent landslides at Tindharia and their control. Indian J Power River ValY Dev 88–194
10. Varnes DJ (1958) Landslide types and processes. In: Landslides and engineering practice, vol 24, pp 20–47

Stability Analysis of Kattipara and Meppady Regions of Hill Soil Slope

P. Aswathi and K. Rangaswamy

Abstract Kerala has been witnessing a large number of slope failure issues nowadays. After 2018 and 2019 floods, due to the landslide activities occurred in Kerala hill regions, people are getting aware of the seriousness of the slope instability problems. Factors such as geomorphology, slope angle, terrain curvature, slope length and steepness, soil type and land use or land cover are considered to affecting slope stability. The present study concentrates on the stability of Kattipara and Meppady regions of hill soil slopes. 2018 Karincholamala landslide comes under Kattippara region, and 2019 Puthumala landslide occurred at Meppadi region. The study area faces several slope failure issues. Rainfall is the primary triggering factor of slope failures in the region. Human-made activities of deforestation, constructing high rise buildings at slopes and mining explorations take part in making slopes more vulnerable to failure. Disturbances of soil due to vehicle motions, quarrying, construction activities, and earthquakes can affect the dynamic stability of slopes. Before performing the stability analysis under static and dynamic loading conditions, strength properties of soils collected from those regions are determined after conducting the experiments. SLOPE/W software using Limit equilibrium and pseudo static methods are used to obtain the factor of safety of slopes. Results are indicating that the soil slopes are about to fail under static as well as pseudo static conditions if the soils along the slopes are submerged below the piezometric line.

Keywords Soil strength · Stability of slope · Statics and dynamics

1 Introduction

Kerala is highly vulnerable to climatic variations and natural disasters, including land slides and geostructural failures. Landslides are becoming the most common natural hazard in recent days of Kerala state. Massive damages to soil structures occurred due to the landslides triggered in the monsoon seasons of 2018 and 2019 years. The

P. Aswathi · K. Rangaswamy (✉)
Department of Civil Engineering, NIT Calicut, Kozhikode 673601, Kerala, India
e-mail: ranga@nitc.ac.in

T. G. Sitharam et al. (eds.), *Local Site Effects and Ground Failures*, Lecture Notes in Civil Engineering 117, https://doi.org/10.1007/978-981-15-9984-2_18

infrastructures are entirely damaged, those including roads, railways, bridges, power supply utilities, and communication networks located nearby the steep slopes. Many people have died or injured and also lost the shelter during the past occurred landslide events. Kerala experiences several types of landslides include rock falls, rock slips, slumps, creeps, debris flows and in a few cases, rotational kinds of slides. Among those, the most prevalent and disastrous type of mass movements observed in Kerala is the 'debris flows.' Majority of downslope slides that have occurred in the Kerala state is due to the strength loss of saturated soil mass. The saturated overburden soil mass containing debris material is ranging in size from soil particles to huge boulders destroying and carrying with it everything that is lying in its path [1].

Recent literature on landslide investigations reveals the significance of analyzing and stabilization of soil slopes in order to mitigate the failures. Arca and Lorenzo [2] present the spatial distribution of Factor of Safety (FS) using Limit Equilibrium Method (LEM) and GIS techniques to improve the landslide mapping at Philippines mountain slopes. The accuracy of the model was validated through actual field observations and also by overlying the landslide inventory. It was found that 78% of inventoried landslides are in low to very low stability zones. This technique provides better insight on the landslide susceptibility throughout the study area and can be used directly for engineering design purposes. Wang et al. [3] analyzes the stability of the Duanjiagou landslide on the Bazhong to Guangan Expressway K134–K135 segment in China. The soil samples were collected from bore holes of the slided zone and carried out direct shear and residual shear tests. Stability was analyzed by the landslide force transmission and numerical simulation methods using Geostudio. The results showed that the safety factor obtained was in good agreement with the field observation situation.

Montrasio et al. [4] examines the efficacy of SLIP model used to derive a simplified method to estimate multiple seasonal cycles of the mean degree of saturation of soil and to carry out the time varying stability analysis of a test site slope. It uses available climatic data, such as air temperature and rainfall depth, and is validated through the comparison with long-term field measurements on a slope in Canneto Pavese, northern Italy. This model used to obtain the factor of safeties of slopes those subjected to excess rain fall. Gutiérrez-Martín et al. [5] developed an innovative analytical model using the limit equilibrium method supported by a geographic information system (GIS). This model is especially useful for predicting the risk of landslides in scenarios of heavy unpredictable rainfall. To validate the model, it is applied to a real life example of a landslide which resulted in human and material losses. Pandit et al. [6] performed a slope stability analyses on Himalayan landslide using limit equilibrium and finite element methods. It was resulting that the slope is just marginally stable, and decided to improve the slope by the application of soil nails and/or pile reinforcement at critical locations along the slope. It also describes the design methodology and the significance of pile reinforcements as a novel technique for slope stabilization.

Present study concentrates on analyzing the stability of Kattipara and Meppady regions of hill soil slopes in Kerala. 2018 Karincholamala landslide comes under Kattippara region in Kozhikode district, and 2019 Puthumala landslide occurred at Meppadi region in Wayanadu district. The areas like Wayanad and Kozhikode districts are prone to deep-seated landslides. However, Idukki and Kottayam areas are susceptible to shallow landslides, which is mainly due to the excessive amount of rainfall [7]. Soil slopes in this study area failed due to excessive rain and other human-made factors like deforestation, mining, construction of high rise buildings along the hill slopes, etc. Rainfall is the primary triggering factor of slope failure in this region. The daily or the threshold rainfall alone cannot be used as the determinant for landslide because antecedent rainfall increases the permeability of the soil, and subsequent storm events may trigger an avalanche or landslide [8]. Due to the excess rainfall, the soil becomes fully saturated and water level reaches the ground surface. Here, the main triggering factor from rainfall is pore pressure build-up which causes to reduce the effective stress. Hence in the numerical model, pore pressure coefficient is considered to one and accordingly the pore pressure heads along the failure surface were calculated to obtain the effective stresses. Factor of safety against sliding is calculated based on effective stresses. The present study used the limit equilibrium numerical method of Morgenstern–price analysis model which one inbuilt in the Slope/W software.

2 Stability Analysis Methods

2.1 Limit Equilibrium Method

For calculating slope stability, various numerical modeling techniques are available now. But this softwares require various geotechnical parameters for stability computations. Hence it's important to study the geotechnical characteristics of soil slopes such as shear strength, grain size distribution, specific gravity, and Atterberg's limits. Various methods used for slope stability analysis include Limit Equilibrium Methods, Numerical methods, Artificial Neural Methods, and Limit Analysis. Among all, the limit equilibrium method is most commonly used due to its reliability and convenience. It considered the principle of static forces and moments. It requires the input of soil shear strength parameters such as cohesion and angle of internal friction including density and slope geometry. Present study used SLOPE/W software using Limit equilibrium method to obtain factor of safety of slopes under static loading. Factor of safety is calculated as

$$FS = \frac{M_r}{M_o} = \frac{\Sigma c \Delta l + \Sigma (W \cos \alpha) \tan \phi}{W \sin \alpha} \quad (1)$$

2.2 Pseudo Static Method

If the soil slopes are susceptible to continuous disturbances due to heavy traffic, quarrying, construction activities, machine induced vibrations and earthquakes, the slopes are needed to be analysed for dynamic stability. Dynamic slope stability could be carried out by using either time history or pseudo static analyses. Present study used pseudo static analysis to get critical acceleration i.e., defined as the ground motion necessary to begin the process of slope failure. In this method, the pseudo static forces in horizontal (F_h) and vertical (F_v) directions are to be applied in different magnitudes to examine the stability of slopes. Pseudo static analysis involves simply adding a permanent body forces i.e., pseudo static coefficients (k_h and k_v) representing the earthquake shaking to a static limit equilibrium analysis [9]. Here the factor of safety can be found out by the equation of

$$F_{SD} = \frac{cl_{ab} + ((W - F_v)\cos\alpha - F_h \sin\alpha)\tan\phi}{(W - F_v)\sin\alpha + F_h \cos\alpha} \tag{2}$$

where $F_h = \frac{W}{g}a_h = k_h W$ *and* $F_v = \frac{W}{g}a_v = k_v W$.

A common approach to utilizing pseudo static analysis is simply to run the limit equilibrium analysis using different values of pseudo static coefficients. The pseudo static coefficient is called as yield coefficient if the factor of safety becomes unity. In the simplest form, any ground acceleration due to inertia that exceeds the product of gravitational force and pseudo static factor is causing to failure of slope under dynamic loads.

3 Study Area

Two study areas namely Kattippara and Meppady hill slopes are chosen for the present study. Kattippara is a small Panchayath (21.29 km^2) in Kozhikkode district consisting of steep hill slopes. It is located about 10 km from Thamarassery town and lies between North 11°21′40″ and East 76°0′35″. The slopes are covered almost with earth material. Kattippara has a generally cool humid climate with a very hot season extending from March to May. The average annual rainfall is more than 3500 mm and it is the highest rainfall in this region which is greater than average rainfall of 3266 mm in Kozhikode city region. Another study area Meppady is a Panchayath (198.65 km^2) in Wayanad district and it shares its southern boundary with Kozhikode district. It's an extremely scenic hill station on the State Highway between Kozhikode and Ooty. The nearest city is Kalpetta and is the headquarters of the Wayanad revenue district. Meppady lies in the coordinates of North 11°34′0″ and East 76°9′0″. It's about 7000 ft above average mean sea level.

3.1 Geology and Geomorphology of the Study Area

Kozhikode district has divided in to three geological belts viz., (i) linear NW–SE trending gneissic belt, along the middle extending from north to south, (ii) a charnockite belt occupying areas in the northeast and south, extending to the adjacent districts and also occurring as pockets within the gneissic terrain, and (iii) a narrow coastal belt. In which Kattippara region consists of Hornblende-biotite gneiss of the Migmatite complex of Archaean period extending from north to south and is well foliated. These medium grained, foliated, banded rock types occur within the migmatites and associated retrograded charnockites. It consists of alternate layers rich in hornblende or biotite. Rocks are having bands of coarse to medium grained light grey to pink granites. The rocks are with highly deformed admixtures of contorted bands and enclaves of pyroxene granulite, calcium granulite, and hornblende-biotite granulite. NW–SE trending dolerite dykes are also present in this region and these dykes are about 10–20 m wide.

Kattippara region mainly consists of denudation hills and valley flats. They are exposed as a group of massive hills with resistant rock bodies and rounded summits and are formed due to differential erosion and weathering. Denudation hills are identified in the satellite imagery by their massive size and dome to elliptical shape. They appear as dark green in color in the satellite imagery. These hills are covered with big boulders and sparse vegetation in contrast to structural hills. This landform, in general, act as high runoff zone, due to its moderate to steep slope (5–25°). Rubber and coconut plantation are the main cultivations in the productive slopes of the region (District survey Report of Minor Minerals, Kozhikode District, 2016). Wayanad district was broadly divided into four geological domains viz., (i) Peninsula Gneissic complex in the north and central part, (ii) Migmatite complex in the south central part, (iii) Charnockite group in the south, and (iv) Wayanad Group in the north. Meppady region comes under charnockite group which comprises the charnockite forming hilly terrain in the south and southeast. Pyroxene granulite and banded magnetite quartzite are occurring as narrow bands within charnockite. Geomorphology of the region consists of structural cum denudation hills along with dissected plateau in the northern part and erosional remnants in the southern part (District survey Report of Minor Minerals, Wayanad District, 2016).

4 Sampling and Geometry of Soil Slope

A sufficient number of undisturbed soil samples were collected from the top and middle part of the soil slopes in Kattippara and Meppady region of hill soil slopes. The soil samples were sealed in the metal core cutters by using paraffin wax and transported to the laboratory without disturbing the structure of soil. Location details of soil collection and slope geometrical features are tabulated in Tables 1 and 2.

Table 1 Location details and slope geometry of Kattippara hill soil slopes

Soil slope at	Soil collected at	Location of soil slope		Slope details	
		Latitude	Longitude	height (m)	Angle (°)
Karinchola	Top Middle	N11°28.046′ N11°28.558′	E075°55.849′ E075°55.836′	20	25
Calvary	Top Middle	N11°28.046′ N11°28.010′	E075°55.256′ E075°55.246′	15	20
Poovanchola	Top Middle	N11°28.787′ N11°28.756′	E075°55.813′ E075°55.806′	10	16
Kualamala-Triveni	Top Middle	N11°28.893′ N11°28.891′	E075°55.789′ E075°55.774′	10	20
Kualamala-Triveni (near water fall)	Top Middle	N11°28.947′ N11°28.936′	E075°55.894′ E075°55.888′	20	25
Theyyathumpara	Top Middle	N11°27.612′ N11°27.585′	E075°55.811′ E075°55.797′	20	27
Mavullapoyil	Top Middle	N11°29.362′ N11°29.375′	E75°54.979′ E075°54.973′	15	18
Cherachankdy	Top Middle	N11°28.816′ N11°28.876′	E075°55.037′ E075°55.041′	25	20
Kelanmoola	Top Middle	N11°28.386′ N11°28.371′	E075°56.202′ E075°56.210′	30	20

Table 2 Location details and slope geometry of Meppady hill soil slopes

Soil slope at	Soil collected at	Location of soil slope		Slope details	
		Latitude	Longitude	height (m)	Angle (°)
Mummykunnu	Top Middle	N11°30.896′ N11°30.900′	E076°08.039′ E076°08.032′	8	40
Kalladi-Puthumala Roadside	Top Middle	N11°30.611′ N11°30.622′	E076°08.240′ E076°08.202′	6	45
Panchami-Kalpatta Roadside	Top Middle	N11°33.788′ N11°33.797′	E076°07.752′ E076°07.715′	6	70
Poothakolli	Top Middle	N11°34.670′ N11°33.660′	E076°06.484′ E076°08.596′	5	60
Thazhe Arrapett	Top Middle	N11°33.238′ N11°32.821′	E076°09.039′ E076°09.443′	6	75

5 Experimental Program

Initially the undisturbed soil samples collected from study area are tested for natural moisture content and bulk unit weight of soil mass. Further the experiments were carried out to determine the particle size distribution, Atterberg's limits in the fine grained soils, and also shear strength parameters. Shear strength parameters of cohesion and friction angles of all the soil samples were determined after conducting the direct shear box tests. All the experimental tests were followed by the test procedures according to Indian standard codes. All the basic properties of soil samples determined from the tests are summarized in Tables 3 and 4.

Results are indicating that the Kattippara soils have the water content in the range of 15–31% with average unit weight of 15–18 kN/m^3. Meppady soils exhibit the water content in the range of 30–40% with average density of 12–17 kN/m^3. However, these water contents are increasing toward saturated values during heavy rains and causing the slopes under failure due to occurring of undrained condition along the failure surfaces. Based on particle size distribution, it is understood that the Kattippara soils are purely cohesionless and Meppady soils are predominantly cohesive soils. Soils are belonging to low to medium plastic consistency. The cohesion strength values are in the range of 0 to 28.5 kPa for Kattippara soil samples and 0 to 40 kPa for Meppady soil samples. Cohesion helps to bind the soil particles tightly and there is an inverse correlation between friction angle and cohesion with landslide potentiality [10]. Friction angles of soil samples were found in-between 18 and 34°. Bonding of the soil particles due to cohesion is influenced and loosened by the presence of lubricating agent (water and ice particles) and ensure the soil mass to collapse.

6 Stability Analysis of Soil Slopes

Slope stability analysis is performed by using numerical program called SLOPE/W feature built in Geostudio 2019 software. Input parameters required for a slope stability analysis includes (i) geometry of slope—description of the stratigraphy and shapes of potential slip surfaces, (ii) soil strength parameters used to describe the capacity of soil, and (iii) unit weight of soil including piezometric surface or pore-water pressures. SLOPE/W uses the analytical methods in solving the stability of soil slopes are: Ordinary (Fellenius) method, Bishop simplified method, Spencer method, Janbu simplified method, Morgenstern–price method, generalized limit equilibrium and finite element methods. Except Morgenstern–price and generalized limit equilibrium methods, all the other methods do not consider interslice shear forces in their analysis. Morgenstern-Price and General limit equilibrium methods use both normal and shear forces between the slices and satisfy the equilibrium condition of both the forces and moments [11].

6.1 Factor of Safety with Static Loading Conditions

In the present study, factor of safety of soil slopes in the regions of Kattippara and Meppady were found out at two extreme conditions of fully dry and saturated states. Typical contours on static stability analysis of two soil slopes in Kattippara and

Table 3 Basic properties of soil samples collected from Kattippara soil slopes

Soil sample at	Natural state of soil		Grain size distribution			Atterberg's limits			Shear strength parameters	
	Density (kN/m^3)	Water content (%)	Gravel (%)	Sand (%)	Silt + clay (%)	LL (%)	PL (%)	PI	C (kPa)	ϕ (degrees)
Karinchola										
Top	16	22.8	3	78	19	NA	NA	0	10	27.3
Middle	15.1	21.5	2	62	36	31	21	10	0	30
Calvary										
Top	16.42	18.6	2	68	30	25	16.2	8.8	3	24.8
Middle	16.02	15.4	2	70	28	NA	NA	0	0	21.7
Poovanchola										
Top	17.16	18	3	53	44	54	33.5	20.5	18	29.6
Middle	15	22	3	54	43	26	18	8	10	6.5
Kualamala-Triveni										
Top	15.8	23.43	1	62	37	38	23	15	20	21.2
Middle	15.26	19.45	2	73	25	34	19.8	14.2	17	18.1
Kualamala-Triveni (near water fall)										
Top	15.46	27.6	1	47	52	68	35.7	32.3	15.8	26.4
Middle	15.16	21.51	1	56	43	29	19	10	2	25.7
Theyyathumpara										
Top	17.69	18.74	3	88	9	NA	NA	0	0	32.5
Middle	17.27	15.64	6	82	12	NA	NA	0	28.5	13.35
Mavullapoyil										
Top	16.44	31	14	53	33	53	35.8	17.2	12	31

(continued)

Table 3 (continued)

Soil sample at	Natural state of soil		Grain size distribution			Atterberg's limits			Shear strength parameters	
	Density (kN/m^3)	Water content (%)	Gravel (%)	Sand (%)	Silt + clay (%)	LL (%)	PL (%)	PI	C (kPa)	ϕ (degrees)
Middle	15	25.8	10	46	44	61	37.4	23.5	11	34
Cherachankandy										
Top	17.45	28.41	3	59	38	46	32	14	20	28.4
Middle	17.37	21.98	0	47	53	50	33.6	16.4	22	32.6
Kelanmoola										
Top	16.88	26.7	2	56	42	50	28.4	21.6	20	32.6
Middle	16.11	19.3	4	64	32	37	22.2	14.8	11	26.6

Table 4 Basic properties of soil samples collected from Meppady hill soil slopes

Soil sample at	Natural state of soil		Grain size distribution			Atterberg's Limits			Shear strength parameters	
	Density (kN/m^3)	Water content (%)	Gravel (%)	Sand (%)	Silt + clay (%)	LL (%)	PL (%)	PI	C (kPa)	Φ (degrees)
Kalladi-Puthumala roadside										
Top	14.42	30.6	6	46	48	67	46.4	20.6	17	30
Middle	14.40	36.8	14	41	45	58	42.75	15.25	0	31
Mummykunnu										
Top	16.9	31.52	0	25.5	74.5	64	40	24	30	22
Middle	17.3	36.48	3	31	66	58	38.46	19.54	25	20
Panchami-Kalpetta roadside										
Top	12.15	30.8	5	28	67	60	35.8	24.2	26	15
Middle	12.50	32.2	4	28	68	52	30.6	21.4	15	27
Poothakolli										
Top	15.11	32.5	0	14	86	66	45	21	40	23
Middle	15.21	35	0	16	84	59	41.2	17.8	21	22
Thazhe Arrapetta										
Top	11.77	28.71	0	20	80	63	43.5	19.5	42	26
Middle	12.89	30.66	0	22	78	56	37.6	18.4	15	26.5

Meppady regions are shown in Figs. 1 and 2. The analysis is extended to all other soil slopes and the factor of safety against slope stability for all the soil slopes under static loading with dry and saturated water conditions was reported in Table 5. Results are indicating that the existing soil slopes are to be failed during rainy season due to excess rainfall and development of continuous pore pressures along the failure surfaces of slopes. The factor of safeties of soil slopes determined is in the range of 0.56–1.5 under saturation during rainfall seasons. Even the slopes are safe against sliding ($FS_L > 1.5$) during dry season without water content, the slopes become unstable with excessive rainfall and pore pressure developments. A few slopes are under margin of stable with factor of safety of 1.5. It is observed that the slide at Karinchola of Kattippara and the slide at Kalladi-Puthumala of roadside of Meppady regions are having lowest factor of safety of 0.45 and 0.28, respectively. Poothakolli slide in Meppady region is safe under static loading with factor of safety around 2.3. However, even the slopes are stable under only rainfall conditions, but they may fail with man-made activities of mine explorations, continuous heavy traffic and

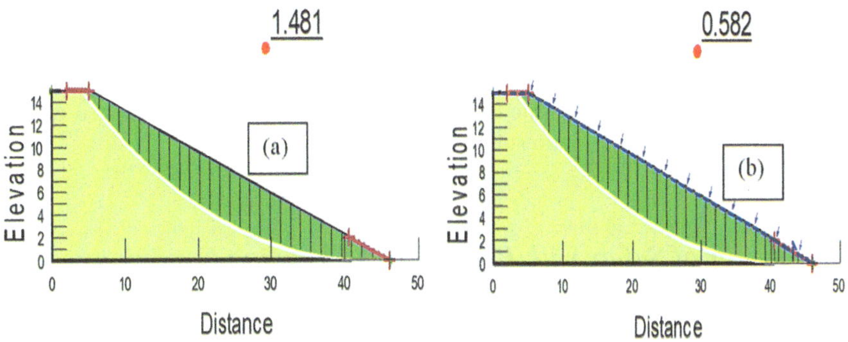

Fig. 1 Typical contours on stability analysis of Calvary hill soil slopes in Kattippara. **a** Dry condition and **b** saturated condition

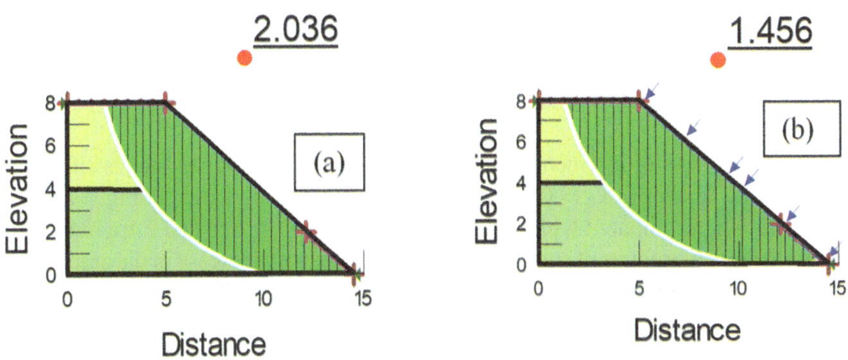

Fig. 2 Typical contours on stability analysis of Mummykunnu soil slopes in Meppady. **a** Dry condition and **b** saturated condition

Table 5 Factor of safety against stability of soil slopes under static loading

Soil slopes in Kattippara region at	Factor of safety against stability		Soil slopes in Meppadi region at	Factor of safety against stability	
	Dry condition	Saturated condition		Dry condition	Saturated condition
Calvary	1.481	0.582	Mummykunnu	2.036	1.456
Karinchola	1.438	0.459	Kalladi-Puthumala Roadside	1.401	0.28
Poovanchola	2.978	1.549	Panchami-Kalpatta Roadside	2.003	1.211
Kualamala-Triveni	2.335	1.502	Thazhe Arrapetta	2.346	1.46
Kualamala-Triveni (Water fall)	1.567	0.627	Poothakolli	2.823	2.233
Theyyathumpara	1.464	0.569			
Mavullapoyil	3.046	1.555			
Cherachankandy	2.736	1.578			
Kelanmoola	1.922	0.866			

industrial vibrations, and seismic events. So the slopes are to be examined for the stability against various pseudo static forces.

6.2 *Factor of Safety with Pseudo Static Loading Conditions*

Slope stability analysis is carried out for all the slopes by varying the pseudo static coefficients, k_h values, in the range of 0.1–0.8. Vertical pseudo static coefficient is considered to zero since it reduces both the driving and resisting forces. A typical contour on pseudo static slope stability of Karinchola, and Kalladi-Puthumala hill soil slopes during excess rainfall and subjected to pseudo static coefficient of 0.2 under dynamic loading is shown in Fig. 3. It is extended to all the other soil slopes and the reduction of factor of safety of Kattippara and Meppady soil slopes by varying pseudo static coefficients in dry and excess rainfall conditions and are shown in Figs. 4, 5, 6 and 7.

Factor of safety of Karinchola and Kalladi-Puthumala hill soil slopes subjected pseudo static loading at $k_h = 0.2$ are about 0.25 and 0.16 as shown in Fig. 3 and are still lesser when compared with static loading. Figures 3, 4, 5, 6 and 7 are indicating that the factors of safeties of soil slopes are decreasing with increase in the amplitudes of horizontal pseudo static coefficients. It is concluding that the soil slopes even under dry conditions fail if the intensity of pseudo static coefficients exceeds above 0.3–0.4. The saturated soil slopes under excess rainfall conditions are completely unstable if the slopes are susceptible to dynamic loading even at pseudo static coefficient of 0.2.

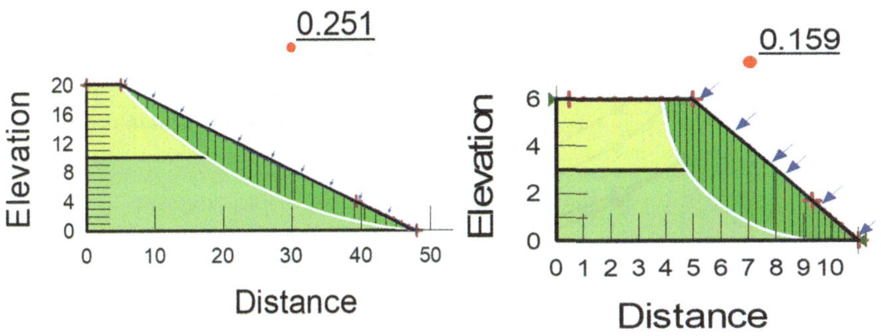

Fig. 3 Typical contours on pseudo static stability analysis with $k_h = 0.2$: **a** Karinchola, and **b** Kalladi-Puthumala hill soil slopes during excess rainfall

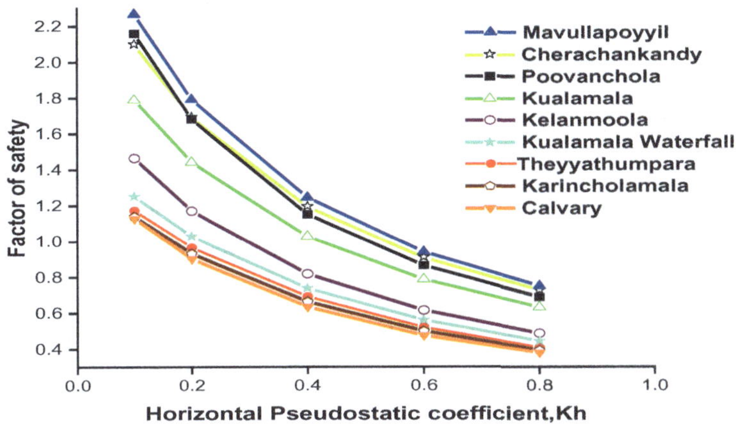

Fig. 4 Reduction of factor of safety of Kattippara soil slopes by varying pseudo static coefficients—dry weather condition

7 Conclusions and Recommendations

A series of experiments were conducted on collected soil samples from the study area of soil slopes and the parameters required to give as input into the numerical program are determined. SLOPE/W software using limit equilibrium static and pseudo static methods are performed to obtain the factor of safety of soil slopes. Present study reveals that from the geometrical features of the slope height and slope angle affect the factor of safety adversely. Meppady soil slopes are having steeper slope angles and show less factor of safety compared to flatter slopes of same height and soil properties. Kattippara soil slopes are flatter when compared with Meppady soil slopes and Kattipara soil slopes are having slightly higher factor of safeties. The major observations are: (i) The existing soil slopes fail during rainy season due to excess rainfall and development of continuous pore pressures along the failure surfaces of

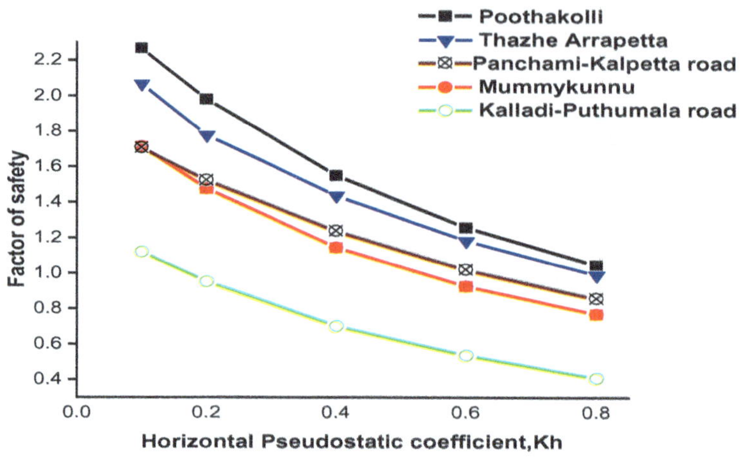

Fig. 5 Reduction of factor of safety of Meppady soil slopes by varying pseudo static coefficients—dry weather condition

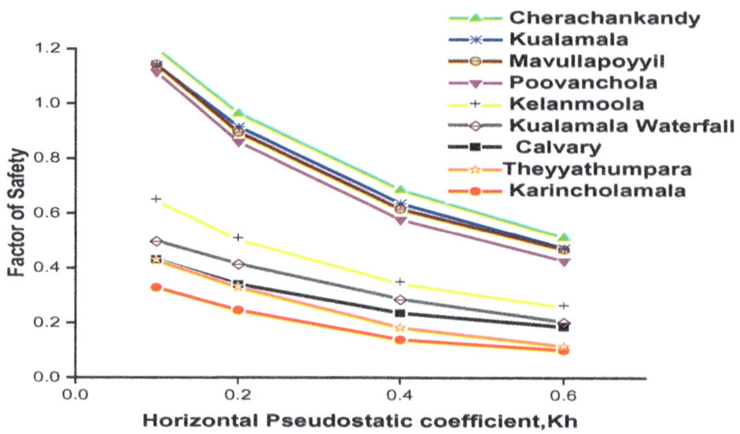

Fig. 6 Reduction of factor of safety of Kattippara soil slopes by varying pseudo static coefficients—excess rainfall condition

slopes. Even a few slopes are under margin of stable with factor of safety of 1.5, but they fail completely under the dynamic influences with increasing the pseudo static coefficients. Soil slopes are susceptible to sliding under dynamic influences even at dry weather conditions. (ii) Slides at Karinchola of Kattippara and the slide at Kalladi-Puthumala of roadside of Meppady regions are having lowest factor of safety of 0.45 and 0.28 under static loading. These factor of safeties are reduced to still lower values 0.25 and 0.16, respectively, with the pseudostatic loading coefficient of 0.2., and (iii) Soil slopes even under dry conditions fail if the intensity of pseudo static coefficients exceeds above 0.3–0.4. The saturated soil slopes under excess

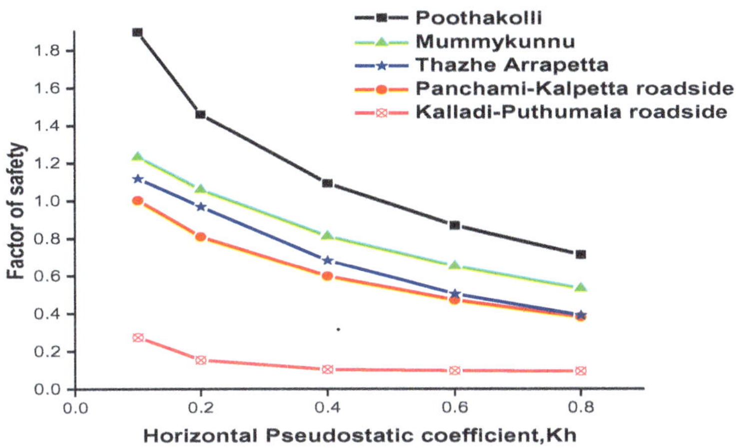

Fig. 7 Reduction of factor of safety of Meppady soil slopes by varying pseudo static coefficients—excess rainfall condition

rainfall conditions are completely unstable if the slopes are susceptible to dynamic loading even at pseudo static coefficient of 0.2. Hence the existing soil slopes are to be stabilized by using various techniques including lime column intrusions, soil nailing, piling or using any innovative chemical grouting methods.

References

1. Thampi PK, Mathai J, Sankar G, and Sidharthan S (1998) Evaluation study in terms of landslide mitigation in parts of Western Ghats, Kerala. Research report submitted to the Ministry of Agriculture, Government of India. Centre for Earth Science Studies, Government of Kerala, Thiruvananthapuram, India
2. Arca MCQ, Lorenzo GA (2018) Landslide hazard mapping using limit equilibrium method with GIS application of roadway traversing mountain slopes: the case of Kitaotao Bukidnon, Philippines. J Nepal Geol Soc 55:93–101
3. Wang X, Xia T, Zhang L, Gao M, Cheng K (2020) Field geological investigations and stability analysis of Duanjiagou landslide. ISPRS Int J Geo-Inf 9:23. https://doi.org/10.3390/ijgi9010023
4. Montrasio L, Valentino R, Meisina C (2018) Soil saturation and stability analysis of a test site slope using the shallow landslide instability prediction (SLIP) Model. Geotech Geol Eng 2018(36):2331–2342. https://doi.org/10.1007/s10706-018-0465-3
5. Gutiérrez-Martín A, Herrada MÁ, Yenes JI, Castedo R (2019) Development and validation of the terrain stability model for assessing landslide instability during heavy rain infiltration. Nat Hazards Earth Syst Sci 19(721–736):2019. https://doi.org/10.5194/nhess-19-721-2019
6. Pandit K, Sarkar S, Samanta M, Sharma M (2016) Stability analysis and design of slope reinforcement techniques for a Himalayan landslide. In: Conference proceedings on recent advances in rock engineering. https://doi.org/10.2991/rare-16.2016.16
7. Kuriakose SL, Sankar G, Muraleedharan C (2009) History of landslide susceptibility and a chorology of landslide-prone areas in the Western Ghats of Kerala in India. Environ Geol 57:1553–1568

8. Rahardjo H, Li XL, Toll DG, Leong EC (2001) The effect of antecedent rainfall on slope stability. Geotech Geol Eng 19:371–399
9. Jibson RW (2011) Methods for assessing the stability of slopes during earthquakes—a retrospective. Eng Geol 122: 43–50
10. Mandal S (2015) Assessing cohesion, friction angle and slope instability in the Shivkhola watershed of Darjiling Himalaya. Res J Earth Sci 3
11. Karhn J (2004) Stability modeling with slope/W: an engineering methodology, 1st edn, 17 pp

Numerical Studies on the Behavior of Slope Reinforced with Soil Nails

Akhil Pandey, Sagar Jaiswal, and Vinay Bhushan Chauhan

Abstract Soil nailing is commonly used to stabilize cut slopes and earth retaining structures, embankments and sometimes to reduce the lateral earth pressure on retaining wall. This is an in-situ reinforcement technique that uses passive rigid bars usually made up of steel that can withstand tensile forces, shearing forces, and bending moments. In the present study, an attempt has been made to analyze the behavior of soil slope reinforced with soil nails by numerical simulation under seismic loading using a finite element numerical modeling tool, OptumG2. Two-dimensional models of the slope are considered (with and without nails) and subjected to static and dynamic loading to investigate the possible modes of failure (base failure, slope failure, and toe failure). The corresponding factor of safety values for unreinforced soil slope is calculated using the strength reduction method (SRM) at different slope angles (β) at 30°, 40°, 50°. Based upon the obtained critical slope surface in the case of the unreinforced slope, soil nails are provisioned in the slope and the same is analyzed under gravity and seismic loading. A detailed investigation of reinforced soil slope has been conducted considering the key factors governing the factor of safety of reinforced slope namely length of the soil nails, the number of soil nails, spacing of soil nails (s), the inclination of soil nails (θ) in different slope angles (β) under seismic loading with a peak horizontal ground acceleration = 0.12 g. The present study also investigates the possible failure patterns under static and seismic conditions, along with the analysis of the development of internal reactions in soil nails such as bending moment and shear force.

Keywords Soil nailing · Finite element method · OptumG2 · Seismic loading

A. Pandey · S. Jaiswal · V. B. Chauhan (✉)
Civil Engineering Department, Madan Mohan Malaviya University of Technology, Gorakhpur 273010, India
e-mail: chauhan.vinaybhushan@gmail.com

A. Pandey
e-mail: akhilpandey812@gmail.com

S. Jaiswal
e-mail: sagarjaiswal.work@gmail.com

T. G. Sitharam et al. (eds.), *Local Site Effects and Ground Failures*, Lecture Notes in Civil Engineering 117, https://doi.org/10.1007/978-981-15-9984-2_19

1 Introduction

The downward falling or sliding of a soil mass from a steep slope may occur in regions like hilly areas, offshore and onshore environment, zones of heavy rainfall, etc. The slip of slope can cause natural damages and sometime cause human losses also [5]. It can be controlled by the use of proper slope stabilization techniques such as soil nailing, etc. Soil nailing is one of the recent in-situ methods for soil slope stabilization and can be used for the future expansion of construction work in the sloping region applicable to hard to stiff fine-grained soils, silty clays, sandy clays and very dense granular soils [1–3]. Soil slope stability analysis is an important component in the design and construction of various geotechnical structures such as rail or road embankment, road widening under existing bridge end, and temporary and permanent excavation in those problematic regions [7]. This technique consists of the passive reinforcement of existing ground by installing closely spaced solid steel bars (i.e., nails) although hallow-system bars are also available. They are passive inclusions, which improve the shearing resistance of the soil. The holes with chosen diameters are drilled at the desired position to install the solid bars and grout or shotcrete is used to avoid caving in the drilled holes before installing the nails [4]. Centralizers, made of synthetic materials are installed to ensure a minimum thickness of grout which completely covers the nail bar. The nail-head which is a square-shaped concrete structure provides bearing strength and transfer bearing loads from the soil mass to soil nails. Temporary and permanent facing is also provided to ensure the support to the exposed soil and it also acts as a bearing surface for the bearing plate [4]. The nails can be arranged in many ways such as steeply inclined nail arrangement, sub-horizontal nail arrangement, and hybrid nail arrangement. Steeply inclined nail arrangement generally satisfies the design objectives of sustaining the earth's pressure upon the liquefaction of the loose fill. Sub-horizontal nail arrangement is effective in sliding failure in cut slopes.

Two types of regions are generated in the soil nail system, an active region which consists of the soil mass above the failure surface which shows the movement of soil masses and passive region which consists of embedded part of soil nail behind the failure surface which doesn't show any movement of soil masses. During the slope failure, the active region tends to deform which results in axial displacement along with soil nails that are placed across the slip plane. This results in the development of tensile forces in soil nail in the passive zone which resists the deformation of the active zone. This tension force results in an increment of the normal force coming on a slip plane and reduces the driving shear force. The soil nails are embedded in the passive region through which it resists the pull-out of the nail from slope through friction between nails and soil [4].

However, shear force and bending moment carrying capacity of the soil nails are conventionally unnoticed in the design procedure [13]. Hence, an attempt is made to examine the generation of these internal stresses, and bending moment and shear force resistance generated in the soil nails are also evaluated for all the cases of the reinforced soil slope in the present study. Detailed analysis of the performance of soil slope reinforced with soil nails under seismic loading is still quite fewer in number. Most of the literature on soil nailed system emphasizes only the static condition. Few studies have been performed for the slope stability of soil nails reinforced slope and its failure mechanism under seismic condition [16]. In the present work, a two-dimensional model of soil slope is considered with and without nails, subjected to gravity as well as dynamic loading and the corresponding minimum factor of safety values for soil slope is calculated by the strength reduction method at different slope angles (β) at 30°, 40°, 50°. Variation of development of maximum bending moment and shear force is also analyzed with the different failure mechanisms.

2 Methodology

In this present work, the finite element method is adopted for numerical simulation of soil slope reinforced with and without soil nails under static and seismic conditions. Once the soil parameter is adopted, soil slope is modeled using the Optum G2 numerical tool [10]. The shear reduction method is used to calculate the factor of safety in all the condition [9] and then the analysis is performed by using soil nails of constant diameter at different nail inclination angle (θ) and spacing (s). Different failure mechanism and effect of bending moment are also analyzed for the different conditions for finding some beneficial conclusion for slope stability of soil using nails.

3 Numerical Modeling

Analysis of soil slope by wedge method is complicated as well as time-consuming, moreover detailed analysis of various soil surface interactions cannot be done through it as this analysis involves many assumptions related to inter-slice force and slip surface, which are not true.

Hence numerical modeling using finite element analysis is preferred because of its simple and efficient approach. A finite element method is a widely adopted method for the analysis of slope stability because no assumptions of inter-slice force and slip surface are required [7]. In this study numerical modeling of soil reinforced with nails under seismic condition is conducted for various parameters to evaluate its performance by finite element analysis. All the variation in the considered parameters for the analysis of reinforced slope is as shown in Table 1.

The Mohr–Coulomb constitutive model is considered for the soil which can simulate perfectly plastic material conditions with the development of irreversible strains.

The strength reduction method using the finite element method is used to obtain the factor of safety for slopes [8]. Designing of soil nails is performed by using FHWA guidelines [6] and the calculation of the diameter of soil nail is performed

Table 1 Slope and soil nail parameters

Spacing of soil nail (s)	Soil slope angle (β)	Soil nail inclination (θ)
0.5 m, 1 m, 2 m	30°	10°, 20°, 30°, 40°
	40°	10°, 20°, 30°, 40°
	50°	10°, 20°, 30°, 40°

by the adopted horizontal and vertical spacing of soil nails [6] and subsequently, the length of soil nails is calculated for the slope.

The soil slopes are reinforced using high yield strength deformed (HYSD) bars nails at four different nail inclinations of 10°, 20°, 30°, and 40°. The boundary conditions are equivalent to the fixed bottom, and the roller at the vertical end of the soil mass is provisioned.

A diagrammatic representation including complete details of the geometric configuration of soil slope with nailed element is shown in Fig. 1.

The accuracy of numerical simulation is governed by the optimum number of total elements considered in the mesh, therefore a sensitivity analysis is carried out to predict the optimum number of total elements in the domain of the numerical mesh considered in the present study [11, 12, 15, 16]. Variation of the total number of elements from 1000 to 4000 with the increment of 1000 elements in each iteration is done, it is noted that a further increase in the total number of elements beyond 2000 elements in the mesh does not improve the factor of safety of the nail soil slope.

The material properties adopted for the numerical simulation are shown in Table 2.

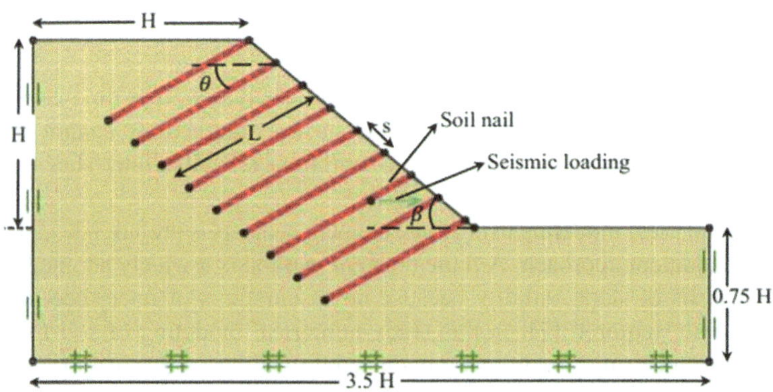

Fig. 1 Numerical Model of soil slope with soil nailing under the seismic condition

Table 2 Material properties [14]

Properties	Sand	Soil nail
Young modulus (E), MPa	20	2.6×10^4
Poisson ratio (υ)	0.2	–
Cohesion (c) (kPa)	1.4	–
Friction angle (∅), degree	30	–
Dry unit weight (γ_d), kN/m^3	17	–
Lateral strength ($F_{L\ max}$), kN/m	-	42

4 Results and Discussion

The soil slope models are analyzed under static and seismic loading conditions. In the case of the seismic condition, a horizontal seismic coefficient of 0.12 g and the factor of safety is determined for the different slopes without soil nails having inclination ranging from 30° to 50°. The factor of safety obtained from this analysis is shown in Fig. 2. The obtained factors of safety are calculated for the lower bound (LB) and upper bound (UB) analysis and presented with the average of factors of safety and error bars in Fig. 2, whose upper and lower tips show the factors of safety in the lower bound and upper bound analysis. A very insignificant difference in UB and LB factors of safety strengthens the fact that proper selection of the number of elements is chosen for carrying out the present study. It is indicated that with the

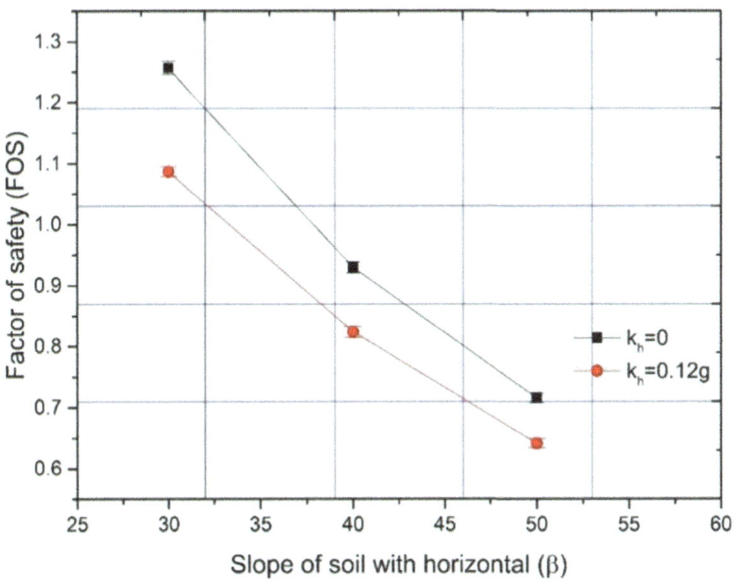

Fig. 2 Variation of the factor of safety under the static and seismic condition

increase in the soil slope, the factor of safety decreases, which is self-explanatory as steeper slopes are unstable. Furthermore, with the application of seismic forces, FOS of a given slope is found to be lower compared to the static case (under gravity loading), and this reduction in the FOS was found to be 11.4, 9.5, and 8.6 for slope having an inclination of 30°, 40°, and 50°.

Furthermore, analysis is also performed for the slope stabilized with soil nails with similar designing parameters and the factors of safety are examined with the variation in spacing and inclination of the soil nails, and slope inclination from the horizontal.

At 0.5 m spacing of soil nails, factor of safety of the soil slope increases with the increase in soil slope inclination of 30° and 40° and remains constant for 50° slope angle.

In case of soil nails spacing of 1 m, the factor of safety increases with the variation of inclination of soil nails at 30° soil slope, for 40° slope angle factor of safety increase up to $\theta = 30°$ but reduce at $\theta = 40°$, and for 50° soil slope angle factor of safety decreases with increase in soil nail inclination up to 30° and decrease at 40° soil nail inclination. At 2 m spacing of soil nails factor of safety increase with the increment in soil nail inclination for 30° and 50° soil slope angle but for 40° soil slope angle factor of safety decreases up to 30° soil nails inclination and show a little increment at $\theta = 40°$.

The variation of a factor of safety with the variation of soil slope angle, spacing of soil nails, and soil nails inclination is shown in Figs. 3, 4, 5.

Analysis of failure patterns observed from the analysis is performed to understand the generation of failure mechanism under seismic loading and the variation of the same due to change in the inclination of the soil slope, soil nails' inclination and spacing. From the observed failure pattern from each analysis, a significant failure pattern is shown in Figs. 6, 7, and 8 that the failure surface obtained from all the simulations performed the present study for a wide range of different slope angles shows two distinct failure surfaces. Failure patterns for different cases are shown in. For a mild slope of 30°, the same failure patterns are obtained with the variation and spacing of soil nails slip surface originating from the top of the slope crest and progresses toward the bottom of the slope with huge soil mass except for 2 m spacing and 40° soil nail inclination. In this case, the top surface of soil slope fails with the small amount of soil mass. For 40° soil slope, failure surface originates from the top of the soil slope surface and moves toward the bottom for 0.5 m and 1 m spacing with the variation of soil nails inclination from 10° to 40° but for 2 m spacing, failure surface shows a different pattern which involves the movement of the soil surface from the surface of soil slope with small soil mass. Failure pattern shows a vivid behavior for β=50°, the only top surface near about the soil nails fails with a small amount of soil mass along with them for the variation of spacing and inclination of soil nails.

Internal stresses developed in the soil nails i.e., bending moment and shear force in nails also aid in the stability of the reinforced slope. In view of the above, maximum bending moment and shear force developed in the nails for all the cases (variation of slope angle, spacing, and inclination of soil nails) of the reinforced slope are

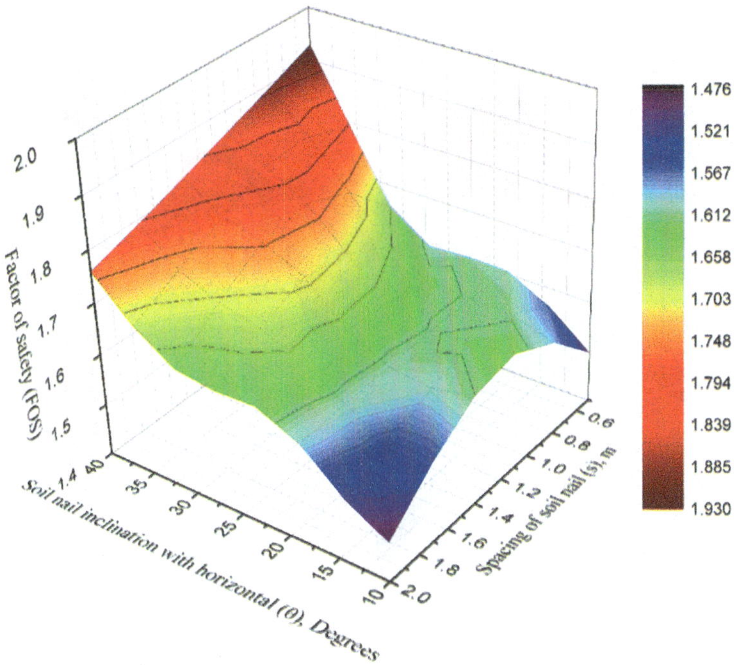

Fig. 3 Variation of a factor of safety with soil nail inclination (θ)attheslope$(\beta) = 30°$

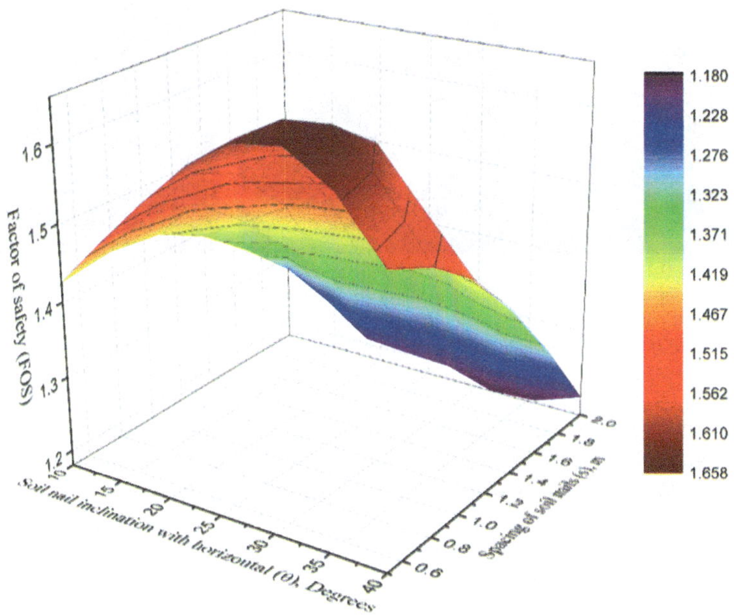

Fig. 4 Variation of a factor of safety with soil nail inclination (θ)attheslope$(\beta) = 40°$

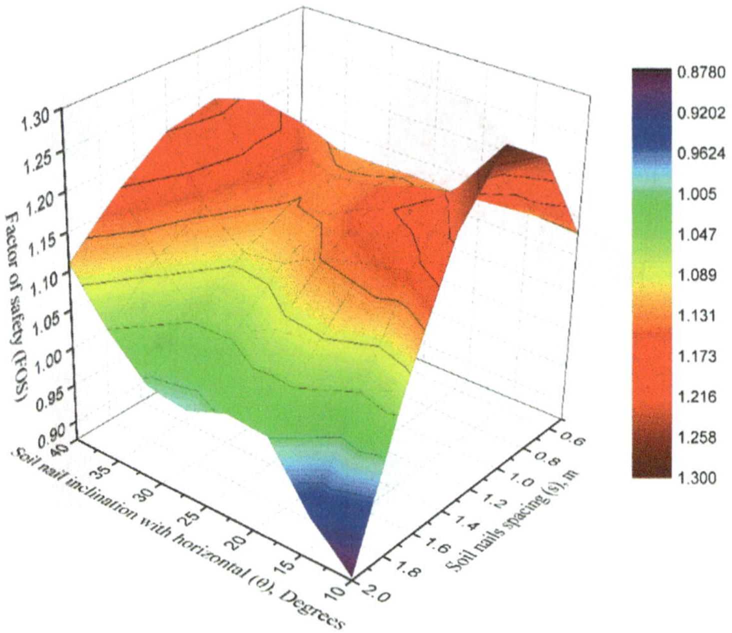

Fig. 5 Variation of a factor of safety with soil nail inclination (θ)attheslope$(\beta) = 50°$

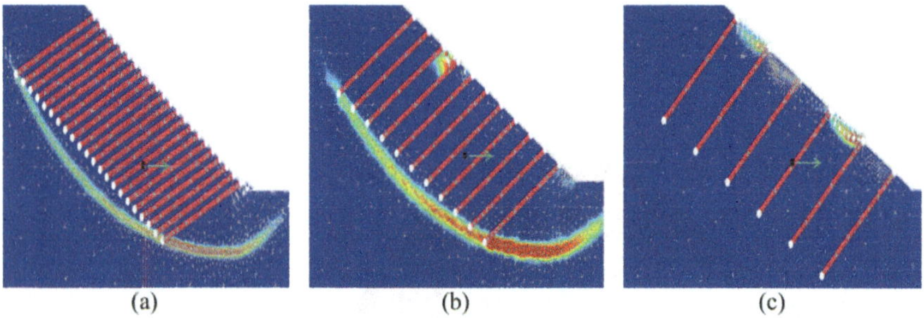

Fig. 6 Failure pattern of soil slope under the seismic condition for slope angle $(\beta) = 30°$ at **a** θ=10° and $s = 0.5$ m; **b** θ=30°and $s = 1.0$ m; and **c** θ=40° and $s = 2$ m

analyzed and summarized in Table 3. In Table 3, Observed magnitude of bending moment and shear force for a given slope with a provided inclination and spacing of nails is shown, where first numeric value signifies the magnitude of maximum bending moment/shear force and the alphabet (U, M, and L) in the parentheses denote the location of the nail for the mentioned internal reaction, where L, M, and U signify the lower ($z = 0$–2 m), middle ($z = 2$–6 m) and upper ($z = 6$–8 m) section of soil slope, z specifies the distance measured from the top horizontal surface of the slope.

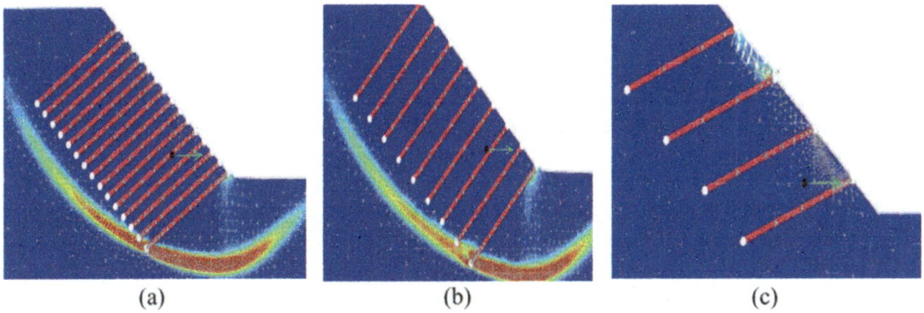

Fig. 7 Failure pattern of soil slope under the seismic condition for slope angle (β) = 40° at **a** θ = 10° and s = 0.5 m; **b** θ = 30° and s = 1.0 m; and **c** θ = 40° and s = 2 m

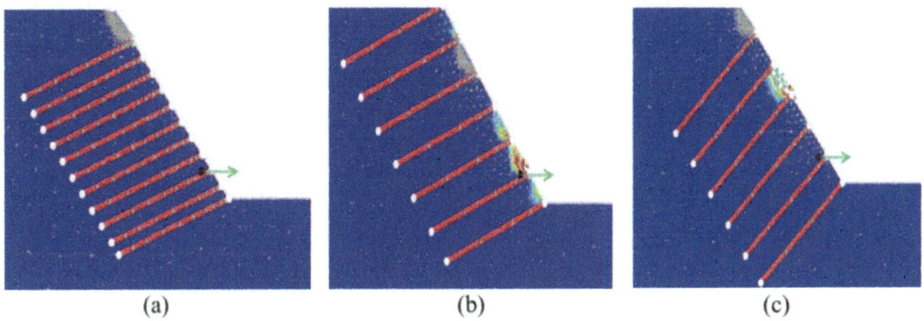

Fig. 8 Failure pattern of soil slope under the seismic condition for slope angle (β) = 50° at **a** θ = 10° and s = 0.5 m; **b** θ = 30° and s = 1.0 m; and **c** θ= 40° and s = 2 m

It is found that spacing and inclination of soil nails plays a vital role in slope stability. At 30° slope angle, maximum bending moment and shear force are generated at θ = 10° for 0.5 m, 1 m, and 2 m soil nail spacing and minimum bending moment and shear force are obtained at θ = 40° and 20° for 0.5 m, 1 m, and 2 m spacing of soil nails, respectively. At 40° slope angle, maximum bending moment and shear force are obtained at θ = 10° and 20° for 0.5 m, 1 m, and 2 m spacing of soil nails and minimum bending moment and shear force are obtained at θ = 40° and 30° for 0.5 m, 1 m, and 2 m spacing of soil nails, respectively. At 50° slope angle, maximum bending moment and shear force are obtained at θ = 40° and 10° for 0.5 m, 1 m, and 2 m spacing of soil nails and minimum bending moment and shear force are obtained at θ= 20° and 40° for 0.5 m, 1 m, and 2 m spacing of soil nails, respectively, as shown in Table 3. It is also noticed that bending moment and shear force developed in the soil nails show a different variation for different slope angles with the variation of spacing and inclination angle of soil nails. In the case of 30° slope bending moment and shear force go on decreasing with the increment in the inclination of soil nails for 10° to 40° and spacing from 0.5 m to 2 m. For 40°, decrement in bending moment and shear force is noticed with the increment of soil nail inclination at 0.5 m and 1 m but at 2 m spacing increase in internal stresses is noticed. At 50° soil slope bending

Table 3 Maximum bending moment and shear force developed in the soil nails for different reinforced slopes

Soil slope (β) →		30°		40°		50°	
Spacing of soil nails (s)	Inclination of soil nail (θ)	Maximum bending moment (kN m)	Maximum shear force (kN)	Maximum bending moment (kN m)	Maximum shear force (kN)	Maximum bending moment (kN m)	Maximum shear force (kN)
0.5 (m)	10°	30.15 (M)	23.49 (M)	33.51 (L)	23.26 (L)	11.95 (U)	8.58 (U)
	20°	23.89 (L)	15.69 (L)	19.00 (L)	18.41 (L)	12.42 (U)	8.69 (U)
	30°	21.98 (L)	14.65 (L)	9.91 (M)	10.18 (L)	15.87 (U)	9.98 (U)
	40°	18.80 (L)	13.92 (L)	11.73 (U)	9.18 (U)	22.74 (U)	14.37 (U)
1 (m)	10°	38.72 (M)	29.55 (M)	39.24 (L)	25.87 (L)	30.28 (M)	22.88 (U)
	20°	32.97 (U)	17.95 (U)	24.01 (L)	16.28 (L)	32.03 (U)	24.24 (U)
	30°	28.38 (U)	17.39 (U)	24.87 (M)	16.78 (M)	34.78 (M)	26.67 (M)
	40°	15.81 (L)	14.84 (L)	23.65 (M)	14.93 (M)	48.45 (M)	30.78 (M)
2 (m)	10°	37.76 (M)	27.53 (M)	36.51 (M)	24.19 (M)	53.49 (M)	38.58 (M)
	20°	26.44 (L)	24.15 (L)	44.56 (M)	27.59 (M)	64.01 (M)	45.97 (M)
	30°	36.16 (L)	22.30 (L)	37.26 (M)	20.51 (M)	52.80 (M)	42.33 (M)
	40°	33.17 (L)	21.51 (L)	43.30 (M)	32.19 (M)	38.01 (M)	27.16 (M)

moment and shear force increase with an increase in soil nail inclination at 0.5 m and 1 m soil nail spacing, at 2 m spacing of soil nails bending moment and shear increases up to 20° soil nail inclination but later on decreases with increase in soil nail inclination.

5 Conclusion

A comprehensive finite element numerical analysis for soil slope reinforced with soil nails at a various inclination (θ) and spacing (s) of soil nails is performed. It is noticed that the inclination (θ) of soil nails and spacing (s) of soil nails play a vital role in the slope stability of sand. On the basis of present work following conclusions are drawn:

1. It is found that keeping the diameter of soil nails fixed, the inclination of 30° gave the maximum factor of safety with respect to 10°, 20° and 40° for the slope of 30° under horizontal seismic coefficient of 0.12 g at the spacing(s) of 0.5 m. The increment of 39.5% in the factor of safety would occur by using soil nails.
2. For the 40° slope, the maximum factor of safety can be obtained at soil nail inclination (θ) of 30° at 0.5 m spacing(s). The increment of 50.27% in the factor of safety can be obtained by using soil nails of given parameters under k_h = 0.12 g

3. In the case of the slope angle of 50°, the increment of 42.9% can be obtained under $k_h = 0.12$ g at soil nail inclination (θ) of 30° at 0.5 m spacing(s).
4. Two different failure patterns are observed during this study, for soil slope 30° and 40° slip surface found from model testing is in a form of circular slip surface that originates from the slope crest and progresses towards the bottom of the slope face and for 50° soil slope only top surface of soil slope fail with the movement of the small amount of soil mass.
5. The maximum bending moment is obtained at 1 m spacing and 10° soil nail inclination for 30° slope angle. For 40° and 50° slope angle maximum bending and shear force is obtained 2 m spacing and 20° soil nail inclination.

References

1. Alsubal S, Indra SH, Babanigida NM (2017) A typical design of soil nailing system for stabilizing a soil slope: case study. Indian J Sci Technol 10(4). https://doi.org/10.17485/ijst/2017/v10i4/110891
2. Azzam W. Sobhey MA (2019) Utilization of soil nailing technique to improve sand slopes under seismic loading. In: International conference on advances in structural and geotechnical engineering. Hurghada, Egypt, pp 0–10
3. Babu GLS, Singh VP (2008) Numerical analysis of performance of soil nail walls in seismic conditions. ISET J Earthquake Technol 31–40
4. Budania R, Arora RP (2016) Soil nailing for slope stabilization : an overview. Int J Eng Sci Comput 6(12):3877–3882
5. Fawaz A, Farah E, Hagechehade F (2014) Slope stability analysis using numerical modelling. Am J Civil Eng 2(3):60–67. https://doi.org/10.11648/j.ajce.20140203.11
6. Federal Highway Administration (FHWA) (2003) Soil nail walls, geotechnical engineering circular no. 7, report no. FHWA0-IF-03–017. Federal Highway Administration
7. Garz6n-Roca J, Capa V, Torrijo FJ, Company J (2019) Designing soil-nailed walls using the amherst wall considering problematic issues during execution and service life. Am Soc Civil Eng 19:1–14. https://doi.org/10.1061/(ASCE)GM.1943-5622.0001453
8. Kannaujiya P, Chauhan VB (2020) Behavior of anchored sheet pile wall. In: Shehata H, Brandl H, Bouassida M, Sorour T (eds) Sustainable thoughts in ground improvement and soil stability, GeoMEast 2019. Springer, Cham, pp 184–195. https://doi.org/10.1007/978-3-030-34184-8_12
9. Ojha R, Chauhan VB (2020) Performance of geosynthetic reinforced segmental retaining walls. In: Shehata H, Brandl H, Bouassida M, Sorour T (eds) Sustainable thoughts in ground improvement and soil stability, GeoMEast 2019. Springer, Cham, pp 196–206. https://doi.org/10.1007/978-3-030-34184-8_13
10. Optum G2 (2017) Optum computational engineering. Copenhagen NV, Denmark
11. Pandey A, Chauhan VB (2020) Evaluation of pull-out capacity of helical anchors in clay using finite element analysis. Geotechnical Special Publication 2020. GSP 317, pp 60–68. https://doi.org/10.1061/9780784482803.007
12. Pandey A, Chauhan VB (2020) Numerical analysis for the evaluation of pull-out capacity of helical anchors in sand. In: Shehata H, Brandl H, Bouassida M, Sorour T (eds) Sustainable thoughts in ground improvement and soil stability, GeoMEast 2019. Springer, Cham, pp 207–218. https://doi.org/10.1007/978-3-030-34184-8_14
13. Rawat P, Chatterjee K (2018) Seismic stability analysis of soil slopes using soil nails. Geotechnical Special Publication, pp 79–87. https://doi.org/10.1061/9780784481486.009
14. Rawat S, Gupta AK (2016) Analysis of a nailed soil slope using limit equilibrium and finite element methods. Int J Geosynth Ground Eng. https://doi.org/10.1007/s40891-016-0076-0

15. Seyed-Kolbadi SM, Sadoghi-Yazdi J, Hariri-Ardebili MA (2019) An improved strength reduction-based slope stability analysis. Geosciences 1–11. https://doi.org/10.3390/geosciences9010055
16. Sengupta A, Giri D (2011) Dynamic analysis of soil-nailed slope. Ground Improve 164:225–234. https://doi.org/10.1680/grim.8.00023

Seismic Stability of Non-homogenous Cohesive Soil by Using Calculus of Variation

Sourav Sarkar and Manash Chakraborty

Abstract In this article, the factor of safety of a cohesive soil slope subjected to seismic load is determined by employing the variational method and pseudostatic analysis. Unlike the conventional limit equilibrium method, there is no requirement to consider any kinematical or static assumption in the variational method. The factor of safety (F) is defined as a functional which is minimized by using the Euler–Lagrangian equation. The (i) transversality and boundary conditions are imposed at the intersection of slip surface, and the slope surface, and (ii) continuity and natural boundary states are forced at the intermediate point of the slip surface. The soil is considered to be completely saturated and loaded under undrained conditions. The cohesion of the soil is assumed to increase linearly with depth. The critical slip surface and consequently critical factor of safety, F_s is being obtained by varying the slope geometry, soil properties, and seismic loadings. The available solutions compare quite well with the convenient solution for the pseudostatic slope stability analysis. The proposed design charts will be quite useful to practicing engineers.

Keywords Variational calculus · Slope stability · Seismic coefficient · Cohesive soil

1 Introduction

Slope stability analysis is one of the most exciting and challenging problems in the geotechnical engineering field. Since the past eight decades, several researchers [4, 9, 11, 17, 28, 34–36] had studied this classical problem thoroughly. Most of the previous studies related to slope stability were accomplished with the help of the limit equilibrium method (LEM). However, in LEM, few assumptions are being considered to assess the stability of slopes. They are as follows: the nature of the interslice forces, the shape of the slip surface, and distribution of the normal stress along the slip surface. Moreover, strain and displacement compatibility are not being considered

S. Sarkar · M. Chakraborty (✉)
Indian Institute of Technology (BHU), Varanasi 221005, India
e-mail: manashchakra.civ@itbhu.ac.in

T. G. Sitharam et al. (eds.), *Local Site Effects and Ground Failures*, Lecture Notes in Civil Engineering 117, https://doi.org/10.1007/978-981-15-9984-2_20

in LEM, and hence, the method suffers severe limitations as discussed by Duncan [10] and Krahn [22]. Several other researchers [3, 12, 18, 23, 27, 37, 38] had applied various other analytical and numerical methods for analyzing the homogenous slopes. Zaki [38] and Lane and Griffiths [23] applied the strength reduction method (SRM) in the framework of the conventional displacement based finite-element method for analyzing the homogenous slope. Yu et al. [37] and Kim et al. [18] used finite-element limit analysis to analyze the homogenous slopes with and without considering the effect of pore water pressure. Michalowski [27] provided stability charts for uniform slopes subjected to pore water pressure and horizontal seismic force by using the kinematic approach of limit analysis. Baker et al. [3] proposed stability charts for homogeneous slopes by applying the variational method and the strength reduction technique considering pseudostatic analysis.

The stability of non-homogeneous slopes was also studied rigorously especially in the past few years [5, 6, 8, 13–16, 20, 21, 24–26, 29]. Gibson and Morgenstern [13] analyzed non-homogeneous slopes considering linearly increasing strength with cohesion at the ground surface (c_{u0}) as zero by using $\phi = 0$ limit equilibrium analysis for the static case. By following his work, Hunter and Schuster [16] extend the problem for $c_{u0} > 0$ and provide solutions. Booker and Davis [5] and Chen et al. [8] used the kinematical approach of limit analysis (LA) to obtain the upper-bound solution for non-homogeneous slope problems. Koppula [20] used the limit equilibrium method to analyze the slope consisting of linearly increasing strength for the static case. Koppula [21] employed the equivalent static force (ESF) method for the seismic case. Leshchinsky and Smith [24], Low [26] evaluated the factor of safety of embankments constructed on the soft clay by adopting LEM. Chai et al. [6] applied the finite-element method (FEM) to assess the failure of an embankment on soft clayey soil-based modified cam clay model. Griffiths and Yu [14] employed Matlab optimization programs to evaluate the critical slip surface and corresponding critical factor of safety of the slope with linearly increasing undrained cohesion with depth and firm stratum at the bottom of the slope. Li et al. [25] applied the variational method to assess the stability of the slope consisting of undrained clay with linearly increasing strength for both static and seismic cases.

After a thorough literature study, it is noticed that very few research works on non-homogeneous soil slopes under seismic forces by using the variational approach had been carried out earlier. The use of the variational approach is more advantageous than the limit equilibrium method as no prior assumption regarding the critical failure surface and the normal stress distributions along the failure surface are needed. The variational method was first introduced by Kopacsy [19] in solving the slope stability problem. Later on, some researchers [1–3, 7, 30–33] carried on a further study on this method and applied it to solve various problems on slope stability.

In this present article, the variational method is implemented to evaluate the critical slip surface and corresponding critical factor of safety (F_s) for any non-homogeneous soil slope subjected to seismic loading. The formulation recommended by Revilla and Castillo [31] for analyzing the homogenous cohesive soil slopes is extended by making generalized for any purely undrained cohesive soil ($\phi_u = 0$) slopes with linearly varying cohesion under seismic conditions. In the present analysis, the effect of pore water pressure is not considered (the provision of pore water pressure is kept for future scope). The critical factor of safety (F_s) is obtained for a different combination of slope angle (β) and cohesion coefficient (λ) of soil corresponding to varying horizontal (k_h) and vertical (k_v) seismic coefficients. The obtained critical slip surfaces are also presented for a few cases.

2 Problem Statement and Mathematical Background

Figure 1 shows a problem containing a rectilinear slope of angle β with non-homogeneous undrained cohesive soil ($\phi_u = 0$) subjected to horizontal and vertical seismic forces within a chosen domain where the cohesion of the soil (c_u) is linearly increasing with depth. The objective is to determine the critical factor of safety (F_s) and the critical slip surface for the given slope. The domain size is kept adequately high so that the slip surface is well accommodated. However, it is to be noted that domain size does not influence the computed solution.

The definition of factor safety is the ratio of the available resisting moment to the actual driving moment. Both these parameters are the function of the slip surface and the slope surface; consequently, they can be represented as functional. The optimization of these functionals is essential for obtaining the critical slip surface of a slope corresponding to the critical factor of safety. The Euler–Lagrangian equation is being applied to optimize the safety functional, as recommended in the theory of calculus of variation. Pseudostatic analysis is being implemented to incorporate the seismic forces.

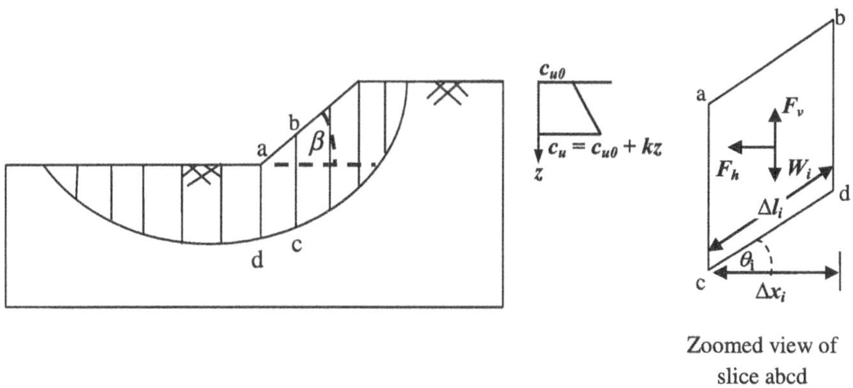

Fig. 1 Schematic diagram of a slope along with slip surface and considered slices

3 Analytical Formulation of the Problem

The linear relationship between undrained shear strength of purely cohesive soil with the depth can be represented as follows:

$$c_u = c_{u0} + kz \tag{1}$$

where c_u is the undrained shear strength of soil at a certain depth z below the ground surface, c_{u0} is the undrained shear strength of soil at the ground surface, and k is the gradient at which undrained shear strength of soil increases with depth z.

In this problem, the variational method is applied in the framework of the ordinary slice method [11] subjected to seismic forces in both horizontal and vertical directions. For any slice, i, as shown in Fig. 1, the equation of static equilibrium along both horizontal and vertical directions are satisfied. Hence, the factor of safety can be formulated as

$$F = \frac{\sum_{i=1}^{n} [c_u \Delta l_i]}{\sum_{i=1}^{n} (W_i \sin\theta_i + F_h \cos\theta_i - F_v \sin\theta_i)} \tag{2}$$

where (i) c_u is the shear strength of soil at a specific depth z.

(ii) W_i, θ_i, Δx_i, Δl_i (=$\Delta x_i/\cos\theta_i$) are the weight, inclination angle, width, and length of the ith slice, respectively; n indicates the total number of slices.

(iii) F_h (=$k_h W_i$) and F_v (=$k_v W_i$) are the horizontal and vertical seismic forces, respectively.

Plugging the expressions of F_h, F_v, and Δl_i in Eq. (2), the factor of safety can be further expressed as

$$F = \frac{\sum_{i=1}^{n} \left[c_u \Delta x_i \sec^2\theta_i\right]}{\sum_{i=1}^{n} W_i[(1 - k_v)\tan\theta_i + k_h]} \tag{3}$$

This formulation is being expressed as the continuous slope surface by following the study of Revilla and Castillo [31] and Baker and Garber [2]

$$F = \frac{\int_{x_0}^{x_3} \left[c_u\left(1 + y_i\prime^2\right)\right] dx}{\int_{x_0}^{x_4} \gamma(f_i - y_i)[(1 - k_v)y_i\prime + k_h] dx} = \frac{\int_{x_0}^{x_3} P(x, y, y\prime) dx}{\int_{x_0}^{x_3} R(x, y, y\prime) dx} \tag{4}$$

where (i) $y_i\prime = \tan\theta_i$, (ii) $W_i = \gamma(f_i - y_i)\Delta x_i$; γ is the unit weight of soil, and f_i and y_i represent the slope and the slip surface, and (iii) x_0 and x_3 are the abscissas of two endpoints where the slip surface intersect with the slope surface.

The rectilinear slope as shown in Fig. 1 can be expressed mathematically with the following three functions within the range of x_0 and x_3:

$$\left.\begin{array}{l} f_1(x) = 0\,(x_0 \le x \le 0) \\ f_2(x) = \frac{h}{h_1}x\,(0 \le x \le h_1) \\ f_3(x) = h\,(h_1 \le x \le x_3) \end{array}\right\} \tag{5}$$

Corresponding to each $f_i(x)$ there would be a distinct $y_i(x)$ in the interval of (x_{i-1}, x_i). The expression of the factor of safety in Eq. (4) takes the following form:

$$F = \frac{\int_{x_0}^{x_1} P_1(x, y, y\prime)dx + \int_{x_1}^{x_2} P_2(x, y, y\prime)dx + \int_{x_2}^{x_3} P_3(x, y, y\prime)dx}{\int_{x_0}^{x_1} R_1(x, y, y\prime)dx + \int_{x_1}^{x_2} R_2(x, y, y\prime)dx + \int_{x_2}^{x_3} R_3(x, y, y\prime)dx} \tag{6}$$

To obtain critical slip surface and corresponding critical factor of safety, F_s the above functionals are required to be minimized. This is being executed by using the Euler–Lagrangian equation as follows:

$$F_s = \frac{\frac{\partial P}{\partial y_i} - \frac{d}{dx}\left(\frac{\partial P}{\partial \prime_i}\right)}{\frac{\partial R}{\partial y_i} - \frac{d}{dx}\left(\frac{\partial R}{\partial y\prime_i}\right)} = \frac{2c_u y\prime\prime_i}{\gamma[(1-k_v)f_i\prime + k_h]};\ (\mathrm{i} = 1, 2, 3) \tag{7}$$

From Eq. 7, the slip surface in the three different ranges, as depicted in Eq. 5, can be obtained. Figure 2 representing all three surfaces is constructed with the assumption that the factor of safety remains to be the same along the slip surface. The three surfaces are

$$y_1 = \frac{k_h}{4(1+\lambda)Nh}x^2 + B_1x + D_1 \tag{8a}$$

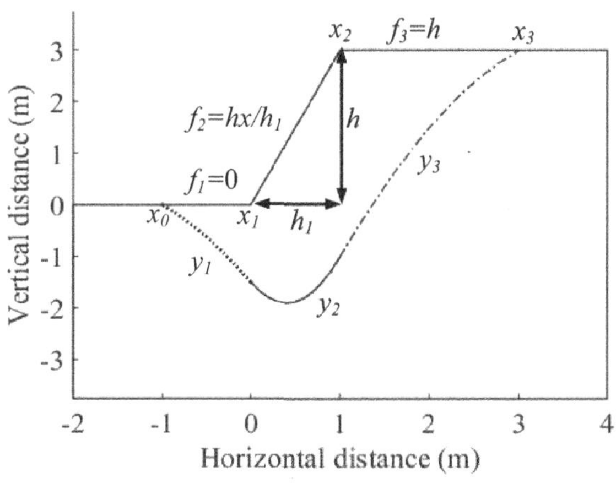

Fig. 2 Slip surface by variational method

$$y_2 = \frac{\left[(1-k_v)\frac{h}{h_1} + k_h\right]}{4(1+\lambda)Nh}x^2 + B_2x + D_2 \tag{8b}$$

$$y_3 = \frac{k_h}{4(1+\lambda)Nh}x^2 + B_3x + D_3 \tag{8c}$$

Here, N (=$c_{u0}/(F_s\gamma h)$) and λ(=kh/c_{u0}) indicate the stability number and the cohesion coefficient, respectively. The equation of each surface contains two integration constants. Therefore, the total number of unknown parameters becomes nine—(i) x_0 and x_3, (ii) F_s, and (iii) six integration constants (B_1, B_2, B_3, D_1, D_2, and D_3). These unknown parameters can be determined by following nine equations.

(a) Two transversality conditions: Specified at two endpoints (x_0 and x_3)

$$[P_{y'} - F_s R_{y'}][f'(x) - y'(x)] + P - F_s R|_{x=x_i} = 0; \text{ where } (i = 0, 3) \tag{9}$$

(b) Two continuity conditions: Specified at the intermediate points x_1 and x_2 (i.e., the intersections of the slip surfaces)

$$y_i(x_i) = y_{i+1}(x_i)\ (i = 1, 2) \tag{10}$$

(c) Two natural boundary conditions: Specified at two intermediate points x_1 and x_2

$$y_i'(x_i) = y'_{i+1}(x_i)\ \ (i = 1, 2) \tag{11}$$

(d) Two boundary conditions: Specified at two end-points (x_0 and x_3)

$$y_1(x_0) = f_1(x_0);\ y_3(x_3) = f_3(x_3) \tag{12}$$

(e) The equation for Factor of Safety, as provided in Eq. (6).

By using these nine equations, nine unknown parameters can be evaluated. The expressions for the unknown parameters are as follows:

$$B_1 = B_2 = -1 - \frac{k_h x_0}{2(1+\lambda)Nh}$$

$$B_3 = 1 - \frac{k_h x_3}{2(1+\lambda)Nh}$$

$$D_1 = D_2 = x_0 + \frac{k_h x_0^2}{4(1+\lambda)Nh}$$

$$D_3 = \frac{(1-k_v)h_1}{4(1+\lambda)N} - 2h_1 + x_0 + \frac{k_h h_1(x_3 - x_0)}{2(1+\lambda)Nh} + \frac{k_h x_0^2}{4(1+\lambda)Nh}$$

$$N = \frac{h(1 - k_v) + k_h(x_3 - x_0)}{4(1 + \lambda)h}$$

According to the literature [14, 25] non-homogeneous soil slope under seismic condition will fail by developing toe failure surface ($x_0 = 0$). Hence in the present analysis, the value of x_0 is considered equal to zero.

$$x_3 = \frac{(1 - k_v)(h_1 + h)}{[(1 - k_v) - k_h]}$$

4 Results and Discussions

In the present study, some typical non-homogeneous slopes are analyzed by using the formulation derived in the previous segment. The solutions are presented in the form of the critical factor of safety (F_s). The height of the slope (h) and the unit weight of soil (γ) are to be maintained identically for all cases. The critical factor of safety is evaluated for different combinations of λ and slope inclination angle (β) corresponding to varying k_h and k_v. The value of β has been varied within the range of 20°–90° at an interval of 5°. Three different values of λ, namely, 0, 1.5, and, 3.0 are chosen for each specific value of β. The magnitude of k_h is considered as 0.0, 0.1, 0.2, and 0.3. For these values of k_h, corresponding magnitudes of k_v are 0, $0.5k_h$, and k_h.

Table 1 depicts the typical result of F_s of non-homogeneous soil slopes for a different combination of λ, β subjected to the different magnitude of k_h and k_v. The

Table 1 F_s for different slopes corresponding to different seismic coefficients

k_h	k_v	λ											
		β = 45°			β = 55°			β = 65°			β = 75°		
		0.0	1.5	3.0	0.0	1.5	3.0	0.0	1.5	3.0	0.0	1.5	3.0
0	0	0.50	1.25	2.0	0.50	1.25	2.0	0.50	1.25	2.0	0.50	1.25	2.0
	0.00	0.48	1.20	1.75	0.47	1.17	1.88	0.46	1.14	1.83	0.44	1.11	1.78
0.10	0.05	0.46	1.14	1.83	0.45	1.12	1.80	0.44	1.09	1.75	0.43	1.06	1.70
	0.10	0.44	1.09	1.92	0.43	1.07	1.72	0.42	1.05	1.68	0.41	1.02	1.64
	0.00	0.43	1.10	1.76	0.42	1.05	1.68	0.40	1.00	1.60	0.37	0.94	1.50
0.20	0.10	0.41	1.02	1.63	0.39	0.98	1.57	0.37	0.93	1.49	0.35	0.88	1.41
	0.20	0.38	0.95	1.52	0.37	0.91	1.46	0.35	0.88	1.40	0.33	0.83	1.33
	0.00	0.37	0.91	1.46	0.34	0.85	1.36	0.31	0.78	1.26	0.29	0.71	1.14
0.30	0.15	0.35	0.87	1.39	0.33	0.82	1.31	0.30	0.76	1.22	0.28	0.70	1.12
	0.30	0.32	0.81	1.29	0.31	0.77	1.23	0.29	0.72	1.16	0.27	0.67	1.08

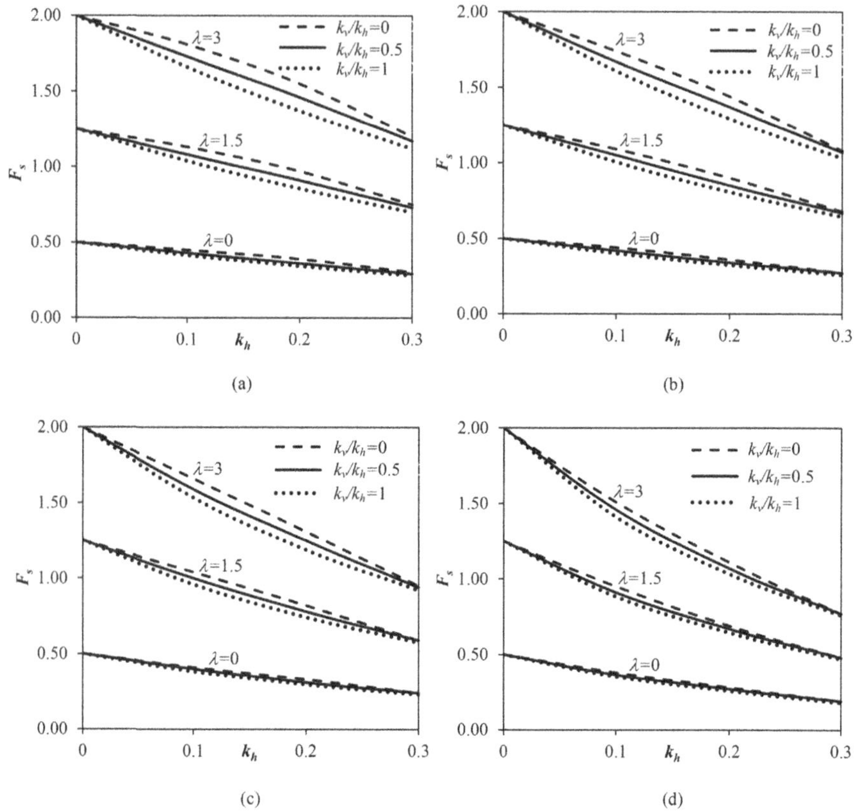

Fig. 3 Variation of critical factor of safety (F_s) with k_h for **a** $\beta = 20\pi$; **b** $\beta = 30\pi$; **c** $\beta = 40\pi$; **d** $\beta = 50\pi$

results are presented for four different values of β, namely, 45°, 55°, 65°, and 75°. Figure 3 shows the variation of F_s with k_h for different values of k_v/k_h. The graphs represent the variation of F_s values for different combination λ values corresponding to four distinct values of β, namely, 20°, 30°, 40°, and 50°. It is clearly noticed from the obtained solutions that the magnitude F_s decreases with the increase in k_h and k_v values. For example, for $\beta = 20°$ and $\lambda = 3$, when the magnitude of k_h increases from 0 to 0.1, F_s reduce by 10.50%, and for the same slope, the reduction of F_s is 66.67% when the magnitude of k_h increases from 0 to 0.3. It is also observed that F_s reduce with (i) increase in β and (ii) decrease in λ. For $k_h = 0.3$, $k_v = 0.5k_h$ and $\beta = 20°$ F_s improves by 75.21% with the increase in λ from 0 to 3. The stability of the slope further improves by reducing β. F_s increases by 32.14% with the decrease of β from 50° to 20° for $\lambda = 3$ and $k_h = k_v = 0.3$.

Figure 4 illustrates the shape of the critical slip surface for β= 20°, 40°, 60°, and 80° with a different combination of k_h. The magnitude of λ and k_v are kept 3 and 0, respectively. The shape of the critical slip surface changes markedly with the

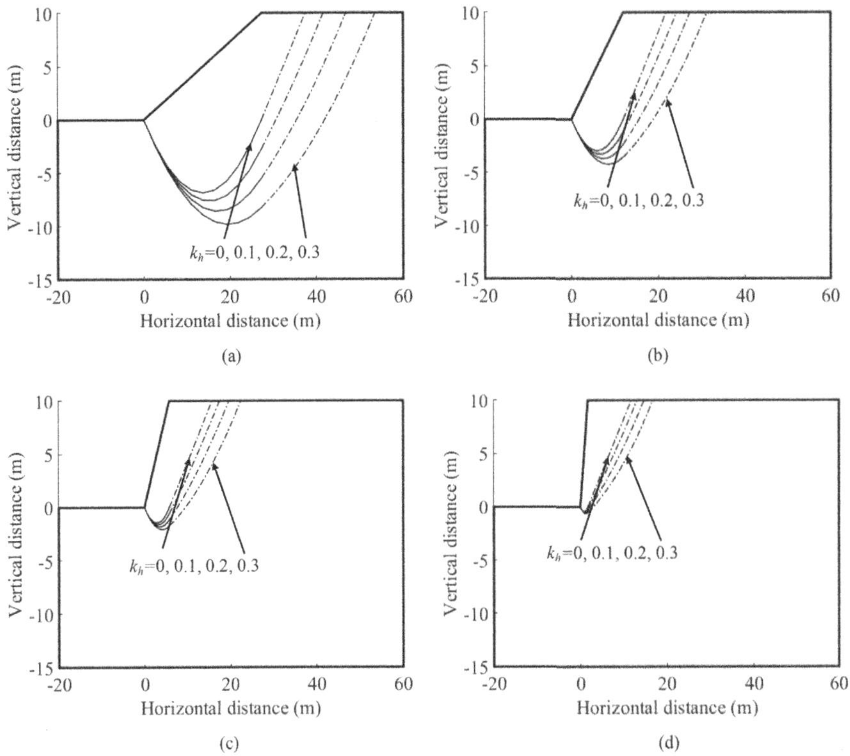

Fig. 4 Form of the critical slip surfaces for **a** $\beta = 20°$, $\lambda = 3$, $k_v = 0$; **b** $\beta = 40°$, $\lambda = 3$, $k_v = 0$; **c** $\beta = 60°$, $\lambda = 3$, $k_v = 0$ **d** $\beta = 80°$, $\lambda = 3$, $k_v = 0$

geometric profile of the slope. The figure also gives the impression that the volume of soil within the critical slip surface increases with the increment of the magnitude of seismic forces.

5 Comparison of Results

Table 2 shows the comparison of the present solution evaluated from the variational method with the available solution provided by Koppula [20] and Li et al. [25] for the static case. The comparison is carried out for different values of cohesion coefficient (λ) corresponding to various β for both static and pseudostatic case. The present solution is compared with the available solution for a particular slope consisting of non-homogeneous purely cohesive soil with slope height, $h = 10$ m, $\gamma = 16$ kN/m^3, and $c_{u0} = 20$ kPa. Table 3 depicts the comparison between the present solution and the result presented by Li et al. [25] for the pseudostatic case. The present analytical solutions are quite agreeable with the available solutions (Table 3).

Table 2 A comparison of F_s values obtained from the present study with the solution of Koppula [20] and Li et al. [25] considering $k_h = k_v = 0$ for $\beta = 45°$

λ	Present study	Koppula [20]	Li et al. [25]
0.5	0.75	1.00	1.00
1	1.00	1.27	1.27
2	1.50	1.79	1.79
3	2.00	2.31	2.31
4	2.50	2.83	2.83
5	3.00	3.34	3.34
10	5.50	5.92	5.92

Table 3 A comparison of F_s values obtained from the present study with the solution of Li et al. [25] considering $k_v = 0$

		$\beta = 60°$		$\beta = 75°$	
k_h	λ	Present study	Li et al. [25]	Present study	Li et al. [25]
0.1	0	0.44	0.57	0.43	0.50
	0.5	0.66	0.74	0.64	0.66
	1	0.88	0.96	0.85	0.78
	1.5	1.10	1.14	1.06	0.89
	2	1.31	1.25	1.28	1.04
	2.5	1.53	1.39	1.49	1.25
	3	1.75	1.79	1.70	1.39
	0	0.38	0.54	0.36	0.45
0.2	0.5	0.57	0.66	0.54	0.57
	1	0.76	0.83	0.72	0.69
	1.5	0.95	0.96	0.90	0.83
	2	1.14	1.14	1.08	0.96
	2.5	1.33	1.25	1.26	1.04
	3	1.52	1.39	1.43	1.25

6 Conclusions

In the present article, pseudostatic analysis of non-homogeneous soil slope is executed by using the variational approach without considering any prior assumption on the shape of the slip surface. Design charts are being presented in the form of the critical factor of safety for different slope geometry (β), soil properties (λ), and seismic forces (k_h and k_v). The factor of safety reduces as the magnitude of k_h and k_v increases, and soil strength decreases. It is also noted that the critical factor of safety decreases significantly with an increment of slope angle for pseudostatic cases. Critical slip surfaces are presented here for a few cases. The present solution has also been well compared with the previous one that instills a further belief in applying the calculus of variation to solve other stability problems. The method can be comfortably used in solving multilayer slope stability problems provided analytical formulation to be changed only.

Acknowledgements The corresponding author acknowledges the support of the "Department of Science and Technology (DST), Government of India" under grant number DST/INSPIRE/04/2016/001692.

References

1. Baker R (2003) Sufficient conditions for existence of physically significant solutions in limiting equilibrium slope stability analysis. Int J Solids Struct 40:3717–3735
2. Baker R, Garber M (1978) Theoretical analysis of the stability of slopes. Geotechnique 28(4):395–411
3. Baker R, Shukha R, Operstein V, Frydman S (2006) Stability charts for pseudostatic slope stability analysis. Soil Dyn Earthq Eng 26(9):813–823
4. Bishop AW (1955) The use of the slip circle in the stability analysis of slopes. Geotechnique 5(1):7–17
5. Booker JR, Davis EH (1972) A note on a plasticity solution to the stability of slopes in inhomogenous clays. Géotechnique 22(3):509–513
6. Chai J, Carter JP (2009) Simulation of the progressive failure of an embankment on soft soil. Comput Geotech 36(6):1024–1038
7. Chen J, Yang Z, Hu R, Zhang H (2016) Study on the seismic active earth pressure by variational limit equilibrium method. Hindawi Publishing Corporation Shock and Vibration, pp 1–11
8. Chen WF, Snitbhan N, Fang HY (1975) Stability of slopes in anisotropic, non-homogeneous soils. Can Geotech J 12(1):146–152
9. Chen ZY, Morgenstern NR (1983) Extensions to the generalized method of slices for stability analysis. Can Geotech J 20(1):104–119
10. Duncan JM (1996) State of the art: limit equilibrium and finite-element analysis of slopes. J Geotech Eng 122(7):577–596
11. Fellenius W (1936) Calculation of stability of earth dam. In: Transactions 2nd congress large dams, Washington, DC, vol 4, pp 445–462
12. Gens A, Hutchinson JN, Cavounidis S (1988) Three-dimensional analysis of slides in cohesive soils. Geotechnique 38(1):1–23
13. Gibson RE, Morgenstern N (1962) A note on the stability of cuttings in normally consolidated clays. Geotechnique 12(3):212–216
14. Griffiths DV, Yu X (2015) Another look at the stability of slopes with linearly increasing undrained strength. Géotechnique 65(10):824–830
15. Hossley A, Leshchinsky B (2019) Stability and failure geometry of slopes with spatially varying undrained shear strength. J Geotech Geoenviron Eng 145(5):06019002
16. Hunter JH, Schuster RL (1968) Stability of simple cuttings in normally consolidated clays. Geotechnique 18(3):372–378
17. Janbu N (1954) Application of composite slip surface for stability analysis. Proc Eur Conf Stab Earth Slopes, Sweden 3:43–49
18. Kim J, Salgado R, Yu HS (1999) Limit analysis of soil slopes subjected to pore-water pressures. J Geotech Geoenviron Eng 125(1):49–58

19. Kopacsy J (1961) Distribution des contraintes a la rupture forme de la surface de glissement et hauteur theorique des talus. In: Proceedings of the 5th international conference on soil mechanics and foundation engineering paris, France, pp 641–650
20. Koppula SD (1984) On stability of slopes in clays with linearly increasing strength. Can Geotech J 21(3):577–581
21. Koppula SD (1984) Pseudo-static analysis of clay slopes subjected to earthquakes. Geotechnique 34(1):71–79
22. Krahn J (2003) The 2001 RM hardy lecture: the limits of limit equilibrium analyses. Can Geotech J 40(3):643–660 (2003)
23. Lane PA, Griffiths DV (2000) Assessment of stability of slopes under drawdown conditions. J Geotech Geoenviron Eng 126(5):443–450
24. Leshchinsky D, Smith DS (1989) Deep-seated failure of a granular embankment over clay. Stability analysis. Soils Found 29(3):105–114
25. Li B, Zhang F, Wang D (2018) Impact of crack on stability of slope with linearly increasing undrained strength. Math Prob Eng 1–11. https://doi.org/10.1155/2018/1096513
26. Low BK (1989) Stability analysis of embankments on soft ground. J Geotech Eng 115(2):211–227 (1989)
27. Michalowski RL (2002) Stability charts for uniform slopes. J Geotech Geoenviron Eng 128(4):351–355
28. Morgenstern NU, Price VE (1965) The analysis of the stability of general slip surfaces. Geotechnique 15(1):79–93
29. Nakase A (1970) Stability of low embankment on cohesive soil stratum. Soils Found 10(4):39–64
30. Narayan CGP (1975) Variational methods in stability analysis of slopes. PhD dissertation submitted to Indian Institute of Technology, Delhi, India
31. Revilla J, Castillo E (1977) The calculus of variations applied to stability of slopes. Geotechnique 27(1):1–11
32. Sarkar S, Chakraborty M (2019) Pseudostatic slope stability analysis in two-layered soil by using variational method. In: Earthquake geotechnical engineering for protection and development of environment and constructions (7th ICEGE, Rome), pp 4857–4864
33. Sarkar S, Chakraborty M (2020) Pseudo-static slope stability analysis for cohesive-frictional soil by using variational method. In: Advances in computer methods and geomechanics. Springer, Singapore , pp 159–171
34. Sarma SK (1973) Stability analysis of embankments and slopes Geotechnique 23(3):423–433
35. Spencer E (1967) A method of analysis of the stability of embankments assuming parallel inter-slice forces. Geotechnique 17(1):11–26
36. Taylor DW (1937) Stability of earth slopes. J Boston Soc. Civ Eng 24(3):197–247
37. Yu HS, Salgado R, Sloan SW, Kim JM (1998) Limit analysis versus limit equilibrium for slope stability. J Geotech Geoenviron Eng 124(1):1–11
38. Zaki A (1999) Slope stability analysis overview. University of Toronto (1999)

Site Characterization Using Satellite Data and Estimation of Seismic Hazard at Ground Surface

Naveen James, Sreevalsa Kolathayar, and T. G. Sitharam

Abstract This paper presents the seismic site characterization carried out for Karnataka (state level) as well as for India (country level) using topographic slope map derived from Digital Elevation Model (DEM) data. Two DEM data, SRTM, and ASTER were used to derive the slope maps. For Karnataka (state level), the slope map was generated from ASTER DEM considering a grid size of 5 × 5 km and for India (country level), the slope map was generated from SRTM DEM considering the grid size of 10 × 10 km. Based on the slope value, every grid point was characterized into various NEHRP site classes, and spatial variation of average shear wave velocity for top 30 m (Vs30) value throughout the study areas is presented in this paper. Peak horizontal acceleration (PHA) at bedrock level was evaluated for the same grid points using deterministic as well as probabilistic methodologies. The amplification factor for every grid point was obtained from the site coefficients corresponding to NEHRP site class. The surface level peak horizontal acceleration (PHA) was then evaluated for every grid point by multiplying bedrock level PHA with the corresponding amplification factor. Spatial variation of seismic hazard at the surface for the state of Karnataka as well as for entire India is presented in this paper.

Keywords Hazard · PSHA · DSHA · DEM · PHA

N. James (✉)
Department of Civil Engineering, Indian Institute of Technology Ropar, Punjab, India
e-mail: naveen.james@iitrpr.ac.in

S. Kolathayar
Department of Civil Engineering, National Institute of Technology Karnataka, Surathkal, India
e-mail: sreevalsa@nitk.edu.in

T. G. Sitharam
Director, Indian Institute of Technology Guwahati, Guwahati, Assam, India
e-mail: sitharam@iisc.ac.in

T. G. Sitharam et al. (eds.), *Local Site Effects and Ground Failures*, Lecture Notes in Civil Engineering 117, https://doi.org/10.1007/978-981-15-9984-2_21

1 Introduction

It is a well-known fact that the characteristics of seismic waves will undergo severe alteration due to impedance contrast while traveling from bedrock to ground surface. Hence, the role seismic site characterization is important in evaluating the intensity of ground shaking at a site due to any future earthquake. Usually, geotechnical or geophysical tests are carried out to characterize the site. For the seismic site characterization of vast regions, topographic gradient data can be very useful. Studies already proved that topographic slope data derived from digital elevation model (DEM) is an index for the seismic site conditions. Several studies have proved that apart from source parameters, the characteristics of surface-level earthquake motion are mainly influenced by engineering properties of overburden soil such as thickness, density, sub-surface geometry. Seismic site characterization aims to categorize the overburden soil strata based on either local geology [1–5] or geotechnical/geophysical test data. The most important parameter needed for the site characterization is the average shear wave velocity for top 30 m depth estimated using geophysical tests such as SASW test, MASW test, suspension logging test, cross-hole test, seismic cone penetration test. The top 30 m is the depth corresponding to the one-quarter wavelength of the period of interest as suggested by Joyner et al. [6]. Many studies have been carried out to correlated geotechnical test data, such as SPT and CPT values with the shear wave velocity [7–9]. Most of the building codes such as [10–12] recommend site characterization based on $\boldsymbol{V}_{s30}$ where

$$V_{s30} = 30 / \sum_{i=1}^{n} \frac{d_i}{V_i} \tag{1}$$

where d_i is the depth of each soil layer, V_i is the corresponding shear wave velocity for each layer and n is the number of soil layers. Soil strata with $V_{s30} > 1500$ m/s are categorized as site class A. Soil strata with V_{s30} ranging from 760 to 1500 m/s, 360–760 m/s and 180–360 m/s are categorized as site class B, C, and D respectively. The strata where the $V_{s30} < 180$ m/s are categorized as site class E.

Wald and Allen [13] presented a technique to derive first-order site-condition maps from topographic gradient using global 30 arc, sec topographic data and VS30 measurements. They found that maps derived from the topographic data were correlating well with other independently derived, regional site-condition maps. Allen and Wald [14] which have examined the use of high-resolution DEMs can be used to recover improved estimates of VS30 beyond that which can be delivered using the original correlations of [13]. James and Sitharam [15] have compared the Vs30 map of Bangalore and Chennai cities, derived from 1 arc-second DEM data with the Vs30 map developed by Sitharam and Anbazhagan [16] and [17] using geotechnical and geophysical tests data. Studies by [15] have found that for both Bangalore and Chennai, the Vs30 maps derived from topographic slope data are comparable with the Vs30 maps developed based on geotechnical and geophysical tests data. Yong et al. [18] have used the geomorphometry for the site characterization using the time averaged shear wave velocity to a depth of 30 m. Grelle et al. (2018) have generated a 3D-topographic surface of East Mountain area (Utah, USA), using the digital elevation model (DEM) for computing Topographic Aggravation Factors (TAF). This paper presents the case studies on site characterization of Karnataka (state level) as well as India (country level) using topographic slope map derived from DEM data, based on the VS30 values. The slope maps of the two regions were derived separately from two different types of DEM (ASTER and SRTM) having different resolutions.

1.1 Digital Elevation Model (DEM)

DEMs are widely used in a large variety of engineering disciplines such as in Hydrology, landslide hazard assessment, and transportation. Two types of DEM are freely available currently, Shuttle Radar Topography Mission (SRTM) and the Advanced Spaceborne Thermal Emission and Reflection Radiometer (ASTER).

Shuttle Radar Topography Mission (SRTM). The Shuttle Radar Topography Mission (SRTM) is a joint venture between the National Geospatial-Intelligence Agency (NGA) and the National Aeronautics and Space Administration (NASA), producing digital topographic data land areas between 60° north and 56° south latitude, with 3-arc second (approximately 90 m) resolution. The data were downloaded from various sources such as Cartosat DEM from NRSC Bhuvan, SRTM from USGS EROS, interpolated seamless SRTM data v4.1 from CGIAR CSI, and void filled data in mountainous terrain from viewfinder panoramas. The absolute vertical accuracy of the elevation data will be 16 m (at 90% confidence). The Advanced Spaceborne Thermal Emission and Reflection Radiometer (ASTER) Global Digital Elevation Model (ASTER GDEM) is a freely available DEM, developed by Ministry of Economy, Trade, and Industry (METI) of Japan in collaboration with the National Aeronautics and Space Administration (NASA) in the United States. The ASTER GDEM covers land surfaces between 83° N and 83° S with a resolution of 1 arc-second (30 m). The estimated vertical accuracy for ASTER GDEM was 20 m at 95% confidence. Since ASTER DEM (Fig. 1a) has higher resolution, it is used to generate the slope map for the state of Karnataka and similarly lower resolution SRTM DEM (Fig. 1b) is used to generate the slope map for entire India. These ASTER DEM and SRTM DEM maps can be freely downloadable from websites.

1.2 Generation of Slope Map

The slope maps of the two study areas were developed in ArcGIS software. The DEMs were first projected to the UTM coordinate system and then reassembled to 5×5 km for Karnataka and 10×10 km for the entire India. Further, using the spatial analyst tool, the slope map for the two study areas was derived from the respective

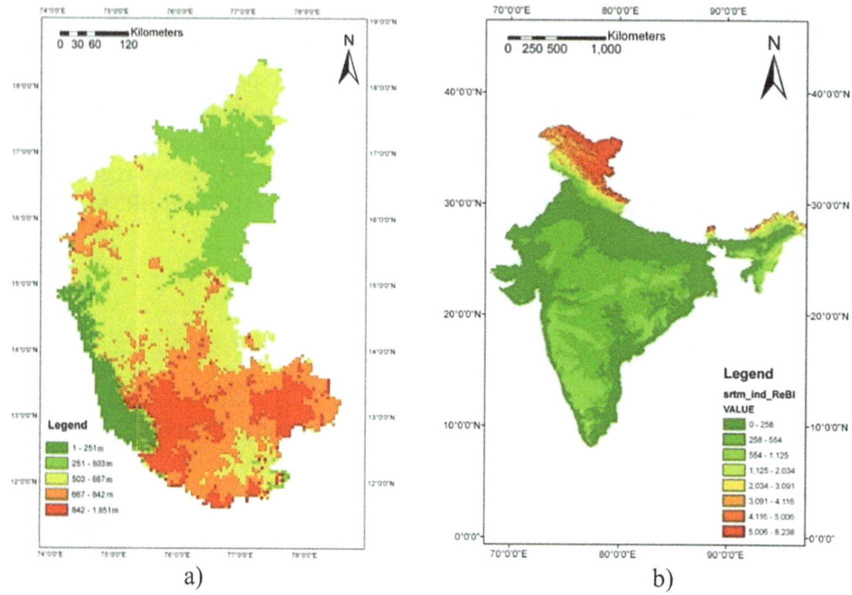

Fig. 1 Digital elevation model **a** Karnataka (ASTER) **b** India (SRTM)

DEM and presented in Fig. 2. Slope maps thus obtained were converted to vector data (points) and then exported in a table form. The geographic co-ordinates and slope values were computed for each grid point considered.

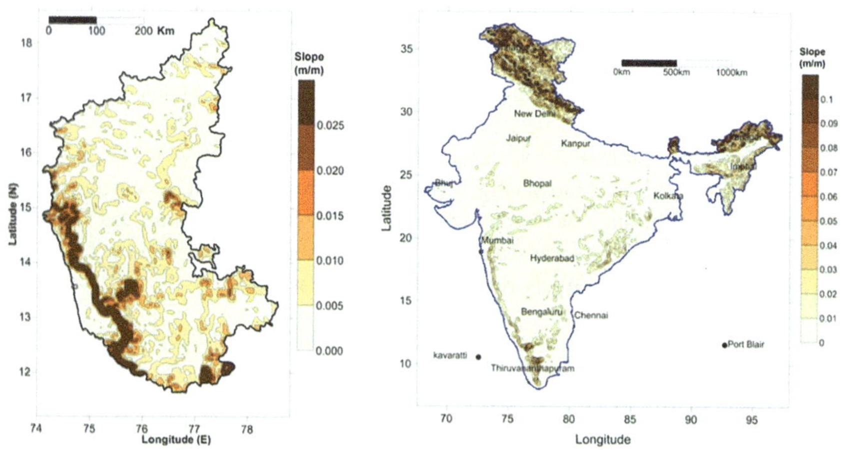

Fig. 2 Slope map of Karnataka and India (gradient values m/m)

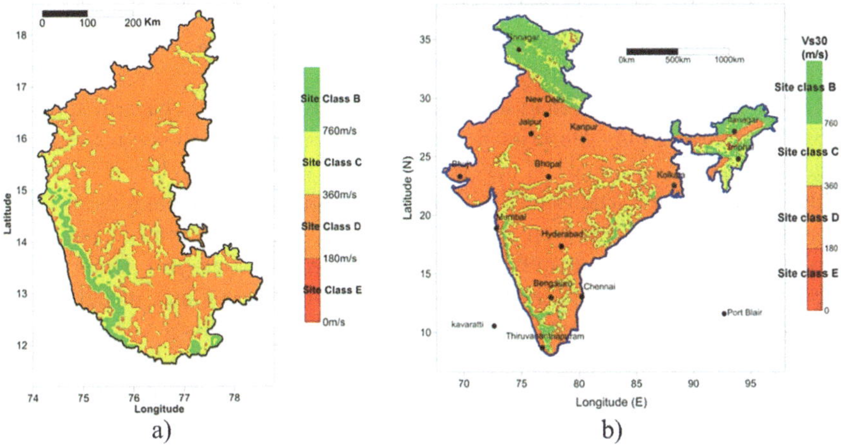

Fig. 3 Spatial distribution of various site classes in Karnataka and India

2 Site Characterization Based on Topographic Slopes

Once slope maps are obtained, site characterization of the two study areas was carried out using methodology suggested by Wald and Allen (2007). Based on the correlation studies conducted for active tectonic and stable continental regions, Wald and Allen (2007) have proposed slope ranges corresponding to each NEHRP site class. Based on the slope range given by Wald and Allen (2007), site class for each grid point was determined. Maps showing the spatial variation in site classes throughout Karnataka and India were developed and presented in Fig. 3. The site classes obtained from various locations in the study areas were then compared with the existing site characterization map for the same location generated for the microzonation purpose by various researchers. Comparing with the studies of [19–22], related to microzonation of Bengaluru, Chennai, New Delhi, and Imphal, the present work identifies similar site class for the above study areas.

3 Evaluation of Seismic Hazard

The conventional seismic hazard analysis provides the peak horizontal acceleration (PHA) at bedrock level. As it is well known that there are lots of variations observed in depth of bedrock, even in a small area of landmass and in many cases, most of the foundations rest at shallow depth and do not extend up to bedrock.

Hence, applying the PHA value at bedrock level obtained from seismic hazard analysis for the earthquake resistant design of shallow founded structures is not the right procedure. So, the PHA at the ground surface has to be evaluated after the characterization of overburden soil mass. As a first step, the PHA at bedrock level for Karnataka as well as entire India for all the grid points where the slope value was evaluated. Both deterministic and probabilistic methodologies were used to evaluate seismic hazard at bedrock.

Deterministic Hazard Analysis. The deterministic seismic hazard approach (DSHA) considers only the critical scenario by assuming the occurrence of the largest magnitude earthquake at the closest proximity of site, thereby giving single-valued ground motion at that site. The deterministic seismic hazard analysis analogies presented by [23] and [24] were used in the present study for Karnataka and India, respectively. In the present study, DSHA was carried out at all grid points (whose slope values are available) in the two study areas using linear source model.

Probabilistic Hazard Analysis. Unlike DSHA, the probabilistic seismic hazard analysis (PSHA) considers various uncertainties on magnitude and distance which are tallied judiciously to get a comprehensive picture of seismic hazard [25]. Similar to DSHA, all the known earthquake events and sources within 300 km from the boundary of study areas (Karnataka as well as India) were compiled and seismicity parameters were evaluated [26]. The PSHA analogy for Karnataka and India presented by [23] and [27], respectively, were used in the present study. Figure 4 describes a general framework for DSHA and PSHA.

Evaluation of PHA at Ground Surface. Non-linear site amplification technique proposed by [27] was adopted in the present study for determining surface-level seismic hazard for the two study areas. The PHA at the ground surface Y_s is obtained by multiplying suitable amplification factor F_s to the bedrock level PHA Y_{br} as per Eq. 2.

$$Y_s = Y_{br} \times F_s \tag{2}$$

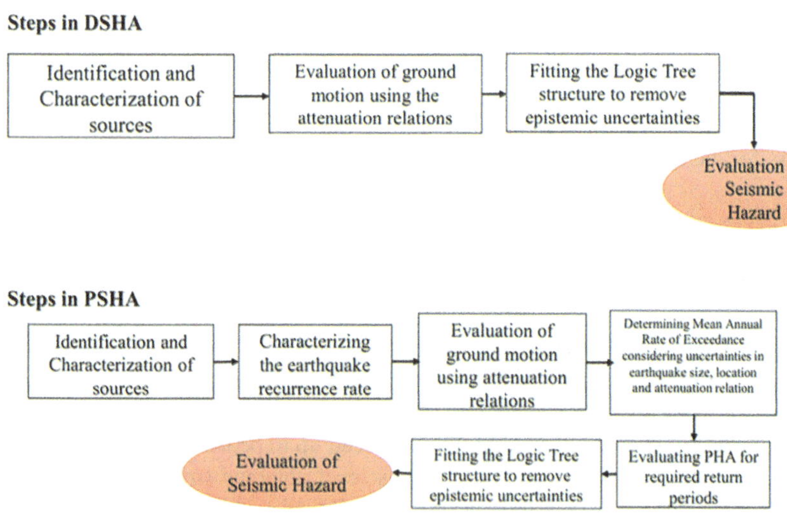

Fig. 4 General methodology for the hazard analysis for DSHA and PSHA

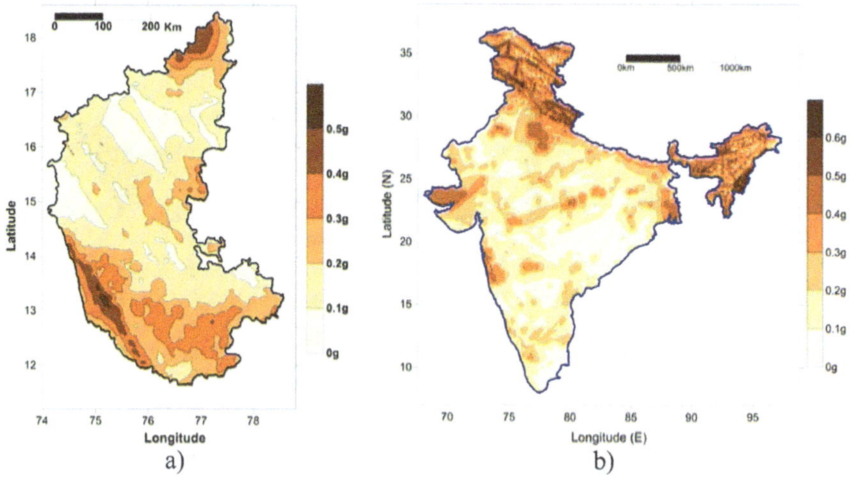

Fig. 5 Spatial variation in PHA value at surface level throughout Karnataka and India (DSHA)

The value of the amplification factor Fs varies with PHA values, and it accounts for the non-linear behavior of soil in amplifying the seismic signals. [28] have developed a regression relationship as per Eq. 3 for determining amplification factor F_s.

$$\ln(F_s) = a_1 Y_{br} + a_2 + \ln \delta_s \tag{3}$$

where a_1 and a_2 are regression coefficients, Y_{br} is the spectral acceleration at rock level, and δ_s is the error term. The values of the regression coefficients a1 and a2 will vary for different site classes and for different time periods. These values were derived based on the statistical simulation of ground motions [28], and they take into account the non-linear site response of soils. For site classes C and D, the amplification factor was found non-linear dependency on bedrock level PHA when compared with site classes A and B. The method adopted for evaluation of the amplification factor (F_s) values consider this effect and the value of F_s varies with the rock level PHA values. Figures 5 and 6 present the spatial variation in PHA value at bedrock throughout Karnataka as well as India from deterministic and probabilistic methodologies, respectively.

4 Conclusions

The seismic hazard at the ground surface is a critical parameter to perform earthquake resistant design of buildings. It is also required to carry out liquefaction analysis for given soil strata. In this paper, the surface level PHA values were evaluated for Karnataka (state level) and India (country level) based on NEHRP site classes by

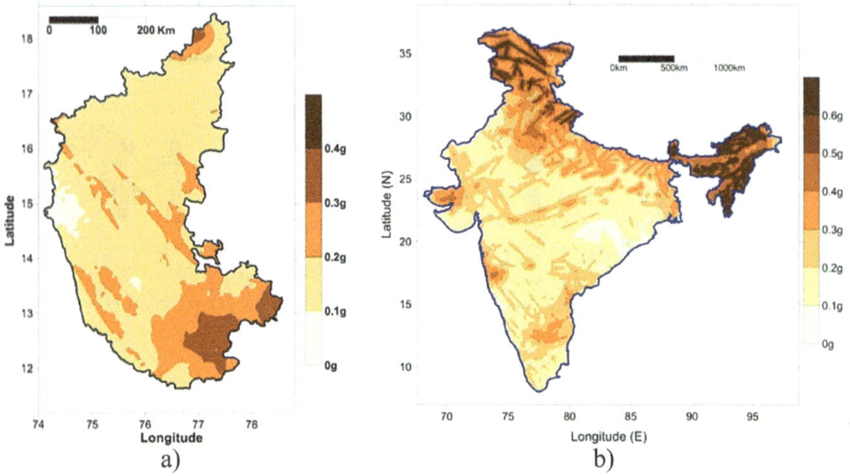

Fig. 6 Spatial variation in PHA value at surface level throughout Karnataka and India (corresponding to 475 years, PSHA)

considering the local site effects. Site characterization study shows that the interior regions of the Karnataka fall in the category of site class D. However, the Western Ghat region of the Karnataka state was identified as site class C to the B-type terrain. Similarly, for India, the site characterization study shows that the majority of central India falls in site class D. The northern and north-eastern regions of India fall in site class C. In the absence of Vs30 values, the site classes can be identified based on local geological conditions. In this paper, the spatial variation of Vs30 throughout Karnataka and India was evaluated from the topographic slope maps. The PHA value was brought to the surface level using appropriate amplification factors. Thus, this method provides a simplified methodology for evaluating the surface-level PHA values. The PHA values obtained based on the DSHA analysis can be taken as the upper bound PHA values.

Thus, this method provides a simplified methodology for evaluating the surface level PHA values. The PHA values obtained based on the DSHA analysis can be taken as the upper bound PHA values. For the state of Karnataka, the very high site amplification (PGA $\leq$ 0.3 g) value is observed for regions close to Bidar in north Karnataka, Mangalore–Udupi in the southwest and Bangalore–Mysore in the southeast. For India, very high site amplification (PGA $\leq$ 0.3 g) is observed for the regions in the north, the northeast, Kutch in the west, and Indo-Gangetic belt in the central to north-central India. The seismic site characterization based on topographic slope maps is ideal for large landmasses like a state or a country where geotechnical/geophysical test-based site characterization is not physically and economically viable. It is recommended to use the topographic slope data to perform the first level of seismic zonation for identifying various classes. Effect of topography, basin geometry, and soil non-linearity on earthquake motion at bedrock were ignored in the present study. These macro-level surface-level seismic hazard maps are excellent tools for engineers and designers to identify potential hazardous zones while planning to set up various infrastructures.

References

1. Medvedev J (1962) Engineering seismology. Academia Nauk Press Moscow, p 260
2. Borcherdt RD, Gibbs JF (1976) Effects of local geological conditions in the San Francisco Bay region on ground motions and the intensities of the 1906 earthquake. Bull Seismol Soc Am 66(2):467–500
3. Shima E (1978) Seismic microzoning map of Tokyo. Proc Second Inter Conf Microzonation 1:433–443
4. Evernden JF, Thomson JM (1985) Predicting seismic intensities. Evaluating earthquake hazards in the Los Angeles Region an earth-science perspective. US Geol Surv Prof Pap 1360:151–202
5. Midorikawa S (1987) Prediction of isoseismal map in the Kanto plain due to hypothetical earthquake. J Struct Eng 33B:43–48
6. Joyner WB, Warrick RE, Fumal TE (1981) The effect of Quaternary alluvium on strong ground motion in the Coyote Lake, California, earthquake of 1979. Bull Seismol Soc Am 71(4):1333–1349
7. Imai T (1982) Correlation of N value with S wave velocity and shear modulus. Proc 2nd ESOPT, pp 57–72
8. Anbazhagan P, Sitharam TG (2008) Mapping of average shear wave velocity for Bangalore region: a case study. J Environ Eng Geophys 13(2):69–84
9. James N, Sitharam TG, Padmanabhan G, Pillai CS (2014) Seismic microzonation of a nuclear power plant site with detailed geotechnical, geophysical and site effect studies. Nat Hazards 71(1):419–462
10. IBC (2009) International building code. International Code Council, Washington
11. Eurocode-8 (2003) BS-EN 1998-1, Design of structures for earthquake resistance part1: General rules, seismic actions and rules for buildings. European Committee for Standardization, Brussels
12. BSSC (2003) NEHRP recommended provisions for seismic regulations for new buildings and other structures (FEMA 450), Part 1: Provisions. Building Seismic Safety Council for the Federal Emergency Management Agency, Washington, D.C., USA
13. Wald DJ, Allen TI (2007) Topographic slope as a proxy for seismic site conditions and amplification. Bull Seismol Soc Am 97(5):1379–1395
14. Allen TI, Wald DJ (2009) On the use of high-resolution topographic data as a proxy for seismic site conditions (VS30). Bull Seismol Soc Am 99(2A):935–943
15. James N, Sitharam TG (2016) Seismic zonations at micro and macro-level for regions in the Peninsular India. Int J Geotech Earthquake Eng (IJGEE) 7(2):35–63
16. Uma Maheswari R, Boominathan A, Dodagoudar GR (2010) Seismic site classification and site period mapping of chennai city using geophysical and geotechnical data. J Appl Geophys 72(3):152–168
17. Yong A, Hough SE, Iwahashi J, Braverman A (2012) A terrain-based site-conditions map of California with implications for the contiguous United States. Bull Seismol Soc Am 102(1):114–128
18. Anbazhagan P, Sitharam TG (2008) Site characterization and site response studies using shear wave velocity. J Seismol Earthquake Eng 10(2):53–67
19. Boominathan A, Dodagoudar GR, Suganthi A, Uma Maheswari R (2008) Seismic hazard assessment of Chennai city considering local site effects. J Earth Syst Sci 117:853–863
20. Hanumantharao C, Ramana GV (2008) Dynamic soil properties for microzonation of Delhi, India. J Earth Syst Sci 117:719–730

21. Pallav K, Raghukanth STG, Singh KD (2015) Estimation of seismic site coefficient and seismic microzonation of Imphal City, India. Using the probabilistic approach. Acta Geophysica 63(5):1339–1367
22. Sitharam TG, Naveen James KS, Vipin K, Ganesha Raj (2012) A study on seismicity and seismic hazard for Karnataka State. J Earth Syst Sci 121(2):475–490
23. Kolathayar S, Sitharam TG, Vipin KS (2012) Deterministic seismic hazard macrozonation of India. J Earth Syst Sci 121(5):1351–1364
24. Kramer SL (1996) Geotechnical earthquake engineering. Pearson Education Pvt. Ltd, Reprinted 2003, Delhi, India
25. Gutenberg B, Richter CF (1944) Frequency of earthquakes in California. Bull Seismol Soc Am 34(4):185–188
26. Sitharam TG, Kolathayar S, James N (2015) Probabilistic assessment of surface level seismic hazard in India using topographic gradient as a proxy for site condition. Geosci Front 6(6):847–859
27. Raghu Kanth STG, Iyengar RN (2007) Estimation of seismic spectral acceleration in Peninsular India. J Earth Syst Sci 116(3):199–214

GIS-Based Landslide Hazard Zonation and Risk Studies Using MCDM

Ankit Tyagi, Reet Kamal Tiwari, and Naveen James

Abstract In India, landslides are the most frequently occurring disaster in the regions of the Himalayas and the Western Ghats. They are mainly triggered either by rainfall or earthquake or the combination of both, causing severe damage to human life and infrastructure. This study presents a comprehensive use of the multi-criteria decision-making (MCDM) method in landslide risk assessment for the Tehri area in the state of Uttarakhand, India. The Tehri area is situated in the Lesser Himalaya of Garhwal hills which lies in zone IV of seismic zoning map of India. Because of the large-scale slope instability in the area, it has received the special attention of the researchers. In the recent past,—many landslide hazards and risk zonation is carried out for different regions in the Uttarakhand state. However, limited work is done considering temporal factors such as seismic ground shaking, rainfall, and seismic amplification at surface level. The DEM data is used to produce topographic characteristics such as slope, aspect, and relative relief. DEM data is also used for the detailed drainage analysis which includes topographic wetness index (TWI), stream power index (SPI), drainage buffer, and reservoir buffer. Seismic hazard analysis is performed using the deterministic methodology to estimate the peak horizontal acceleration. The amplification factor is calculated using the non-linear site amplification method. In this study, the analytical hierarchy process (AHP) is used to evaluate the landslide hazard index which is used to generate landslide hazard zonation (LHZ) map. Further, the landslide vulnerability assessment is done for the study area. The vulnerability map of the study area is derived in terms of landuse/landcover (LULC) using remote sensing data of Landsat 8 which can provide useful information that helps people to understand the risk of living in an area.

Keywords PHA · AHP · GIS · Landslide

A. Tyagi (✉) · R. K. Tiwari · N. James
Department of Civil Engineering, Indian Institute of Technology Ropar, Rupnagar 140001, India
e-mail: annkittyagi@gmail.com

R. K. Tiwari
e-mail: reetkamal@iitrppr.ac.in

N. James
e-mail: naveen.james@iitrpr.ac.in

T. G. Sitharam et al. (eds.), *Local Site Effects and Ground Failures*, Lecture Notes in Civil Engineering 117, https://doi.org/10.1007/978-981-15-9984-2_22

1 Introduction

Landslide is defined as a displacement of slopes, forming earth material by fall, topple, slide, or flow under the gravity [1]. Landslides are a major threat to both lives and property and damages natural resources in the Himalayas. Naithani [2] reported that every year more than 200 people die, and property damage is more than US$ 1 billion in the Himalayas because of landslides which is approximately 30% of worldwide landslide losses. The Himalayas are highly susceptible to landslides due to their varying slope, complex geology, and are triggered by heavy rainfall, earthquakes, floods, etc. There is a need to carry out Landslide Hazard zonation (LHZ) to identify these landslide-prone areas. Landslide Hazard Zonation is a method of classifying different parts of an area according to their landslide potential [3]. In the last few decades, landslide hazard and risk assessment have become a topic of great interest for engineering professionals. It started with field-based studies of hazard zonation with manual data integration [4] which have some disadvantages, such as small area cover, poor integration capacity, and manual overlay. Recently remote sensing and geographic information systems (GIS) technology have attracted great attention with their brilliant spatial data processing capacity. With their help, it is possible to efficiently gather, visualize, operate, and combine different types of data such as aspect, geology, LULC, slope, etc. of a selected zone that is used to generate LHZ [5] map. In this study, the LHZ map is generated for the Tehri Garhwal region of Uttarakhand with the help of satellite data and GIS technology. The vulnerability map is derived from the LULC classification of multispectral data. Using the AHP technique, vulnerable regions and hazard zones were combined to obtain the risk zone of the area.

2 Study Area

The Tehri area is located at latitude 30° 22′ 40″ N and longitude 78° 28′ 50″ E in the Lesser Himalaya of Garhwal hills in Uttarakhand, India. The area of study covers about 1500 km^2 area which falls in a part of Bhagirathi and Bhilangna river basins. These two rivers confluence to form a huge reservoir at Tehri Dam where landslides are very common. A total of 195 landslides were recorded in this area from field observations, image interpretation, and historical data (Fig. 1).

3 Data Used

For deriving the causative factors of a landslide, various types of remote sensing data, earthquake data, and ancillary data are used as listed in Table 1.

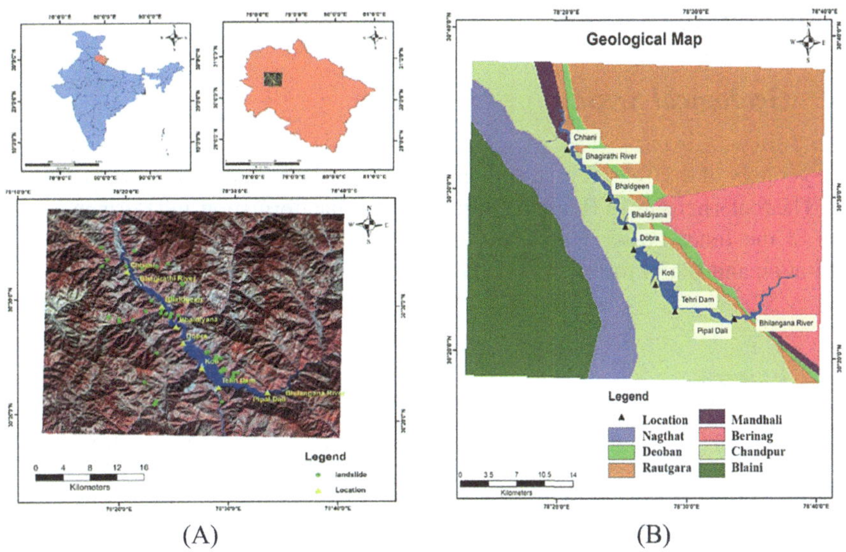

Fig. 1 **a** Location and landslide inventory map, **b** geological Map of the study area

Table 1 Data used

Data type	Sensors	Resolution	Derivative
Multispectral data	Landsat 8	30 m	LULC Landslide inventory
DEM	Cartosat DEM	30 m	Slope Aspect Relative relief Curvature TWI SPI Drainage density
Ancillary data	Published geology map.	1:250,000	Digitized geology map
	A report on Uttarakhand soils.	1:250,000	Digitized soil map
	Toposheet 53 J/7 NW	1:50,000	Digitized base map

4 Terrain Parameters

4.1 *Geological Setting*

The geology of the Tehri region defined here is mostly based on the work of Valdiya [6]. The rocks here are of Rautgara formation, Deoban formation and Berinag formation of the Inner Lesser Himalayas, and the rocks Chandpur formation, Nagthat formation and Blaini formation of outer Lesser Himalayas are found here.

4.2 *Soil Type*

Soil categories of the reservoir area consist of alluvial soil, sandy soil, clayey soil [7]. Soils of different categories have different influences on landslides. Clayey soil is comparatively more stable because of thick green vegetation cover. Recently deposited alluvium soil and boulders are unstable as they are less compacted and have more moisture. As compared to alluvial soil, sandy loamy soil is more stable.

4.3 *Primary Topographic Attributes*

Topographic map and DEM data are used to derive primary factors of topography. Slope angle is varying from 0° to 75°. The slope angle is classified into five classes based on its influence to cause landslide (Figs. 2, 3). It is widely accepted that area having low slope angle is less prone, whereas areas with a higher slope angle have more chance of landslides [8].

The aspect of a slope governs the Sun ray's effect which is related to temperature and other climatic conditions (Fig. 4). Aspect as a great influence on the terrain can be seen in the Himalayan region as south direction slopes are forested and moist, whereas north direction slopes are glaciated and dry. The Himalayan terrain slopes that are facing south have more chances of landslides [9]. Relative relief is another important DEM derivative, which is the difference in maximum and minimum points of elevation in an area. Slope curvature is another important factor as water flowing in and out of the area is controlled by it [10]. Calculation of curvature is done parallel to the direction of the maximum slope in which, a positive profile specifies that the terrain is concave and a negative value specifies that the terrain is convex and a value of zero specifies that the terrain is level [11]. Acceleration of flow across the terrain is affected by curvature. Therefore, a curvature map is created, indicating both profiles of concave and convex.

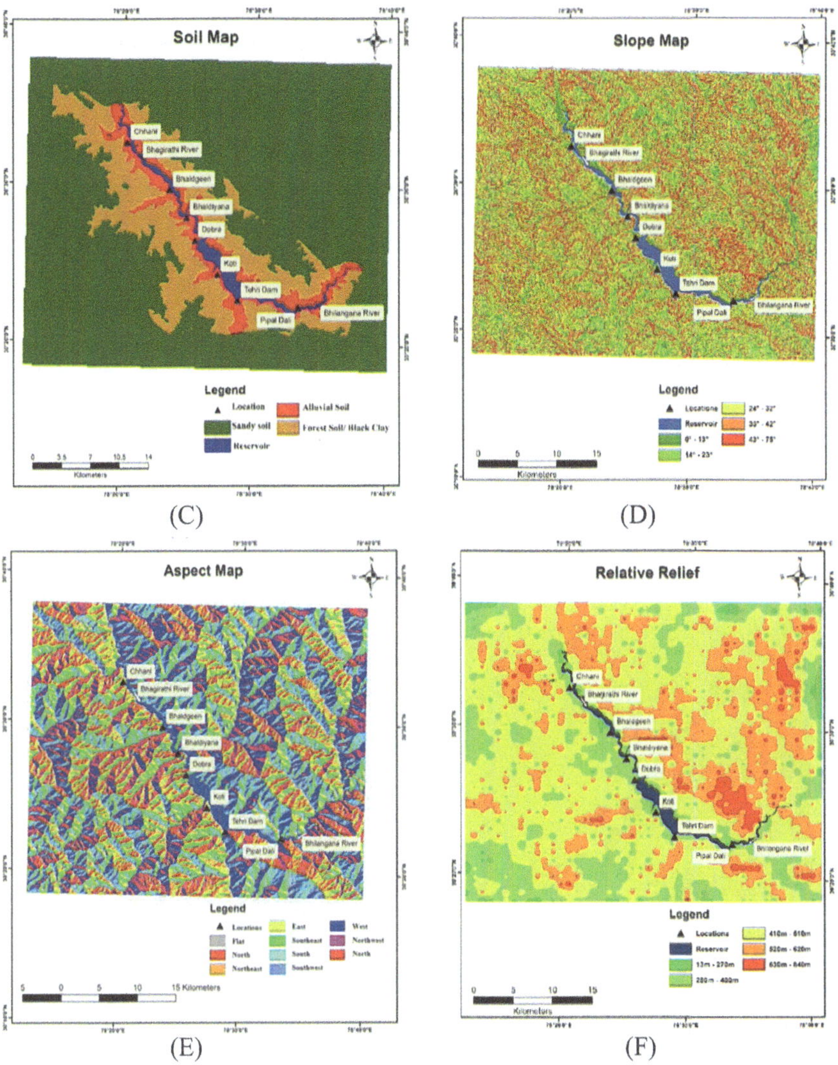

Fig. 2 **c** Soil map, **d** slope map, **e** aspect map, and **f** relative relief map of the Tehri area

4.4 Secondary Topographic Attributes

For the LHZ mapping, two secondary topographic units, i.e., TWI and SPI are used. TWI indicates the accumulation of flow of the terrain TWI [12]. The following formula is used to calculate

$$\mathrm{TWI} = \ln \frac{\mathrm{CA}}{\tan slp} \tag{1}$$

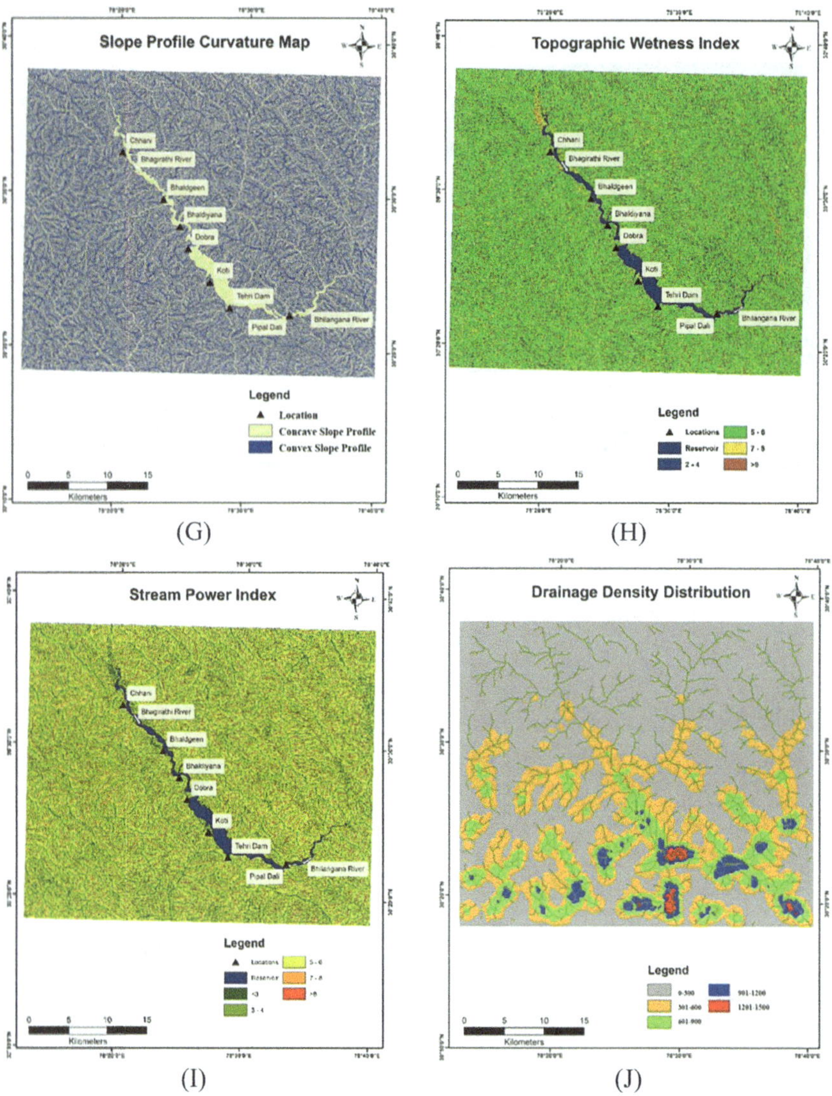

Fig. 3 **g** Curvature map, **h** TWI map, **i** SPI map, and **j** Drainage density map of the Tehri area

here CA represents the catchment area and *slp* represents the slope gradient of the area. As the area increases and slope decreases, soil moisture content and TWI increases. SPI is another important factor which corresponds to the erosive power of the stream. The following formula is used to calculate

$$\text{SPI} = \ln(\text{CA} \times \tan slp) \quad (2)$$

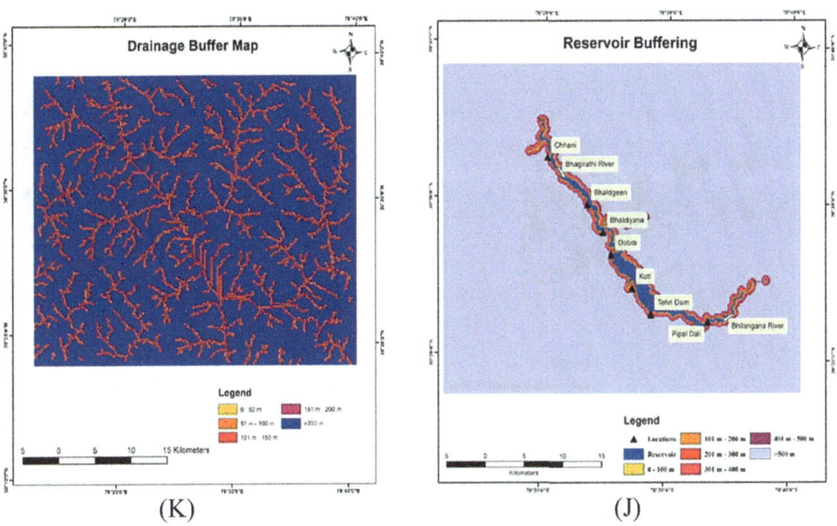

Fig. 4 **k** Drainage buffer map and **j** reservoir buffer map of the Tehri area

As the area and slope increases, erosion and SPI increases. Many landslides are triggered during the monsoon season because of the high stream density of the Tehri reservoir area. A high value of SPI indicates more possibility of a landslide. SPI values have been divided into five categories by applying a natural break classifier.

4.5 Distance/Buffer Layers

The Arc hydro tool of ESRI GIS package is used to derive a drainage map from the DEM. As the drainage density increases, the possibility of landslide occurrence also increases. In hilly regions, unstable slopes are created because of the continuous erosion of its banks by streams that are very prone to failures. Drainage buffer map, drainage density map, and reservoir map are generated as drainage pattern has a great influence on landslides.

4.6 Seismicity

Uttarakhand has been repeatedly shaken by earthquakes in the past. Uttarkashi suffered one of India's deadliest earthquakes in 1991 that killed nearly 730 people and affected over three lakhs. Another major earthquake hit the Chamoli district of Uttarakhand in 1999 where 103 people died. Hence in this study, within 500 km of radius from the study area, events of earthquakes were taken from the USGS website

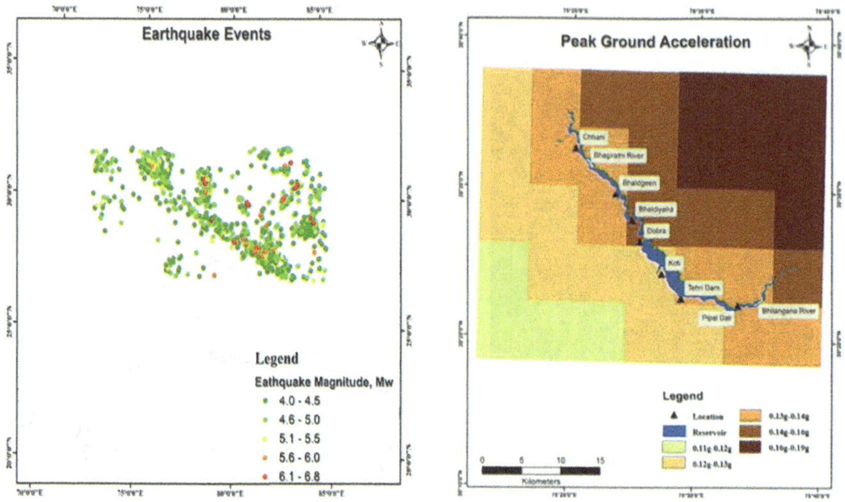

(K) (L)

Fig. 5 **k** Decluttered earthquake event and **l** peak ground acceleration map of the Tehri area

(Fig. 5). About 950 seismic events were found in the 500 km vicinity of the study area from 1950 to 2019 of magnitude 4 and above. Hence, it can be stated that the Uttarakhand in the Himalayas is situated in a highly seismic zone. A deterministic method is used in this study for seismic hazard assessment. A relationship for the Himalayas in India for peak ground accelerations which was developed by Sharma [13] is used to calculate PHA. Following is the general formula which shows the relationship between ground motion and parameters like magnitude, distance, etc.

$$\log(y) = b_1 + b_2M + b_3\log\left[\sqrt[2]{R_{JB}^2 + b_4^2}\right] + b_5S + b_6H \quad (3)$$

here, y is acceleration, S is kept as 1 as the study area is rocky, H varies from 0 to 1, M is the magnitude of the earthquake, R_{JB} is Joyner distance, and b_1, b_2, b_3, b_4, b_5, and b_6 are the regression coefficients. For the study area, a map of PHA is prepared at the bedrock level (Fig. 5). To calculate PHA at the surface level, amplification technique is used here [14]. The PHA at the surface is calculated for slopes having more than or equal to 10°. The surface with a slope angle less than 10° are neglected considering them to be flat [15]. The surface with slope angle 10° and above comes under B-type site class [16], for which the velocity of shear wave is more than 760 m/s for top 30 m overburden (V_{s30}) [17]. The following relationship is used to calculate PHA at the surface level.

$$Y_s = Y_{br} + F_s \quad (4)$$

Here, Y_s is surface-level PHA, F_s is the amplification factor, and Y_{br} is PHA at bedrock, where amplification factor is calculated using the following relationship proposed by Raghu Kanth and Iyengar in 2007.

$$\ln F_S = a_1 Y_{br} + a_2 + \ln(\delta_s) \tag{5}$$

Here, a_1 and a_2 are site class regression coefficients, Y_{br} is the spectral acceleration at the bedrock level, and δ_s is the error term [18].

5 Methodology

An inventory map for the landslide is prepared with the help of satellite data, field observation and historical information available. Many landslides were found in the close vicinity of the reservoir. In this study, 12 causative factors are used to prepare LHZ map. According to Yalcin [19], it is necessary to be sure that any selected factor is operational, complete, uniform, measurable, and non-redundant in GIS Study. For synthesizing weights of these causative factors, the AHP method is used. It is a method to derive ratio by paired comparisons. AHP has many applications in the selection of the sites and doing LHZ [20]. AHP is an MCDM technique invented by Saaty [21] where both objective as well as subjective factors can be considered while taking decisions [22]. AHP breakdowns difficult decision-making problems into a pyramid of factors and alternatives. Weights are assigned to factors and alternatives on a nine-point scale (Table 2) by pair-wise comparison between them. In this study, the comparison is done between all twelve causative factors and their respective classes. Matrix is formed with different causative factors and their respective classes [23]. For comparison, weights are provided which are subjective. In this study, weights are given based on a field visit and past works of literature. From the matrix, Eigenvalues and Eigenvectors are obtained which represent the influence of that factor or class for causing a landslide.

As these weights are given subjective, hence it is important to check the consistency of the weight provided. To check consistency, Saaty (1980) has defined two

Table 2 The priorities scale between two causative factors in AHP according to Saaty

Preference	Degree
1	Equal
3	Moderate
5	Strong
7	Very strong
9	Extreme
2, 4, 6, 8	In-between
Reciprocals	Reverse

Table 3 Random consistency index prepared by Saaty

N	1	2	3	4	5	6	7	8	9	10	11	12	13	14	15
RI	0	0	0.57	0.90	1.11	1.23	1.31	1.42	1.44	1.48	1.52	1.54	1.55	1.58	1.58

terms, i.e., consistency index (CI) and consistency ratio (CR). The following formula is used to calculate the consistency index:

$$\mathrm{CI} = \frac{\lambda_{\max} - N}{N - 1} \tag{6}$$

Here, N is the number of factors or classes and $\lambda_{\max}$ is the maximum eigenvalue. After calculating the CI, Random index is obtained from the table prepared by Saaty. From the ratio of CI and RI, the consistency ratio is calculated.

Table 3 of the Random index was prepared by Satty in 1980 by doing random sampling. CR value greater than 0.1 shows inconsistency. While CR value 0 shows that it is perfectly consistent. The following are the matrix prepared for all the causative factors and their respective classes (Table 4).

The next step was the calculation of LHI. It was computed by multiplying weights with respective classes and then using the following formula:

$$\mathrm{LHI} = \sum_{j=1}^{N} \text{weight of factor}\left(W_j\right) \times \text{weight of factor classes}\left(W_{ij}\right) \tag{7}$$

where W_{ij} denotes weight of ith class of causative factor J. LHI map generated is again divided into five classes as very low, low, moderate, high, and very high.

6 Landslide Vulnerability Map

Landslide vulnerability map in the form of LULC is prepared from multispectral Landsat 8 data. LULC map is generated by a supervised classification technique. The maximum likelihood classification algorithm uses the spectral signature of the training class to classify the whole image into five classes, i.e., reservoir, settlement, dense forest, scrub forest, and agriculture land. Among them settlement or built-up areas are more vulnerable than other areas.

Table 4 Scores obtained by performing AHP

Factors and classes	1	2	3	4	5	6	7	8	9	10	11	12	13	Normalized eigen
Factors														
Geology	1													0.05455
LULC	1/2	1												0.03570
Soil	1/3	1	1											0.03637
Slope	4	5	4	1										0.13450
Aspect	1/4	1/3	1/3	1/5	1									0.02222
Relative relief	3	4	4	1/3	3	1								0.10070
Curvature	1/4	1/4	1/3	1/6	1/2	1/3	1							0.01762
TWI	1/2	2	1	1/4	2	1/3	4	1						0.04065
SPI	1/3	1/2	½	1/5	2	¼	2	1/3	1					0.02454
Drainage density	4	4	4	1	5	3	5	4	5	1				0.13061
Drainage buffer	4	4	4	1	4	1/3	5	4	5	1	1			0.11822
Reservoir buffer	4	5	4	3	5	4	5	5	5	3	3	1		0.20538
PHA	3	4	4	1/3	3	½	5	3	4	1/3	1/4	1/3	1	0.07889
CR = 0.0785, maximum eigenvalue = 14.4615														
Geology														
Blaini formation	1													0.07565
Nagthat formation	3	1												0.24704
Chandpur formation	4	2	1											0.33496
Berinag formation	1	1/4	1/5	1										0.07176
Deoban formation	1	1/4	1/5	1	1									0.06555
Mandhali formation	1/3	1/5	1/5	1/3	1/2	1								0.03877
Rautgara formation	4	1/3	1/3	3	3	4	1							0.16623
CR = 0.0788, maximum eigenvalue = 7.62														
LULC														
Dense forest	1													0.07173
Scrub forest	3	1												0.15323
Agriculture	4	2	1											0.24412
Settlement	5	4	3	1										0.53092

(continued)

Table 4 (continued)

LULC

CR = 0.077, maximum eigenvalue = 4.209

Soil cover

Alluvial soil	1			0.638
Forest soil	1/5	1		0.101
Sandy soil	1/3	3	1	0.262

CR = 0.021, maximum eigenvalue = 3.030

Slope

0°–13°	1					0.04308
14°–23°	2	1				0.06682
24°–32°	4	3	1			0.14245
33°–42°	6	5	3	1		0.29466
43°–75°	8	6	4	2	1	0.45299

CR = 0.051, maximum eigenvalue = 5.230

Aspect

North	1									0.04161
Northwest	3	1								0.05852
West	3	2	1							0.08241
Southwest	4	3	3	1						0.15122
South	5	4	4	3	1					0.30804
Southeast	4	3	3	3	1/3	1				0.21129
East	2	2	1	1/3	1/4	1/3	1			0.06652
Northeast	2	1	½	1/3	1/4	1/4	1/2	1		0.05234
Flat	1/3	1/3	1/3	1/4	1/5	1/5	1/3	1/3	1	0.02712

CR = 0.026, maximum eigenvalue = 9.302

Relative relief

13–270 m	1					0.04674
280–400 m	2	1				0.07483
410–510 m	3	2	1			0.11732
520–620 m	5	4	3	1		0.25099
630–840 m	8	6	5	3	1	0.51012

CR = 0.0538, maximum eigenvalue = 5.239

Profile curvature

Concave	1		0.66666
Convex	1/2	1	0.33333

(continued)

Table 4 (continued)

Profile curvature							
CR = 0, maximum eigenvalue = 2.00							
TWI							
2–4	1						0.07600
5–6	3	1					0.14751
7–8	4	3	1				0.27356
>9	5	4	2	1			0.50293
CR = 0.06988, maximum eigenvalue = 4.1887							
SPI							
<3	1						0.06232
3–4	2	1					0.09846
5–6	3	2	1				0.16083
7–8	4	3	2	1			0.26192
>8	5	4	3	2	1		0.41646
CR = 0.0248, maximum eigenvalue = 5.1103							
Drainage density							
0–300	1						0.06232
301–600	2	1					0.09846
601–900	3	2	1				0.16083
901–1200	4	3	2	1			0.26192
1201-1500	5	4	3	2	1		0.41646
CR = 0.0248, maximum eigenvalue = 5.1103							
Drainage buffer							
0–50 m	1						0.48637
51–100 m	1/3	1					0.23394
101–150 m	1/4	1/3	1				0.15493
151–200 m	1/5	¼	1/2	1			0.08045
>200 m	1/7	1/6	1/4	1/3	1		0.04432
CR = 0.0781, maximum eigenvalue = 5.347							
Reservoir buffering							
0–100 m	1						0.38348
101–200 m	1/3	1					0.25485
201–300 m	1/4	1/2	1				0.16406
301–400 m	1/5	1/3	1/2	1			0.10303
401–500 m	1/6	1/4	1/3	½	1		0.06494
>500 m	1/9	1/5	1/4	1/3	1/2	1	0.02964

(continued)

Table 4 (continued)

Reservoir buffering						
CR = 0.0923, maximum eigenvalue = 6.568						
PHA						
0.11–0.12 g	1					0.06232
0.12–0.13 g	2	1				0.09846
0.13–0.14 g	3	2	1			0.16083
0.14–0.16 g	4	3	2	1		0.26192
0.16–0.19 g	5	4	3	2	1	0.41646
CR = 0.0248, maximum eigenvalue = 5.1103						

7 Results

AHP technique is used to give weightage to different causative factors for preparing an LHZ map for the study area. Each factor of the raster map is given weight value. Weighted raster maps were integrated resulting in an LHI map, where the high value of LHI means that the grid is lying on a more critical landslide zone and the low value of LHI means that the grid is lying on a less critical landslide zone. Hence, the LHI map is classified into the following five categories: Very low, low, moderate, high, and very high (Fig. 6). Entire region is divided into five zones of susceptibility, i.e., 1% of the entire area is found in very low class, 59% in low class, 30% in moderate class, 9% in high class, and 2% in very high. Vulnerable areas in the form of built-up

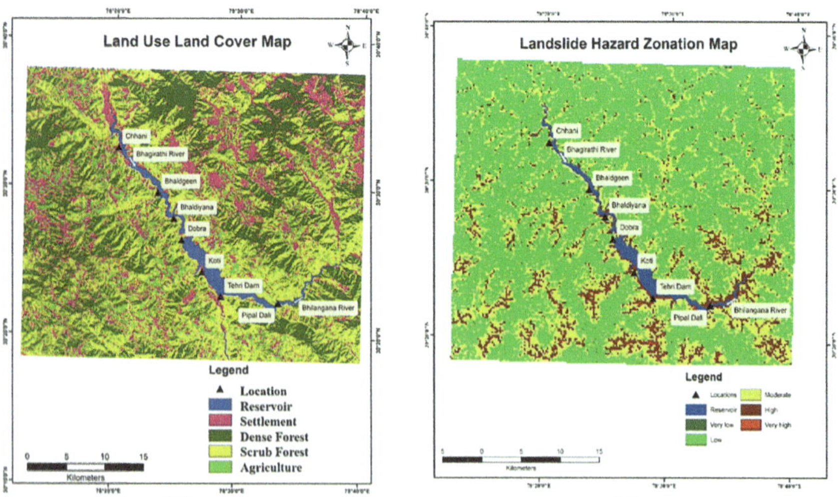

(M) (N)

Fig. 6 **m** Vulnerability Map and **n** landslide hazard zonation map of the Tehri area

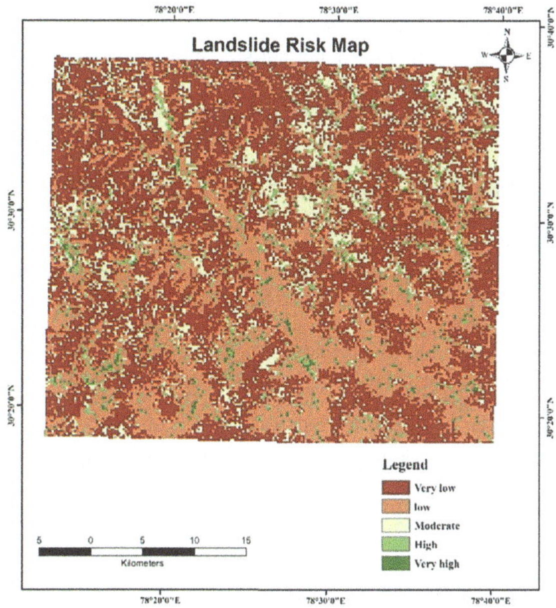

Fig. 7 Landslide Risk map of the Tehri area

areas are found out using LULC map (Fig. 6) and a landslide risk map (Fig. 7) is prepared from LHZ map and landslide vulnerability map.

8 Conclusions

The present study uses the ability of AHP in predicting areas prone to landslides. This method is effective in allocating weights to different factors and their respective classes based on the relative importance of different factors, which depends on the choice of expert and skilled knowledge. Weights are used to classify study areas into zones and a hazard map is prepared. Further, LULC classification is applied to the study area and vulnerable areas are identified. A risk map is generated by overlaying hazard and vulnerability maps. Recognizing zones with a high level of risk can help in increasing security in urban planning by applying safety and maintenance actions.

References

1. Varnes DJ (1978) Slope movement types and processes. In: Schuster RL, Krlzek RJ (eds) landslides: analysis and control. Transportation Research Board, Nat. Acad. Scl., Washington, D.C., pp 11–33
2. Naithani AK (1999) The Himalayan landslides. Employment News, vol. XXIII no. 47, 20–26 February, 1–2

3. Varnes DJ (1984) Landslide hazard zonation: a review of principles and practice. UNESCO, Paris, pp 1–63
4. Anbalagan R (1992) Landslide hazard evaluation and zonation mapping in mountainous terrain. Eng Geol 32:269–277
5. Gupta RP, Joshi BC (1990) Landslide Hazard Zonation using the GIS approach—a case study from the Ramganga Catchment, Himalayas. Eng Geol 28:119–131
6. Valdiya KS (1980) Geology of Kumaun lesser Himalaya. Wadia Institute of Himalayan Geology, Dehradun. Interim. Report, 291
7. Kumar R, Anbalagan R (2015) Landslide susceptibility zonation of Tehri reservoir rim region using binary logistic regression model. Curr Sci 1662–1672
8. Gupta RP, Saha AK, Arora MK, Kumar A (1999) Landslide hazard zonation in a part of Bhagirathy Valley, Garhwal Himalayas, using integrated Remote Sensing & GIS. J Him Geol 20:71–85
9. Nagarajan R, Mukherjee A, Roy A, Khire MV (1998) Temporal remote sensing data and GIS application in landslide hazard zonation of part of Western Ghat, India. Int J Remote Sens 19:573–585
10. Ayalew L, Yamagishi H (2005) The application of GIS-based logistic regression for landslide susceptibility mapping in the Kakuda–Yahiko Mountains, Central Japan. Geomorphology 65:15–31
11. ESRI FAQ (2012) What is the Jenks optimization method? http://support.esri.com/en/knowledgebase/techarticles/detail/26442
12. Wilson JP, Gallant JC (2000) Terrain analysis: principles and applications. Wiley & Sons Ltd., New York, USA, pp 1–27
13. Sharma ML (1998) Attenuation relationship for estimation of peak ground horizontal acceleration using data from strong-motion arrays in India. Bull Seismol Soc Am 88(4):1063–1069
14. Stewart JP, Liu AH, Choi Y (2003) Amplification factors for spectral acceleration in tectonically active regions. Bull Seismol Soc Am 93:332–352
15. James N, Sitharam TG (2017) Multicriteria based landslide hazard assessment and vulnerability studies for the state of Sikkim, India. In: 16th world conference on earthquake, Santiago Chile
16. Wald DJ, Allen TI (2007) Topographic slope as a proxy for seismic site conditions and amplification. Bull Seismol Soc Am 97:1379–1395
17. BSSC (2003) NEHRP recommended provisions for seismic regulations for new buildings and other structures (FEMA 450), Part 1: Provisions. Building Seismic Safety Council for the Federal Emergency Management Agency, Washington, D.C., USA
18. Raghu Kanth STG, Iyengar RN (2007) Estimation of seismic spectral acceleration in peninsular India. J Earth Syst Sci 116:199–214
19. Yalcin A (2008) GIS-based landslide susceptibility mapping using analytical hierarchy process and bivariate statistics in Ardesen (Turkey): comparisons of results and confirmations. Catena 72:1–12
20. Ayalew L, Yamagishi H, Marui H, Kanno T (2005) Landslides in Sado Island of Japan: Part II. GIS-based susceptibility mapping with comparisons of results from two methods and verifications. Eng Geol 81:432–445
21. Saaty TL (1980) The analytic hierarchy process: planning, priority setting, resource allocation. McGraw-Hill International Book Company, New York
22. Feizizadeh B, Blaschke T (2013) GIS-multicriteria decision analysis for landslide susceptibility mapping: comparing three methods for the Urmia lake basin, Iran. Nat Hazards 65:2105–2128
23. Saaty TL (1977) A scaling method for priorities in hierarchical structures. J Math Psycho 15:234–281

A Comparative Study on Landslide Susceptibility Mapping Using AHP and Frequency Ratio Approach

Malemnganbi Lourembam Chanu and O. Bakimchandra

Abstract Landslide is a naturally occurring phenomenon in most of the mountainous regions of the world. Manipur, being a landlocked hilly state, is continuously facing the problems of landslides in the rainy seasons and the economic conditions are highly affected due to blockages of the highways of the region which served as the lifelines. So it becomes very important to check the problems caused due to this natural disaster. In this particular study, an attempt is being made for developing the landslide susceptibility mapping of the region using two GIS-based landslide susceptibility approaches—Analytic hierarchy approach and Frequency ratio approach. Eight causative factors Land Use Land Cover (LULC), Normalized Difference Vegetation Index (NDVI), slope, aspect, curvature, elevation, rainfall, and soil types are considered in the study. The output landslide susceptibility maps developed by the two different approaches have been compared and validation of both the models have been done using landslide locations of the region. Both the models show good accuracy but the Frequency Ratio shows higher accuracy when compared to the AHP approach.

Keywords Landslide · Analytic hierarchy process (AHP) · Frequency ratio · Geographical information system (GIS) · Success rate curve

1 Introduction

Landslides are one of the natural phenomena prevalent in the mountainous region of the world. If there is any downward movement caused by the influence of gravity, it is termed a landslide [1]. Landslides are one of the major natural geological hazards prevalent in Manipur, a border state of India. Sometimes landslide is triggered due to

M. L. Chanu (✉) · O. Bakimchandra
Department of Civil Engineering, National Institute of Technology, Imphal 795004, Manipur, India
e-mail: malemnganbilc@gmail.com

O. Bakimchandra
e-mail: bakim143@gmail.com

T. G. Sitharam et al. (eds.), *Local Site Effects and Ground Failures*, Lecture Notes in Civil Engineering 117, https://doi.org/10.1007/978-981-15-9984-2_23

anthropogenic activities such as large-scale deforestation, agricultural practices on slopes, wildfires, etc. ([2, 3]). Each and every year, many casualties, death, and huge economic loss have been reported from various parts of the world and nearly 17% of fatalities from natural hazards are from landslides worldwide [4]. Generally, landslides are caused by the combination of geo-environmental factors [1, 5]. Landslide occurrences depend on the complex interactions among a large number of causative factors. These factors can be categorized into static and dynamic factors (rainfall and earthquakes) [6]. Static factors or primary factors are basic surface-related characteristics that are related to sliding [7]. Proper understanding of the causative factors is necessary in order to develop reliable landslide susceptibility maps. The identification of susceptible landslide zones is a must for proper planning and construction purposes and also for the safety of the inhabitants.

Through proper surveying and analysis by scientific studies, it is possible to assess and predict the landslide susceptible regions and the damage can be reduced through proper planning purposes and using the prediction models. Landslide susceptibility is defined as the probability of landslide occurrence based on the conditions of the local terrain [8]. A landslide susceptibility map shows the landslide potential areas of the future by considering the factors that contributed to past landslides [9]. The spatial prediction of landslide hazards is one important field of geoscientific research in which statistical classification rules have been applied [10]. Multi-temporal information is particularly important since future landslide hazards partly depend on the scarps of past events [11] and should therefore be modeled conditionally.

In this particular study, an effort is being constructed to produce the landslide susceptibility map of the region using the Frequency Ratio Approach and Analytic Hierarchy Process. Eight causative factors are taken into consideration as topographic data—elevation, slope, aspect, curvature, soil information, Normalized difference Vegetation Index (NDVI), Tropical Rainfall Measuring Mission (TRMM) precipitation data. Landslide location data, collected from field survey and various departments, has been employed to develop the spatial relationship with the various causative factors. The landslide causative factors are integrated into the GIS platform using the frequency ratio approach and AHP to generate the final Landslide Susceptibility Map. The maps were validated and the accuracy of the models has been checked.

2 Study Area

Manipur is a state lying at the northeastern border of India with a total area of 22,327 km^2. The region is geographically divided into two broad regions, viz., the hill and the valley regions. The average elevation of the valley is about 790 m above the sea level and that of the hill's ranges from 1500 to 1800 m. Around 90% of the total area constitutes the hilly regions and the remaining 10% constitutes the valley region. Specially, the hilly districts are highly affected in the rainy months. The state has

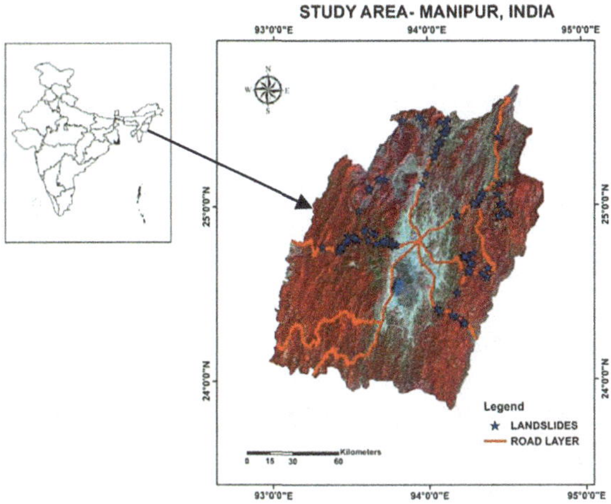

Fig. 1 Study area, Manipur, India

to face huge losses in economy due to blockages of Highways- 37, 2, 102, and 202 during the monsoons and rainy seasons every year (Fig. 1).

In general, the causes of landslides in Manipur are mainly due to weak lithology consisting of complex structural features like highly jointed rock formation; unstable slope and high intensity rainfall during the rainy seasons. Landslides occurring in the areas are both natural and anthropogenic. Landslides are quite frequent and past records show that landslides mostly occur during the rainy seasons in the month of June–September in the region. The damage caused by landslides is huge and the state has to face huge economic losses.

3 Data Description and Methodology

Eight factors have been considered for the preparation of the landslide susceptibility map of the region. The acquisition of the conditioning factors involves a wide variety of data sources. LANDSAT 8 imageries downloaded from the USGS have been used to generate the LULC map of the region, using the algorithm of SAM (Spectral Angle Mapper) and is also used to generate the vegetation index, Normalized difference Vegetation Index (NDVI). SRTM (Shuttle Radar Topographic Mission) DEM data of 30 m resolution has been used to generate various topographic layers—slope, aspect and curvature of the area. And soil map from the National Bureau of Soil Survey and Land Use Planning (NBSSLUP) has been used to obtain the soil type (Fig. 2).

Historical landslides locations of past field visits collected from Earth Science Department, Manipur University and Manipur Science and Technology Council (MASTEC) and also landslides locations taken during reconnaissance field survey have been used for model validation. Rainfall data from (Tropical Rainfall Measuring

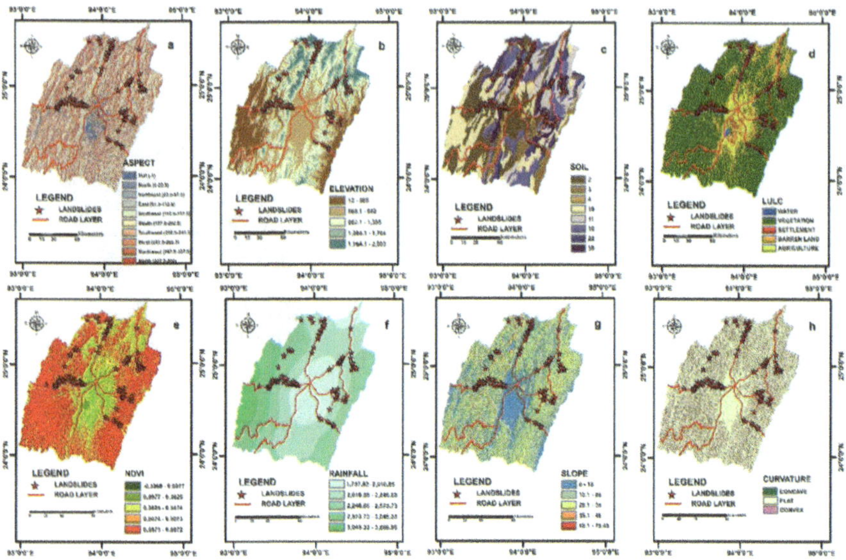

Fig. 2 Landslide conditioning factors **a** Aspect, **b** Elevation, **c** Soil type, **d** LULC, **e** NDVI, **f** Rainfall, **g** Slope, **h** Curvature

Mission) TRMM is used to generate the annual rainfall layer, for the year 2017, using the Geostatistical-Kriging interpolation technique (Fig. 3).

Different data have been derived and prepared from various sources. LANDSAT 8 images are obtained to derive the LULC and the spectral index NDVI. SRTM DEM data is used for the derivation of various topographic parameters like slope, aspect, curvature, elevation map, and rainfall map is prepared using Kriging Interpolation technique from TRMM data.

4 Assigning Weights for Each Factor and Classes Using AHP

The Analytic Hierarchy Process (AHP) is a semi-quantitative multi-criteria decision-making process of measurement through pair-wise comparisons of causative factors and to derive priority scales. It has four levels, viz., defining problem, determination of goals and alternatives, construction of pair-wise comparison matrix, determining weights and obtaining overall priority. AHP converts complex decision-making problem into a hierarchy of factors and alternatives. Weights of each factor and alternatives are assigned on a nine-point ordinal scale (Table 1) by pair-wise comparison between them. Both subjective and objective factors are to be considered in the decision-making process [12].

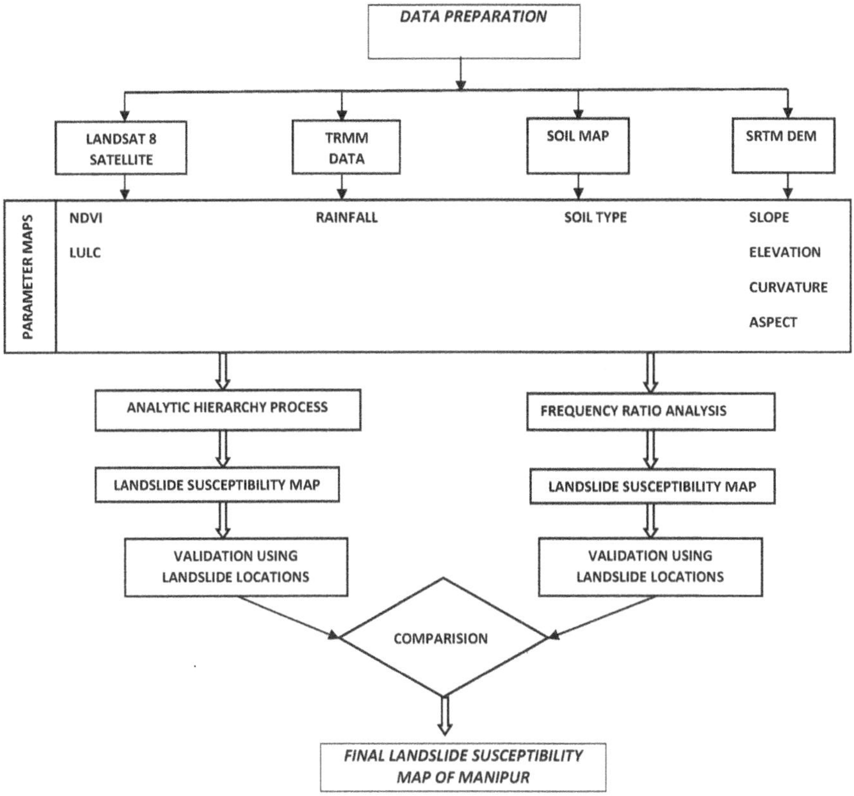

Fig. 3 Methodological framework

Table 1 Ordinal scale representing preference of judgment [13]

Importance scale	Degree of preference	Definition
1	Equal	Equal importance
3	Moderate	One factor moderately favored over other
5	Strongly	Judgment strongly favors one over other
7	Very strongly	One factor very strongly favored over other
9	Extreme importance	One factor favored over other in highest degree
2, 4, 6, 8	Intermediate	Compensation between weights 1,3, 5,7, and 9
Reciprocals	Opposite	Refers inverse comparison

In the present study, the relative value of each pair of the factors and factor classes was determined by the frequency of landslides on each class. Matrix calculation is used to derive the weights of factor/class weights in terms of the Eigen vector. One important aspect of the AHP is the calculation of Consistency Index (CI) and Consistency Ratio (CR).

Table 2 Random consistency index (RI) [14]

N	1	2	3	4	5	6	7	8	9	10	11	12	13	14	15
RI	0	0	0.58	0.9	1.12	1.24	1.32	1.41	1.45	1.49	1.51	1.53	1.56	1.57	1.59

CR is calculated as

$$\text{CR} = \frac{CI}{RI} \tag{1}$$

where RI stands for random index (Table 2) (Satty 1980) based on the number of random samples. CR value of 0.1 is the maximum threshold of consistency of the matrix. CR value >0.1 is thought to be inconsistent whereas value 0 indicates a perfectly consistent comparison result.

AHP method has its advantages in weighting/rating of factors and their classes along with some deficiencies. Pair-wise comparison can provide a simple and acceptable decision rule. Some landslide conditioning factors have a certain degree of dependency in influencing landslides but AHP considers each factor in the hierarchy as an independent entity. Overall AHP multi-criteria decision-making provides a very simple and flexible decision-making.

Landslide susceptibility index (LSI) was computed as given below

$$\text{LSI} = \sum_{j=1}^{N} \text{Weight of factor (Wj)} \times \text{Weight of factor classes} \left(\text{W}_{\text{ij}}\right) \tag{2}$$

where W_{ij} = weight of *i*th class of factor j.

Lastly, the LSI map was classified into five classes as very low, low, moderate, high, and very high susceptibility classes of susceptibility using Natural Jenk's break classification.

4.1 Frequency Ratio Approach for Landslide Susceptibility

Frequency ratio approach is a bi-variate statistical approach of landslide susceptibility assessment based on observed relationships between landslide distribution and landslide causative factors. Generally, it is necessary to assume that landslide occurrence is determined by landslide-related factors to predict landslides, and that future landslides will occur under the same conditions as past landslides. Using this assumption, the relationship between landslides occurring in an area and the landslide-related factors can be distinguished from the relationship between landslides not occurring in an area and the landslide-related factors.

The frequency ratio is used to represent the distinction quantitatively. The frequency ratio is the ratio of the area where landslides occurred to the total study area, and also, is the ratio of the probabilities of a landslide occurrence to a non-occurrence for a given factor's attribute. Therefore, the greater the ratio above unity, the stronger the relationship between landslide occurrence and the given factor's attribute, and the lower the ratio below unity, the lesser the relationship between landslide occurrence and the given factor's attribute.

$$\text{Frequency ratio, } FR = \frac{\%target\ occurence\ in\ each\ sub\ category}{\%category\ of\ an\ independent\ factor} = \frac{(points\ in\ factorclass/total\ points)}{(factor\ class\ area/total\ area)} \quad (3)$$

$$\mathrm{PR} = |\mathrm{RFmax} - \mathrm{RFmin}|/|\mathrm{RFmax} - \mathrm{RFmin}|\mathrm{min} \quad (4)$$

Table 5 has been used to give weights to each and every factor and the Relative Frequency (RF) from Table 3 has been used to give weights to each class of every factor.

5 Results

The relative frequency obtained from Table 3 has been used to assign weights to the factor classes of the eight conditioning factors and the prediction rate from Table 4 has been used to assign weights to the conditioning factors (Table 5).

The final Landslide Susceptibility Index Map (LSI) for Frequency Ratio Approach is found out by summing up all the weighted factors.

$$\mathrm{LSI} = \sum \mathrm{F_{ij}} * \mathrm{W_{ij}} \quad (5)$$

where $\mathrm{F_{ij}}$ = Weighted factors and $\mathrm{W_{ij}}$ = weights obtained from Prediction rate.

Landslide Susceptibility Index (LSI) for Analytic Hierarchy Process (AHP) [15] was computed using Eq. 2.

Lastly, the LSI maps have been classified into five classes as—no landslide, low, moderate, high, and very high susceptibility classes of susceptibility using Natural Jenk's break classification, Jenkins optimization technique. In the above landslide susceptibility maps produced by Frequency Ratio and AHP Approach, the patterns of the landslide classes are almost similar. To understand the pattern more, the area of each class has been found out which is shown in Table 6.

The above map shows the status of the combination of the two resulting maps obtained from frequency ratio and AHP approach and Table 7 depicts the area falling under each status class (Figs. 4 and 5).

Table 3 Landslide area falling under each class of the factors and their relationship with landslides

Factors	Factor classes	No. of points	% of points	Class area	% of class area	Ratio (+)	Relative frequency
Slope angle	1	6300	6.730769231	7,799,085	31.448641	0.21402417	0.04
	2	26,100	27.88461538	8,859,387	35.724150	0.78055364	0.14
	3	51,300	54.80769231	6,203,538	25.014837	2.19100737	0.40
	4	9000	9.615384615	1,658,073	6.685931	1.43815198	0.26
	5	900	0.961538462	279,351	1.126441	0.85360745	0.16
Total		93,600	100	24,799,434	100	5.47734461	1
Aspect	1	0	0	847,866	3.539918	0	0.00
	2	9900	10.57692308	2,751,132	11.486229	0.92083514	0.12
	3	2700	2.884615385	2,769,487	11.562863	0.24947243	0.03
	4	13,500	14.42307692	3,149,691	13.150249	1.09679115	0.14
	5	15,300	16.34615385	3,235,370	13.507967	1.21011208	0.15
	6	17,100	18.26923077	2,861,654	11.947668	1.52910429	0.19
	7	10,800	11.53846154	3,059,620	12.774194	0.90326334	0.11
	8	12,600	13.46153846	3,180,422	13.278554	1.01378046	0.13
	9	11,700	12.5	2,944,193	12.292276	1.01689873	0.13
Total		93,600	100	23,951,569	100.000000	7.94025762	1
Elevation	1	14,400	15.38461538	4,444,324	17.921070	0.85846525	0.17
	2	15,300	16.34615385	8,367,688	33.741446	0.48445327	0.10

(continued)

Table 3 (continued)

Factors	Factor classes	No. of points	% of points	Class area	% of class area	Ratio (+)	Relative frequency
	3	41,400	44.23076923	6,488,959	26.165753	1.69040687	0.34
	4	18,900	20.19230769	4,063,427	16.385160	1.23235343	0.25
	5	3600	3.846153846	1,435,037	5.786571	0.66466887	0.13
Total		93,600	100	24,799,435	100	4.93034769	1
Curvature	1	61,200	65.38461538	11,328,410	45.680117	1.43135832	0.56
	2	4500	4.807692308	2,556,218	10.307566	0.46642361	0.18
	3	27,900	29.80769231	10,914,805	44.012317	0.67725797	0.26
Total		93,600	100	24,799,433	100	2.5750399	1
LULC	1	0	0	115,634	0.468443	0	0.00
	2	0	0	312,288	1.265105	0	0.00
	3	7200	7.692307692	16,649,192	67.447265	0.11404922	0.03
	4	1800	1.923076923	1,633,547	6.617635	0.29059882	0.07
	5	84,600	90.38461538	5,974,094	24.201553	3.73466184	0.90
Total		93,600	100	24,684,755	100	4.13930987	1
Soil	1	4500	4.807692308	6,632,774	26.745720	0.17975558	0.01
	2	5400	5.769230769	2,325,911	9.378906	0.61512833	0.04
	3	5400	5.769230769	258,565	1.042627	5.53336202	0.37
	4	15,300	16.34615385	5,170,820	20.850598	0.78396572	0.05
	5	10,800	11.53846154	3,426,371	13.816355	0.83513067	0.06
	6	23,400	25	3,696,266	14.904668	1.67732686	0.11

(continued)

Table 3 (continued)

Factors	Factor classes	No. of points	% of points	Class area	% of class area	Ratio (+)	Relative frequency
	7	11,700	12.5	2,080,577	8.389631	1.48993434	0.10
	8	17,100	18.26923077	1,208,101	4.871496	3.75023022	0.25
Total		93,600	100	24,799,385	100	14.8648337	1
NDVI	1	0	0	107,702	0.436190	0	0.00
	2	27,900	29.80769231	3,497,782	14.165912	2.10418447	0.39
	3	39,600	42.30769231	4,530,842	18.349774	2.30562469	0.43
	4	19,800	21.15384615	6,585,861	26.672539	0.79309457	0.15
	5	6300	6.730769231	10,077,056	40.811774	0.16492224	0.03
Total		93,600	100	24,691,541	100.4361899	5.36782597	1
Rain	1	9900	10.57692308	4,693,059	18.92442624	0.55890324	0.15
	2	51,300	54.80769231	8,786,654	35.43155658	1.54686098	0.42
	3	26,100	27.88461538	6,912,463	27.87401483	1.0003803	0.27
	4	6300	6.730769231	2,855,297	11.51378183	0.5845837	0.16
	5	0	0	1,551,477	6.256220525	0	0.00
Total		93,600	100	24,798,950	100	3.69072822	1

Table 4 Prediction rate calculation using the above equation and the information from Table 3

Class	PR
Slope angle	1.874262
Aspect	1
Elevation	1.270137
Curvature	1.945859
LULC	4.685121
Soil	1.870182
NDVI	2.230425
Rain	2.17639

Table 5 Weights obtained from frequency ratio and pair-wise comparison

Factors	Prediction rate using FR	Prediction rate using pairwise comparison
Aspect	100.000	46.914
Elevatiion	127.014	59.588
Soil	187.018	87.738
Slope angle	187.426	87.930
Curvature	194.586	91.289
Rain	217.639	102.104
NDVI	223.043	104.639
LULC	468.512	219.799

Table 6 Area falling under each landslide susceptibility classes

Class	Frequency ratio approach		Analytic hierarchy approach	
	Area (Km^2)	% Area	Area (Km^2)	% Area
No landslide	5189.56	23.36	7317.70	32.94
Low	7513.96	33.83	6308.47	28.40
Moderate	4135.54	18.62	4975.35	22.40
High	2547.22	11.47	2358.88	10.62
Very high	2825.97	12.72	1251.85	5.64

Table 7 The status of combined maps and the area for each status

Value	Count	FR	AHP	Status	Area (km^2)
1	2,223,461	3	3	Moderate	2001.115
2	900,192	4	4	High	810.1728
3	1,592,210	5	4	VH-H	1432.989
4	3,937,685	2	2	Low	3543.917
5	1,931,552	3	2	M-L	1738.397
6	1,139,685	1	2	NL-L	1025.717
7	3,208,101	2	1	L-NL	2887.291
8	1,199,578	2	3	L-M	1079.62
9	1,243,185	5	5	Very high	1118.867
10	4,607,890	1	1	No landslide	4147.101
11	147,620	4	5	H-VH	132.858
12	1,781,954	4	3	H-M	1603.759
13	18,600	1	3	NL-M	16.74
14	125,099	3	4	M-H	112.5891
15	304,576	5	3	VH-M	274.1184
16	314,792	3	1	M-NL	283.3128
17	3479	2	4	L-H	3.1311
18	138	3	5	M-VH	0.1242
19	3	1	4	NL-H	0.0027
20	483	4	2	H-l	0.4347

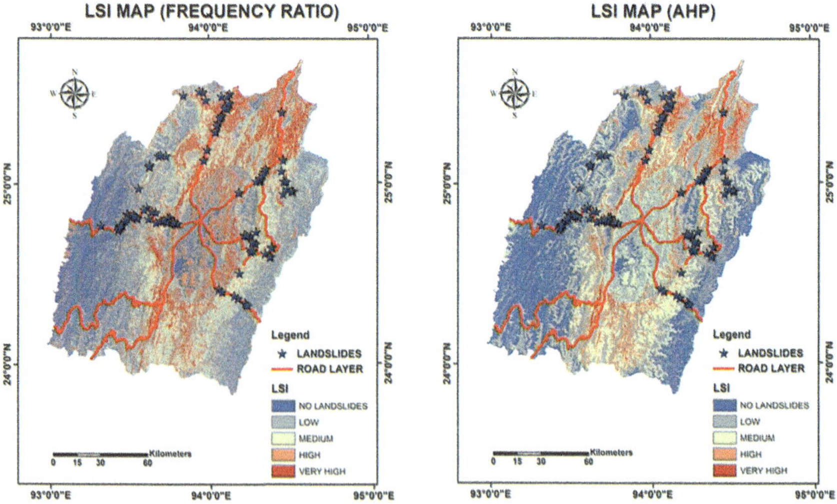

Fig. 4 Landslide susceptibility maps obtained from frequency ratio and AHP approach

6 Validation

Validation is important in landslide susceptibility analysis. Without validation, no interpretation is possible and no support of the method or the input information can be provided [16]. For validation of landslide susceptibility calculation models, it is assumed that landslides are related to spatial information such as topography, LULC, and future landslides will be triggered by specific factors such as rainfall. Two types of plots have frequently been used in landslide susceptibility modeling: the success rate and prediction rate curves [17]. The success rate curve is obtained by varying the decision threshold and plotting the respective sensitivities against the total proportions of the data set classified as landslide. This may be done on independent test data sets (Fig. 6).

Some authors use either test data from the training area but for a different time period for evaluation [17, 18], or data from an adjacent test area [17]. But in this particular study, 75% of the total landslide has been used to train the model which will give the success rate of the model and remaining 25% has been used to test the model which will give the prediction rate of the model and thus accuracy of the model can be checked.

The Area Under Curve (AUC) value for finding accuracy was calculated by using simple trapezium method. AUC value of 0.7512 is obtained from the success rate curve and 0.7314 from the prediction rate curve for AHP approach. So, the model gave an accuracy of 75.12 and 73.14% for success and prediction rate, respectively. Whereas an accuracy of 75.26 and 74.21% have been obtained from the success rate and prediction rate curve for Frequency Ratio Approach.

7 Conclusions

From the present study, we can conclude that both AHP and Frequency Ratio Approach can be conventionally used for the landslide susceptibility analysis for regional scale by combining only some easily available data such as topographic slope, aspect, curvature, elevation which can be derived from DEM and ancillary data such as soil map and rainfall data were also considered for the better results. It can be concluded from the validation that both the model shows acceptable agreement in accordance with the landslide occurrences with the frequency ratio showing a slightly higher accuracy i.e. 75.12% as compared to AHP approach, i.e., 75.26%.

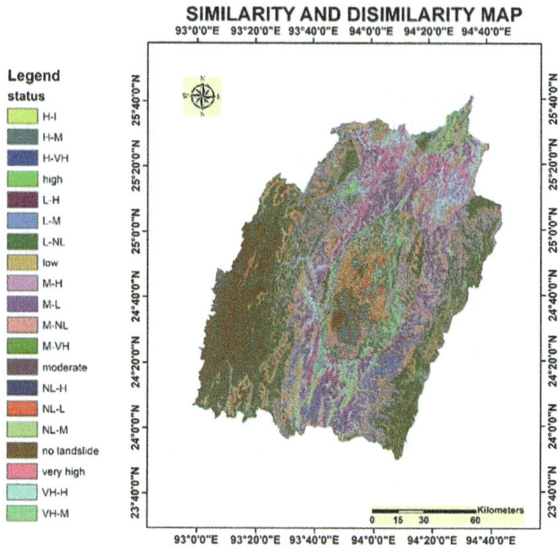

Fig. 5 Combined status map of the similarity and dissimilarity

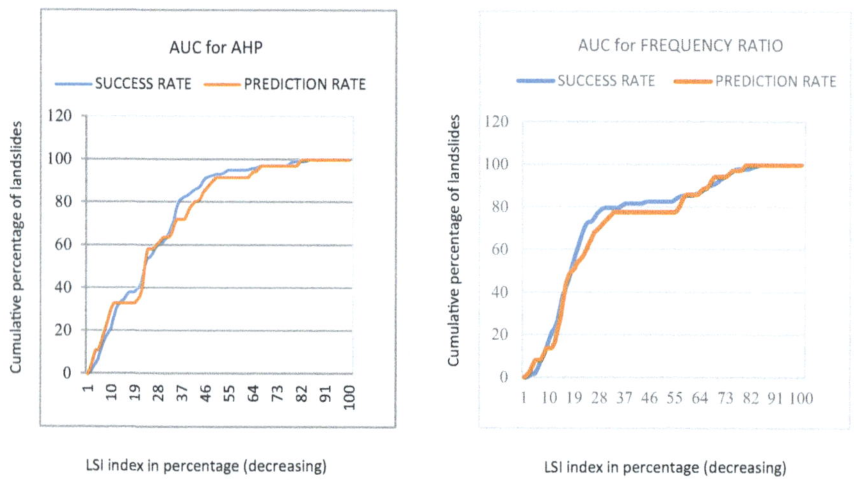

Fig. 6 Cumulative frequency diagrams for AHP approach and frequency ratio, respectively

References

1. Varnes DJ (1984) Landslide hazard zonation: a review of principles and practice, natural hazards 3. Communication landslides IAEG, UNESCO, Paris, France, p 63
2. Sidle RC, Pearce AJ, O'Loughlin CL (1985) Hillslope stability and land use. Water Resources Monograph Series 11, American Geophysical Union, Washington, DC

3. Kuruppuarachchi T, Wyrwoll KH (1992) The role of vegetation clearing in the mass failure of hillslopes: Moresby Ranges, Western Australia. Catena 19, 193–208
4. CRED (2006) Retrieved from international disaster database. https://www.cred.be/emdat
5. Hutchinson, J.N. (1995) Keynote paper: landslide hazard assessment. In: Bell (ed) Landslides, Balkema, Rotterdam, pp 1805–1841
6. Dai FC, Lee CF (2002) Landslide characteristics and slope instability modeling using GIS, Lantau Island, Hong Kong. Geomorphology 42, 213–238
7. Sidle RC, Ochiai H (2006) Landslide processes, prediction, land use. American Geophysical Union, Washington, DC, 1–312
8. Brabb E (1984) Innovative approaches for landslide hazard evaluation. In: IV International symposium landslides, Toronto, pp 307–323
9. Santacana N, Baeza B, Corominas J, de Paz A, Marturia J (2003) A GIS-based multivariate statistical analysis for shallow landslide susceptibility mapping in La Pobla de Lillet Area (Eastern Pyrenees Spain). Nat Hazards 30:281–295
10. Brenning A (2005) Spatial prediction models for landslide hazards: review, comparison and evaluation. Nat Hazards Earth Syst Sci 5:853–862
11. Casadei M, Dietrich WE, Miller NL (2003) Testing a model for predicting the timing and location of shallow landslide initiation in soil-mantled landscapes. Earth Surf Proc Land 28:925–950
12. Yalcin A (2008) GIS-based landslide susceptibility mapping using analytical hierarchy process and bivariate statistics in Ardesen (Turkey): comparisons of results and confirmations. Catena, 72, 1–12
13. Saaty TL (1977) A scaling method for priorities in hierarchical structures. J Math Psycho, 15, 234–281.
14. Saaty TL (1980) The analytic hierarchy process: planning, priority setting, resource allocation. McGraw-Hill International Book Company, New York
15. Saaty T (2008) Decision making with the analytical hierarchy process. Int J Serv Sci 1(1):83–98
16. Chung CJF, Fabbri A (2003) Validation of Spatial Prediction Models for Landslide Hazard Mapping. Nat. Hazards, 30, 451–472
17. Chung CF, Fabbri A (2003) Validation of spatial prediction models for landslide hazard mapping. Nat Hazards 30:451–472
18. Chung CF, Kojima H, Fabbri AG (2002) Stability analysis of prediction models for landslide hazard mapping, in Allison 3–19

Support Vector Machine for Evaluation of Liquefaction Potential Using SPT Data

Dev Kumar Pradhan, Suvendu Kumar Sasmal, Vamsi Alla, and Rabi Narayan Behera

Abstract The geotechnical engineers usually have difficulties in resolving complicated problems that involve a number of influencing parameters. Sometimes, it is also complicated to describe a problem mathematically. This study aims for an alternative algorithm known as Support Vector Machine (SVM), which can be used to solve the classification-type problem. In this study, liquefaction potential is analysed using influencing parameters viz. the Standard Penetration Test (SPT) values, different soil parameters, and depth of water table taken from a borehole database of a certain depth. This borehole database was prepared by collecting borehole data of different sites located in different parts of the country. A deterministic approach has been used to evaluate the liquefaction potential and it is expressed in the form of factor of safety (*FS*). The SVM model developed using the dataset of the liquefaction potential evaluated from a deterministic approach showed an overall accuracy of 96.8%.

Keywords Liquefaction · Standard penetration test · Deterministic approach · Classification · Support vector machines

1 Introduction

If a mass of sand in a saturated condition with a greater void ratio than its critical void ratio is subjected to a suddenly applied shearing stress, as from an earthquake,

---

D. K. Pradhan · S. K. Sasmal · V. Alla · R. N. Behera (✉)
Department of Civil Engineering, National Institute of Technology Rourkela, Rourkela 769008, India
e-mail: rnbehera82@gmail.com

D. K. Pradhan
e-mail: devpradhan.nit@gmail.com

S. K. Sasmal
e-mail: suvendukumarsasmal@gmail.com

V. Alla
e-mail: vamsi.royalcivil@gmail.com

T. G. Sitharam et al. (eds.), *Local Site Effects and Ground Failures*, Lecture Notes in Civil Engineering 117, https://doi.org/10.1007/978-981-15-9984-2_24

heavy blasting, pipe driving, or any other dynamic force, the sand tends to decrease in volume. As a result, the pore water is subjected to a suddenly applied hydrostatic excess pressure, and a portion of the weight of overlying material is transferred from inter-granular pressure to pore water pressure. The effective stress in the soil is thus reduced. Since the shearing strength of granular soil depends on the effective stress, this transfer of pressure causes a sudden decrease in shearing strength. If it is reduced to a value below the applied shearing stress, the mass will fail in shear. This type of failure is sometimes referred to as liquefaction. As such, there is no single major method to study the liquefaction problem. In the present work, this problem is addressed using soft computing technique. One of the recent trends in soft computing across engineering applications is the use of Support Vector Machine (SVM). In recent times, SVM has been used in the field of geotechnical engineers by various researchers. Zhang et al. [12] analyzed the soil concrete interfacial strength using LS-SVM. Zhang et al. [11] used the SVM method for the analysis of geometry factors in rock slope. Ji et al. [4] analyzed the stability of slopes using this technique. Kurnaz and Kaya [5] studied the compression index of clay using the Support Vector Machine. Borthakur and Dey [1] observed the behavior of micropile in clay using the SVM technique. One of the key criteria to forecast the occurrence of the liquefaction is to observe the SPT *N* values of the corresponding soil. The Standard Penetration Tests is generally conducted as per IS-2131:1981 [3]. This test is performed to determine different soil parameters like relative density or density index, angle of internal friction, unconfined compressive strength of clay, ultimate pile load-carrying capacity, ultimate bearing capacity as per shear criteria, and allowable bearing pressure on the basis of settlement criteria. In this test, a split sampler is placed over the soil in a borehole of 55–150 mm diameter and this sampler is driven by the dynamic mechanism of hammer. The weight of hammer is 65 kg and the height is 760 mm. The test is conducted at every 2–5 m interval or at a change of soil strata. The SPT number is defined as the number of blows of hammer required for 300 mm penetration. Usually, test is performed in 3 stages 150 mm each. SPT number is taken as the number of blows required for the last 300 mm of penetration, which means the number of blows required for the first 150 mm penetration is ignored. In this study, a number of SVM classification techniques are presented to predict the liquefaction potential in terms of SPT data. Outcomes corresponding to different kernels are also presented. A comparative analysis among the different considered methods is also discussed, which will be helpful in deciding the method selection for liquefaction analysis.

1.1 Support Vector Machine (SVM)

SVM is one of the most reliable soft computing techniques to predict the dependent variable in terms of the independent variable. SVM was introduced as a classifier by Cortes and Vapnik [2], which was initially popularized in the Neural Information Processing Systems community.

In a linearly separable case, for a data set, a linear hyperplane that can separate two classes can be defined by Eq. (1),

$$f(x) = w \cdot x + b = 0 \tag{1}$$

where $w \in R^n$, $b \in R$ = bias. Separating hyperplane for two classes can also be represented as

$$w \cdot x_i + b \geq 1 \text{ for } y_i = 1 \tag{2}$$

$$w \cdot x_i + b \leq -1 \text{ for } y_i = -1 \tag{3}$$

The above two equations can be written as

$$y_i(w \cdot x_i + b) \geq 1 \tag{4}$$

Sometimes, slack variables ($\xi_i > 0$) are used because of the effects of misclassification during the selection of training data of mixture of classes. So Eq. (4) can be written as

$$y_i(w \cdot x_i + b) \geq 1 - \xi_i \tag{5}$$

$\frac{|1+b|}{\|w\|}$ = perpendicular distance from $w \cdot x_i + b = -1$ from the origin.

Similarly, $\frac{|b-1|}{\|w\|}$ = perpendicular distance from $w \cdot x_i + b = 1$ from the origin. The margin ($\rho(w, b)$) between the planes is simply (Fig. 1)

$$\rho(w, b) = \frac{2}{\|w\|} \tag{6}$$

For nonlinear classification kernel function is used, which calculates the dot product of two vectors.

SVM Classifier: The classifiers used in this study are as follows:

C-SVM: In this type of SVM,

minimization of error function:

$$\frac{1}{2}w^T w + C\sum_{i=1}^{N} \xi_i \tag{7}$$

subjected to

$$y_i\left(w^T Ø(x_i) + b\right) \geq 1 - \xi_i \text{ and } \xi_i \geq 0, i = 1, \ldots, N \tag{8}$$

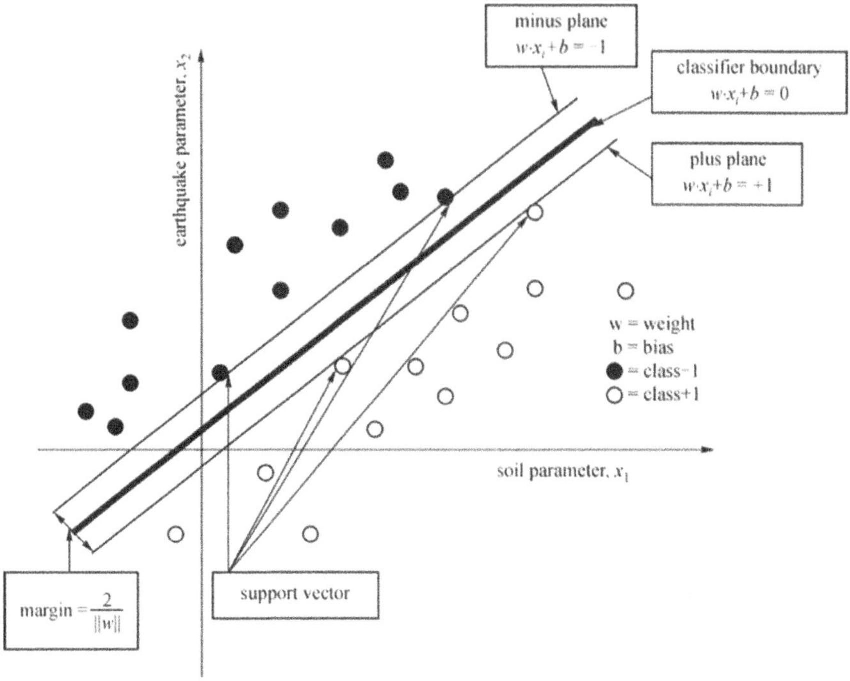

Fig. 1 Support vector with maximum margin [7]

where C = capacity constant, w = vector of coefficients, b = constant, and ξ_i = parameters for handling non-separable data (inputs). The index i labels the N training cases. The kernel ϕ is used for the purpose of transform data from the input (independent) to the feature space. The problem with the parameter C is that it can take any positive value and the other one is that it has no direct interpretation. Therefore, it is hard to choose correctly and one has to resort to cross-validation or direct experimentation to find a suitable value.

nu-SVM: Schölkopf et al. [8] reformulated SVM to take a new regularization parameter *nu*. This parameter is that it is bounded between 0 and 1 and also it has a direct interpretation.

The classification model involves minimization of error function:

$$\frac{1}{2}w^T w - \nu\rho + \frac{1}{N}\sum_{i=1}^{N}\xi_i \tag{9}$$

subjected to the constraints

$$y_i\left(w^T Ø(x_i) + b\right) \geq \rho - \xi_i, \xi_i \geq 0,\ \text{i} = 1, \ldots, \text{N and } \rho \geq 0. \tag{10}$$

Bound constraint c-SVM: In this type of SVM offset b is introduced which simplifies the dual solution with only bound constraints remaining.

Minimize the error function:

$$\frac{1}{2}w^T w + \frac{1}{2}b^2 + \frac{C}{N}\sum_{i=1}^{N}\xi_i \tag{11}$$

subjected to

$$y_i\left(w^T \emptyset(x_i) + b\right) \geq 1 - \xi_i, \xi_i \geq 0 \text{ and i } = 1, \ldots, \text{N} \tag{12}$$

Weston and Watkins multi-class SVM [10]:

Minimization of error function:

$$\frac{1}{2}\sum_{m=1}^{k} w_m^T w_m + C\sum_{i=1}^{l}\sum_{t \neq y_i}\xi_{i,t} \tag{13}$$

subjected to

$$\begin{gathered} y_i\left(w^T \emptyset(x_i) + b\right) \geq t\left(w^T \emptyset(x_i) + b\right) + 2 - \xi_{i,t}, \\ \xi_{i,t} \geq 0, \text{ i } = 1, \ldots, 1, \text{ and } t \in \{1, .., k\}\backslash y_i \end{gathered} \tag{14}$$

the resulting decision function is

$$\arg\max_m f_m(x) = \arg\max_m\left(w_m^T \emptyset(x) + b_m\right) \tag{15}$$

Crammer and Singer's multi-class SVM:

Minimization of error function:

$$\frac{1}{2}\sum_{m=1}^{k} w_m^T w_m + C\sum_{i=1}^{l}\xi_i \tag{16}$$

subjected to

$$w_{y_i}^T \emptyset(x_i) - w_t^T \emptyset(x_i) \geq 1 - \delta_{y_i,t} - \xi_i, \text{ I } = 1, \ldots, 1 \text{ and } t \in \{1, \ldots, k\} \tag{17}$$

where, $\delta_{i,j}$ is the Kronecker delta (defined as 1 for I = j and as 0 otherwise). The resulting decision function is

$$\arg\max_m f_m(x) = \arg\max_m w_m^T \emptyset(x) \tag{18}$$

2 Methodology

2.1 Deterministic Approach

The cyclic stress ratio [9]:

$$CSR = 0.65 \times \frac{a_{\max}}{g} \times \frac{\sigma_v}{\sigma_v'} \times r_d \tag{19}$$

where $a_{\max}$ = maximum horizontal acceleration, σ_{vo}, σ'_{vo} = total and effective overburden stress, r_d is the stress reduction coefficient.

The expressions to estimate the average values of stress reduction factor (r_d) have already been reported in Liao and Whiteman [6].

2.2 SVM Approach

In this technique, the seven input variables considered were the depth of interest, SPT value, bulk density at the depth of interest, saturated density at the same depth, groundwater table, fine contents, and the earthquake moment magnitude (M_w). The output considered for the classification were liquefied soil and non-liquefied soil in the deterministic approach case. The types of SVM classifiers used in this study are C-SVM, nu-SVM, bound constraint c-SVM, Weston Watkins multi-class SVM, and Crammer–Singer multi-class SVM along with different kernel functions to obtain an optimal model.

The kernel functions are used for nonlinear SVM classification. The kernel functions are as follows:

- The linear kernel function is the simplest of kernel functions.

$$k(x, x') = \langle x, x' \rangle \tag{20}$$

- The Gaussian Radial Basis Function (RBF) kernel

$$k(x, x') = \exp\left(-\sigma \left\|x - x'\right\|^2\right) \tag{21}$$

- The polynomial kernel

$$k(x, x') = \left(\text{scale} \cdot \langle x, x' \rangle + \text{offset}\right)^{\text{degree}} \tag{22}$$

- The hyperbolic tangent kernel

$$k(x, x') = \tanh(\text{scale} \cdot \langle x, x' \rangle + \text{offset}) \tag{23}$$

- The Bessel function kernel

$$k(x, x') = \frac{\text{Bessel}^{n}_{(\nu+1)}(\sigma \|x - x'\|)}{(\|x - x'\|)^{-n(\nu+1)}} \tag{24}$$

- The Laplace Radial Basis function kernel

$$k(x, x') = \exp(-\sigma \|x - x'\|) \tag{25}$$

- The ANOVA Radial Basis function kernel

$$k(x, x') = \left(\sum_{k=1}^{n} \exp\left(-\sigma \left(x^k - x'^k\right)^2\right)\right)^d \tag{26}$$

- The linear spline kernel in one dimension

$$k(x, x') = 1 + xx' \min(x, x') - \frac{x + x'}{2}(x, x')^2 + \frac{\left((\min(x, x'))^3\right)}{3} \tag{27}$$

and for multidimensional case

$$k(x, x') = \prod_{k=1}^{n} k\left(x^k, x'^k\right) \tag{28}$$

The optimal model was chosen which gave the least number of misclassification or better accuracy. Training and testing of SVM for classification problem have been carried out using R Studio.

Training and testing performance can be calculated from the following formula:

$$\begin{aligned} &\text{Training performance}\,(\%)\ \text{or testing performance}\,(\%) \\ &= \left(\frac{\text{No of data predicted accurately by SVM}}{\text{Total data}}\right) \times 100. \end{aligned} \tag{29}$$

3 Result and Discussions

In this study, the database of liquefaction potential based on SPT values is prepared by using deterministic approaches. The database consists of 532 cases. The data represents the SPT value data of 5 sites situated in different locations of India. For the study, data were taken from various boreholes of certain depths from the ground surface.

The database of liquefaction potential based on SPT values is prepared by evaluating the factor of safety against liquefaction using 532 SPT values obtained from 5 sites. The factor of safety is the ratio of cyclic resistance ratio (CRR) to cyclic stress ratio (CSR). Liquefaction is predicted to occur if the factor of safety is less than or equal to 1. After evaluation, it is observed that there are 276 liquefied cases and 256 non-liquefied cases.

In this study, SVM is performed in R Studio with the help of the e1071 package and kernlab package. The optimal model is found using different SVM classifiers with the help of different kernels. For finding the optimal model determination of certain parameters which were obtained using trial and error procedure. The details are shown in Table 1.

Out of 532 cases, 372 cases were considered for training and 160 cases were considered for testing the SVM model. In the training data, 196 cases were liquefiable cases and 176 cases were non-liquefiable cases. In testing data, 80 cases were liquefiable cases and 80 cases were non-liquefiable cases. The optimal model is chosen which gave the least number of misclassification or better accuracy.

It can be seen that the optimal model is obtained using c-SVM classifier with radial kernel. In training process, parameters obtained cost = 35 and gamma = 0.49 gave accuracy of 99.19%. When the same parameters were used in the testing process, it showed an accuracy of 91.25%. In the training process, out of 372 cases only 3 cases were misclassified whereas in the testing process it misclassified 14 cases out of 160 cases, achieving an overall accuracy of 96.8%. The summary of the results is presented in Table 2. The Bound constraint c-SVM, Wetson and Watkin SVM and Crammer–Singer SVM have very good accuracy both for training and testing, indicating the fact that these methods have good generalization capability. Irrespective of the different techniques used in the present analysis for liquefaction analysis, the radial and polynomial kernels yield very good results in terms of the accuracy of the model.

4 Conclusions

The liquefaction data, based on SPT data obtained from the deterministic approach, is analyzed to develop the SVM model.

- The optimal model is achieved using c-SVM classifier and radial kernel function using cost parameter = 35 and gamma parameter = 0.49. The trained SVM model correctly classified 369 out of 372 cases of training data (99.19% accuracy) and 146 out of 160 cases of testing data (91.25% accuracy).
- Radial and Polynomial Kernels are found to be more useful in terms of the accuracy of the model, irrespective of the classifier considered.

Table 1 Parameters used to develop SVM model using liquefaction data

Classifier	Kernels	Parameter used	No. of support vectors
C-SVM	Radial	cost = 35 gamma = 0.49	143
	Polynomial	cost = 2 gamma = 0.47 degree = 3	140
	Linear	cost = 1 gamma = 0.142	170
	Sigmoid	cost = 24 gamma = 0.01 coeff = 0	179
nu-SVM	Radial	nu = 0.16 gamma = 0.5	153
	Polynomial	nu = 0.16 gamma = 0.08 degree = 5	127
	Linear	nu = 0.46 gamma = 0.01	174
	Sigmoid	nu = 0.46 gamma = 0.01 coeff = 0	175
Bound constraint c-SVM	Polynomial	cost = 1 degree = 1 scale = 1 offset = 1	169
	Linear	cost = 1	169
	Radial	cost = 1 sigma = 0.153	190
	Hyperbolic tangent	cost = 1 scale = 1 offset = 1	159
	Laplace	cost = 1 sigma = 0.190	218
	Bessel	cost = 1 sigma = 1 order = 1 degree = 1	186
	Anova	cost = 1 sigma = 1 degree = 1	167
Weston–Watkins multi-class SVM	Polynomial	cost = 1 degree = 1 scale = 1 offset = 1	169

(continued)

Table 1 (continued)

Classifier	Kernels	Parameter used	No. of support vectors
	Linear	cost = 1	169
	Radial	cost = 1 sigma = 0.153	190
	Hyperbolic tangent	cost = 1 scale = 1 offset = 1	158
	Laplace	cost = 1 sigma = 0.190	218
	Bessel	cost = 1 sigma = 1 order = 1 degree = 1	186
	Anova	cost = 1 sigma = 1 degree = 1	167
Crammer–Singer multi-class SVM	Polynomial	cost = 1 degree = 1 scale = 1 offset = 1	167
	Linear	cost = 1	166
	Radial	cost = 1 sigma = 0.153	180
	Hyperbolic tangent	cost = 1 scale = 1 offset = 1	120
	Laplace	cost = 1 sigma = 0.190	205
	Bessel	cost = 1 sigma = 1 order = 1 degree = 1	175
	Anova	cost = 1 sigma = 1 degree = 1	160

- The Bound constraint c-SVM, Wetson and Watkin SVM and Crammer–Singer SVM have very good generalization ability.

Table 2 Summary of SVM classification results

Classifier	Kernels	Training accuracy (%)	Testing accuracy (%)	Overall accuracy (%)
c-SVM	Radial	99.19	91.25	96.8
	Polynomial	92.47	85.625	90.41
	Linear	86.56	87.5	86.84
	Sigmoid	83.33	88.75	84.96
nu-SVM	Radial	98.65	91.25	96.42
	Polynomial	93.27	83.125	90.22
	Linear	82.8	88.75	84.58
	Sigmoid	83.06	88.75	84.77
Bound Constraint c-SVM	Linear	53.76	51.875	53.19
	Polynomial	83.33	87.5	84.58
	Radial	87.09	90	87.97
	Laplace	87.36	89.375	87.97
	Bessel	86.56	90	87.6
	Anova	85.48	88.125	87.6
	Hyperbolic tangent	58.87	59.375	59.02
Weston Watkin SVM	Linear	83.33	87.5	84.58
	Polynomial	83.33	87.5	84.58
	Radial	87.09	90	87.97
	Laplace	87.365	89.375	87.97
	Bessel	86.56	90	87.6
	Anova	85.48	88.125	86.27
	Hyperbolic tangent	59.14	56.87	58.45
Crammer Singer SVM	Linear	84.67	86.875	85.33
	Polynomial	83.87	86.875	84.77
	Radial	87.63	90	88.35
	Laplace	89.78	88.75	89.47
	Bessel	86.3	88.125	86.84
	Anova	86.56	88.75	87.22
	Hyperbolic tangent	68.55	71.875	69.54

References

1. Borthakur N, Dey AK (2020) Evaluation of group capacity of micropile in soft clayey soil from experimental analysis using SVM-based prediction model. Int J Geomech 20(3):04020008
2. Cortes C, Vapnik V (1995) Support-vector network. Mach Learn 20
3. IS: 2131:1981(Reaffirmed 2002) Method for standard penetration test for soils (First Revision)
4. Ji J, Zhang C, Gui Y, Lü Q, Kodikara J (2017) New observations on the application of LS-SVM in slope system reliability analysis. J Comput Civil Eng 31(2):06016002
5. Kurnaz TF, Kaya Y (2018) The comparison of the performance of ELM, BRNN, and SVM methods for the prediction of compression index of clays. Arab J Geosci 11(24):770
6. Liao SSC, Whitman RV (1986) Catalogue of liquefaction and non-liquefaction occurrences during earthquakes. Department of Civil Engineering, Massachusetts Institute of Technology
7. Samui P (2013) Liquefaction prediction using support vector machine model based on cone penetration data. Front Struct Civil Eng 7:72–82
8. Schölkopf B, Smola A, Williamson R, Bartlett PL (1998) New support vector algorithms. NeuroCOLT technical report NC-TR-98-031
9. Seed HB, Idriss IM (1971) Simplified procedure for evaluating soil liquefaction potential. J Soil Mech Found Div 97(9):1249–1273
10. Weston J, Watkins C (1999) Multi-class support vector machines. In: Proceedings of ESANN-99
11. Zhang Y, Dai M, Ju Z (2016) Preliminary discussion regarding SVM kernel function selection in the twofold rock slope prediction model. J Comput Civil Eng 30(3):04015031
12. Zhang C, Ji J, Gui Y, Kodikara J, Yang SQ, He L (2016) Evaluation of soil-concrete interface shear strength based on LS-SVM. Geomech Eng 11(3):361–372

Assessment of Seismic Liquefaction of Soils Using Swarm-Assisted Optimization Algorithm

T. Vamsi Nagaraju, Ch. Durga Prasad, Babloo Chaudhary, and B. M. Sunil

Abstract Assessment of liquefaction potential of soils due to the earthquake has been carried out in this research using the nature-inspired Metaheuristic swarm-assisted algorithm (PSO). An assessment has been made on the basis of actual field data from the previous research. The field data consists of 59 sets having variables of total stress of soil ($\neg_0$), effective stress of the soil ($\neg'_0$), percentage fines, mean size of soil particles (D_{50}), standard penetration value (SPT), the equivalent dynamic shear stress ($T_{av}/\neg'_0$), maximum horizontal acceleration at ground surface (a/g) and the earthquake magnitude (M). PSO-based models were developed for both single variable and multivariable linear approaches. The results revealed that for the assessment of liquefaction of soils, the developed PSO models perform good estimations in terms of the errors and convergent solution. And also, with a damping coefficient and varying input variables, there is a significant improvement in the best solution. These developed models can be useful for practicing engineers in the field.

Keywords Liquefaction · SPT · Dynamic shear stress · PSO algorithm

1 Introduction

Civil engineering infrastructure construction in the earthquake influence zone has a major concern about the liquefaction potential of soils in the designing process [1]. Liquefaction of the soil depends on many factors such as soil type, soil state, water table and earthquake intensity [2–4]. In recent times and past many civil infrastructure damages occurred during earthquakes mostly by the liquefaction of soils. The efficient way of determination of the liquefaction potential of soils in the field is Standard Penetration Test (SPT) and Cone Penetration Test (CPT), and in

T. V. Nagaraju (✉) · Ch. D. Prasad
Department of Civil Engineering, S.R.K.R. Engineering College, Bhimavaram 534204, India
e-mail: Varshith.varma@gmail.com; tvnraju@srkrec.edu.in

B. Chaudhary · B. M. Sunil
Department of Civil Engineering, National Institute of Technology Karnataka, Mangalore 575025, India

T. G. Sitharam et al. (eds.), *Local Site Effects and Ground Failures*, Lecture Notes in Civil Engineering 117, https://doi.org/10.1007/978-981-15-9984-2_25

the laboratory is Cyclic Stress Ratio (CSR) for better understanding the liquefaction of soils [5, 6]. Nowadays, many researchers were interested to develop prediction tools to estimate the many geotechnical parameters including the liquefaction potential of soils. From the very past, many researchers have been contributed correlations for liquefaction of soils based on the field SPT and CPT data sets [7, 6]. Later, correlations were developed by using laboratory data. However, these correlations not provided stable and accurate predictions because of having many constraints. To overcome the constraints with promising predictions, artificial intelligence techniques have emerged with complex structures. The gaining importance in artificial intelligence stems from their ability to find the best solutions in an effective and efficient way. In artificial intelligence techniques, few of them are inspired by neurobiological studies and others from nature-inspired algorithms. The prediction models were developed based on sufficient input–output data sets [8, 9]. Further, it keeps generating and regenerating new data. To use artificial intelligence to its fullest potential in a civil engineering discipline, it can optimize material utilization, reduce cumbersome testing, reduce time-consuming tasks, assess properties of materials, and reduce construction costs [8, 9]. The developed prediction models can estimate the behaviour of materials by considering the required data information obtained from the experimental results. In this regard, some research was carried out to develop models for the prediction of liquefaction potential of soils. Over time, only mathematical methods are used for developing correlations and equations for liquefaction potential. These methods have no convergent findings due to the complex structure. This has recently motivated researchers for the application of optimization techniques for predicting the liquefaction potential of soils, and a brief account of results is presented below.

Young-Su and Byung-Tak [10] considered undrained cyclic triaxial testing results of sand were used as input data sets for developing prediction models to predict Cyclic Resistance Ratio (CRR) of soils using Artificial Neural Network (ANN). Hanna et al. [11] adapted General Regression Neural Network (GRNN) models that were developed between the seismic parameters and soil properties. The obtained results were convergent with 94%.

Bagheripour et al. [12] proposed a model to estimate the liquefaction potential using the Genetic Algorithm (GA). The main advantage of this model is that it relates both the Factor Of Safety (FOS) and the probability of liquefaction, which helps in the estimation of liquefaction potential when deterministic approaches are considered. Samuli and Sitharam [13] developed prediction models for liquefaction susceptibility using machine learning techniques such as ANN and Support Vector Machine (SVM). In the case of liquefaction susceptibility prediction, the SVM approach is more capable over the ANN models based on standard penetration data.

This paper presents the prediction models for the assessment of liquefaction potential of soils using nature-inspired Particle swarm Optimization Algorithm (PSO). For developing the estimation models total stress of soil ($\neg_{o}$), effective stress of the soil ($\neg'_{o}$), percentage fines, mean size of soil particles (D_{50}), standard penetration value (SPT), the equivalent dynamic shear stress ($T_{av}/\neg'_{o}$), maximum horizontal acceleration at the ground surface (a/g) and the earthquake magnitude (M) were considered as input variables and the output variable is Liquefaction Potential of Soils (LPS) in the field. Furthermore, the data sets used in this paper were compiled from the previous literature. This paper also explores the importance of PSO in the field of geotechnical engineering.

1.1 Overview of Particle Swarm Optimization

Many optimization algorithms were emerging day by day, but very few optimization algorithms applications got a great response from all the disciplines such as Genetic Algorithm (GA), Particle Swarm Optimization (PSO), Flower Pollination Algorithm (FPA), Bat Algorithm (BA) and Invasive Weed Optimization Algorithm (IWO) [14]. PSO algorithm is nature-inspired algorithm, it was proposed by Kennedy and Eberhart in the year 1995 [15]. PSO has been widely used in many disciplines including civil engineering [8, 9, 14]. The main advantage of PSO is easy for implementation and no requirement of gradient information. PSO mechanism works in a way that a group of birds searching randomly for food in two-dimensional space. The position of each bird is indicated as a solution to a given problem. Initially, these positions are randomly searched. In case if they are not found at an initial position they shift to the next random position where it found food, this is the best solution for that task. In PSO, from the definition, each element has the following properties, let y_i^k denotes the element position vector for finding the solutions and next position for each element finds randomly as

$$y_i^{k+1} = y_i^k + v_i^{k+1} \tag{1}$$

where v_i^{k+1} is the particle velocity vector which processes the optimization. The position of every individual bird is y_i^k and the velocity is v_i^k. The best position of each particle is p_{best} and the overall optimized position is g_{best}. Then the velocity of each organism in the swarm is updated as

$$V_i^{k+1} = \omega V_i^k + a_1 m_1 (p_{best_i} - x_i^k) + a_2 m_2 (g_{best_i} - x_i^k) \tag{2}$$

where, a_1 and a_2 are acceleration constants mainly helps in the movement of a particle in a given iteration; m_1 and m_2 are random numbers; ω is the inertia weight which helps in controlling the behaviour of the object in direction of convergence.

In this study, for the classification of liquefaction potential, single variable and multivariable approaches were adopted. Moreover, the multilinear prediction model equation was developed for predicting the liquefaction potential of soil in the field. For this liquefaction estimation, 8 input variables were considered such as earthquake magnitude (X1), total stress of soil (X2), effective stress of the soil (X3), standard penetration value (X4), maximum horizontal acceleration at the ground surface (X5), the equivalent dynamic shear stress (X6), percentage fines (X7) and mean size of soil particles (X8), these variables were multiplied by respective coefficients such as c_1, c_2, c_3, c_4, c_5, c_6, c_7 and c_8, and additional coefficient c_o was considered. Equation (3) is expressed as the developed prediction multivariable equation for classification of field liquefaction potential.

$$
\begin{aligned}
Liq_{(est)} = c_1 \cdot X1 + c_2 \cdot X2 + c_3 \cdot X3 + c_4 \cdot X4 + c_5 \cdot X5 \\
+ c_6 \cdot X6 + c_7 \cdot X7 + c_8 \cdot X8 + c_o
\end{aligned}
\tag{3}
$$

where c_1, c_2, c_3, c_4, c_5, c_6, c_7 and c_8 are coefficients used to evaluate liquefaction potential classification. For classification of liquefaction potential, if the value is found to be (+1) then say yes (Y) for the possibility of liquefaction in field or in case value is found to be (−1) then say no (N) for the possibility of liquefaction in field.

2 Database Compilation

The data sets were taken from the Tokimatsu and Yoshimi [16] case study report on soil liquefaction. The data sets include the field behaviour of the soil liquefaction during the various earthquakes experienced in the past such as Mino-Owari (1891), Kanto (1923), Tohnankai (1944), Fukui (1948), Niigata (1964), Tokachi-Oki (1968), Miyagiken-Oki (1978), San Francisco (1957), Alaska (1964), San Fernando (1971), Guatemala (1976), and Imperial Valley (1979). Moreover, this case study data have been widely used for developing prediction models based on different approaches [17, 18]. For the classification of these data sets, a MATLAB code was developed to evaluate the data sets. These evaluation processes consist of a series of steps such as objective function, boundary of iterations and acceleration constants. Few data sets were tabulated in Table 1 [17].

3 Results and Discussion

The objective of this section is to evaluate and discuss the prediction models of liquefaction potential developed by using PSO. The results of the PSO models were developed by using seven different input variables as mentioned in the previous section. The data set includes 59 samples. Initially, an attempt has been made to study the influence of individual variables for predicting liquefaction potential in the field. Later, by considering the input variable as mentioned in the previous section were used to develop a linear prediction model for predicting liquefaction potential in the field (X9) using PSO.

Table 1 Data set for liquefaction potential classification

Earthquake magnitude	Total stress (kPa)	Effective stress (kPa)	SPT value, N	a/g	Dynamic shear stress	Fines (%)	D_{50} (mm)	Liquefaction behaviour in field
7.9	186.4	96.1	20.0	0.32	0.36	0.0	0.46	Y
7.9	130.5	81.4	10.0	0.32	0.32	5.0	0.28	Y
7.9	93.2	73.6	5.7	0.2	0.16	20.0	0.20	N
7.9	149.1	100.1	2.0	0.2	0.18	0.33	0.15	N
8.0	89.3	59.8	8.0	0.2	0.19	10.0	0.40	Y
8.0	130.5	81.4	10.0	0.16	0.16	5.0	0.28	Y
7.3	66.7	35.3	7.0	0.35	0.39	35.0	0.13	N
7.3	128.5	63.8	20.0	0.40	0.45	0.0	0.45	N
7.5	130.5	71.6	8.0	0.16	0.17	2.0	0.30	Y
7.5	84.2	46.1	6.0	0.16	0.18	0.0	0.40	Y
6.7	61.8	34.4	5.0	0.12	0.12	5.0	0.70	Y
6.7	74.6	49.1	4.0	0.12	0.10	10.0	0.15	N
5.5	111.8	48.1	6.0	0.19	0.10	30	0.20	Y
8.3	56.9	53.0	10.0	0.16	0.22	10.0	0.20	Y
6.6	72.6	86.3	9.0	0.45	0.29	20.0	0.10	Y
7.5	72.6	28.4	8.0	0.14	0.17	3.0	1.00	Y
6.6	107.9	51.0	31.0	0.6	0.45	11.0	0.12	N
6.6	72.6	46.1	7.0	0.2	0.21	34.0	0.09	Y
6.7	118.7	66.7	10.0	0.10	0.09	0.0	0.60	N
6.7	93.2	68.7	9.0	0.14	0.10	20.0	0.15	N

3.1 Effect of Individual Variables on Susceptibility of Liquefaction Potential in Field

All the input variables X1, X2, X3, X4, X5, X6, X7 and X8 were individually considered for the prediction of field susceptibility of liquefaction using particle swarm optimization. Figure 1a–h shows the classification results of liquefaction susceptibility.

Figure 1a–h shows the relative individual parameters' influence on the liquefaction potential assessment by considering each individual parameter as input variables and liquefaction potential as output variable. All 8 parameters have significant importance in the liquefaction potential. However, for developing predicting model considering individual input parameter exhibits complexity in classification of problem.

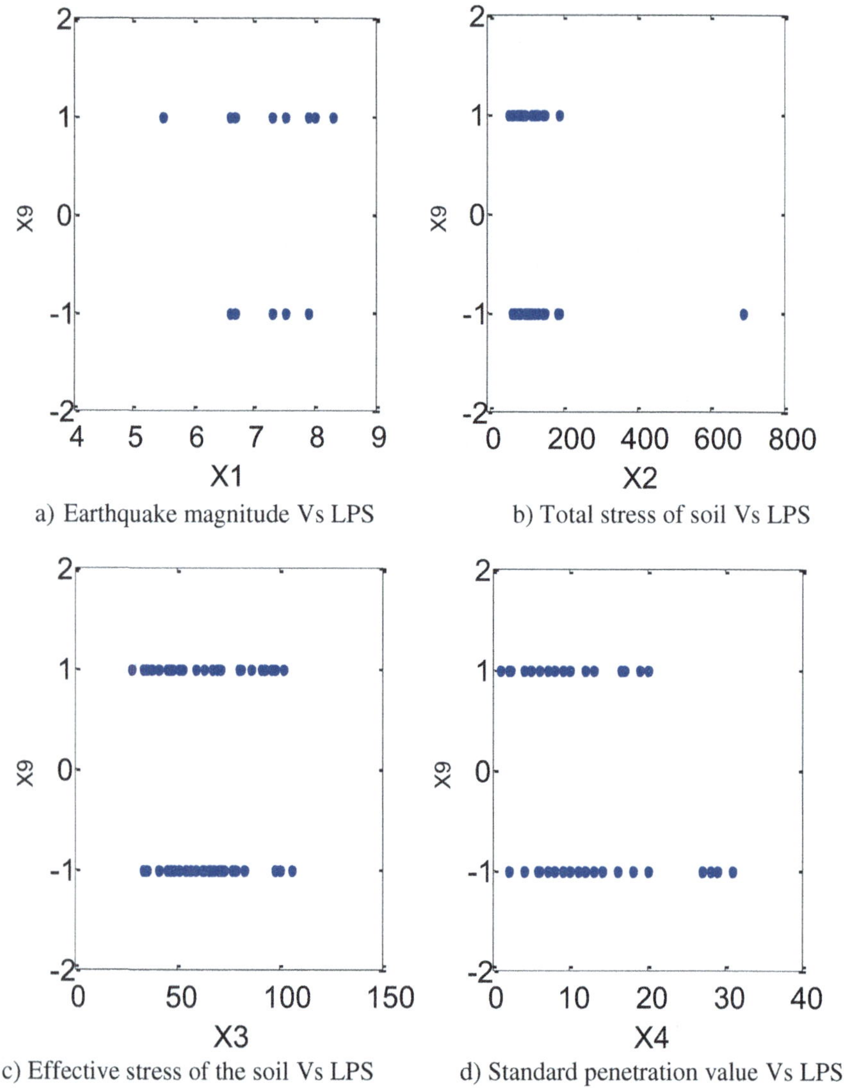

a) Earthquake magnitude Vs LPS
b) Total stress of soil Vs LPS
c) Effective stress of the soil Vs LPS
d) Standard penetration value Vs LPS

Fig. 1 Single variable classification of liquefaction potential based on PSO

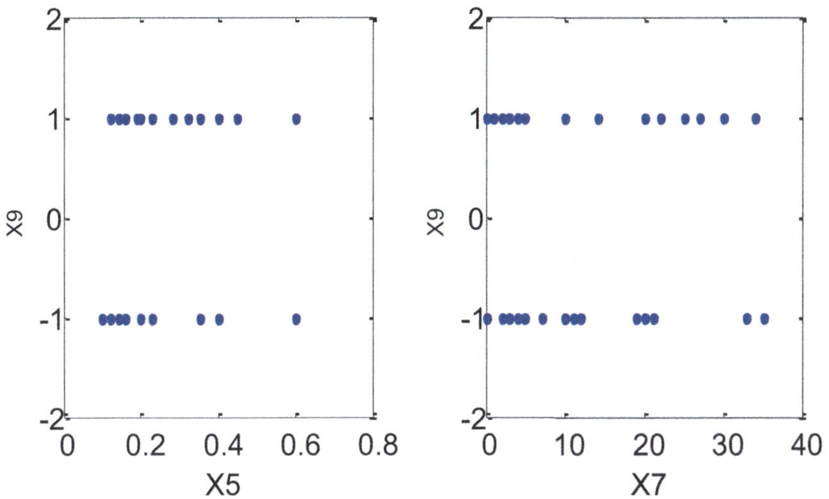

e) a/g Vs LPS f) Equivalent dynamic shear stress Vs LPS

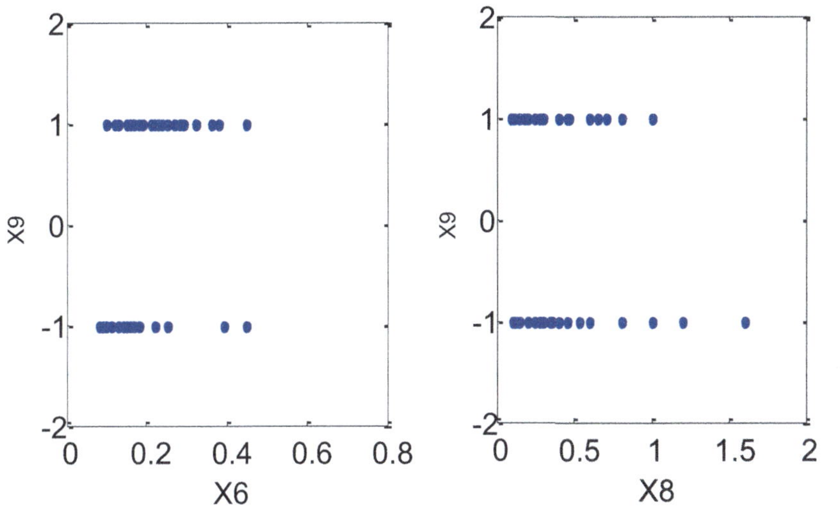

g) Percentage fines Vs LPS h) Mean size of soil particles Vs LPS

Fig. 1 (continued)

3.2 Effect of Multivariable Approach on Susceptibility of Liquefaction Potential in Field

Multivariable linear prediction was developed using PSO. All the input variables as previously mentioned were considered for estimating liquefaction potential in the field. Figure 2 shows the classification of liquefaction potential in the field.

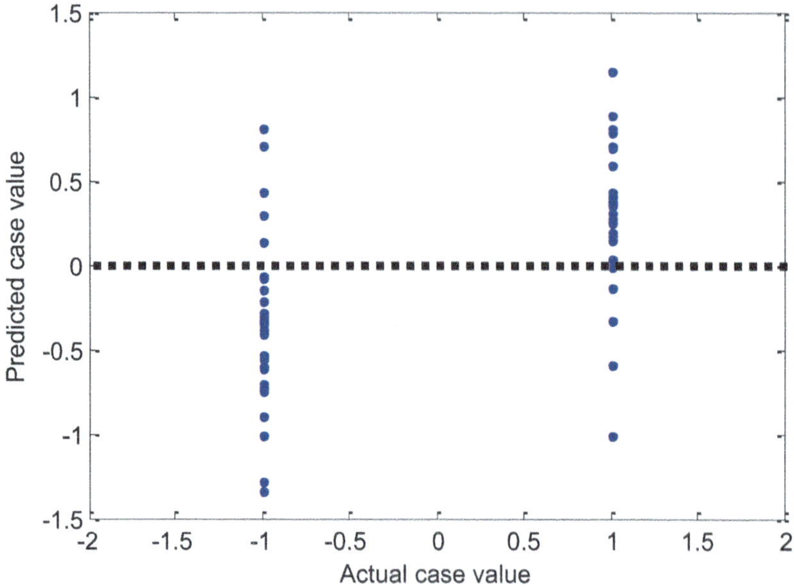

Fig. 2 Multivariable classification based on PSO

The PSO-based linear regression approach results indicate that the PSO model has been successful in predicting the susceptibility of liquefaction in the field. The developed relationship between the input data variables and the output of the liquefaction possibility estimation has 90% convergent predictions.

For predicting the field behaviour of liquefaction of soils using PSO, the Eq. (4) is developed for classification.

$$\begin{aligned} Liq_{(est)} &= 0.5023 \cdot X1 - 0.001 \cdot X2 + 0.00896 \cdot X3 \\ &\quad - 0.08156 \cdot X4 + 0.5365 \cdot X5 + 3.0703 \cdot X6 \\ &\quad - 0.01316 \cdot X7 - 0.12729 \cdot X8 - 3.9882 \end{aligned} \tag{4}$$

The developed equation, as aforementioned, classification of liquefaction is expressed in terms of possibility of filed liquefaction (+1) and not possibility of field liquefaction (−1). The PSO equation provides the best convergent results.

4 Conclusions

In this study, based on the field liquefaction data of Tokimatsu and Yoshimi's [16] case study report on soil liquefaction, an optimization algorithm PSO is used to estimate the behaviour of soil liquefaction in the field. The conclusions are drawn from this computational study as follows:

(1) The obtained classification results indicated that multilinear variable model is best suitable compared to the single variable classification models for estimating liquefaction potential in the field.
(2) Particle swarm optimization can be an alternative method for the classification of liquefaction potential in the field with 90% convergent results.
(3) Although many approaches are in practice for predicting liquefaction potential, the PSO prediction model would be easy to implement and no requirement for gradient information.

References

1. Seed HB (1987) Design problems in soil liquefaction. J Geotech Eng 113(8):827–845
2. Ishibashi I, Sherif MA, Cheng WL (1982) The effects of soil parameters on pore-pressure-rise and liquefaction prediction. Soils Found 22(1):39–48
3. Juang CH, Zhang J, Khoshnevisan S, Gong W (2017) Probabilistic methods for assessing soil liquefaction potential and effect. Geo-Risk 122
4. Ling YH (1992) Liquefaction of granular soils with non-cohesive and cohesive fines. In: Proceedings of the tenth world conference on earthquake engineering, 19–24 July 1992, vol 3. Madrid, Spain. CRC Press, p 1491
5. Schneider JA, Moss RES (2011) Linking cyclic stress and cyclic strain based methods for assessment of cyclic liquefaction triggering in sands. Géotech Lett 1(2):31–36
6. Zhou J, Li E, Wang M, Chen X, Shi X, Jiang L (2019) Feasibility of stochastic gradient boosting approach for evaluating seismic liquefaction potential based on SPT and CPT case histories. J Perfor Construct Facilit 33(3):04019024
7. Seo MW, Olson SM, Sun CG, Oh MH (2012) Evaluation of liquefaction potential index along western coast of South Korea using SPT and CPT. Mar Georesour Geotechnol 30(3):234–260
8. Nagaraju TV, Prasad CD, Raju MJ (2020) Prediction of California bearing ratio using particle swarm optimization. In: Soft computing for problem solving. Springer, Singapore, pp 795–803
9. Nagaraju TV, Prasad CD, Murthy NGK (2020) Invasive weed optimization algorithm for prediction of compression index of lime-treated expansive clays. In: Soft computing for problem solving. Springer, Singapore, pp 317–324
10. Young-Su K, Byung-Tak K (2006) Use of artificial neural networks in the prediction of liquefaction resistance of sands. J Geotech Geoenviron Eng 132(11):1502–1504
11. Hanna AM, Ural D, Saygili G (2007) Neural network model for liquefaction potential in soil deposits using Turkey and Taiwan earthquake data. Soil Dyn Earthquake Eng 27(6):521–540
12. Bagheripour MH, Shooshpasha I, Afzalirad M (2012) A genetic algorithm approach for assessing soil liquefaction potential based on reliability method. J Earth Syst Sci 121(1):45–62
13. Samui P, Sitharam TG (2011) Machine learning modelling for predicting soil liquefaction susceptibility. Nat Hazards Earth Sys Sci 11(1):1–9
14. Yang XS, Gandomi AH, Talatahari S, Alavi AH (eds) (2012) Metaheuristics in water, geotechnical and transport engineering. Newnes

15. Kennedy J, Eberhart R (1995) Particle swarm optimization. IEEE, 0-7803-2768-3/95, pp 1942–1948
16. Tokimatsu, K., & Yoshimi, Y. (1983). Empirical correlation of soil liquefaction based on SPT N-value and fines content. Soils and Foundations, 23(4), 56–74.
17. Goh AT (1994) Seismic liquefaction potential assessed by neural networks. J Geotech Eng 120(9):1467–1480
18. Sen G, Akyol E (2010) A genetic-algorithm approach for assessing the liquefaction potential of sandy soils. Nat Hazards Earth Syst Sci 10(4)

Liquefaction Potential Evaluation: Necessity of Developing Energy Correction Factor for SPT N-value

S. M. Ali Jawaid

Abstract The liquefaction analyses proposed by various researchers such as Seed et al. (1971) and Seed and Idriss (1985) are based on Cyclic Stress Ratio (CSR) and Cyclic Resistance Ratio (CRR). The Cyclic Resistance Ratio (CRR) is dependent on corrected SPT blow count N_{60}. Since hammer energy efficiency in India is different in comparison to the USA, it is necessary to develop an energy correction factor for India as the same was carried out by China and Japan. In this paper, the I.S. code procedure of liquefaction evaluation as well as energy correction factor developed for various countries were reviewed, and it is found that there is a need to develop one such energy correction factor for India.

Keywords Liquefaction · Earthquake · SPT energy correction

1 Introduction

About 54% of land in India is vulnerable to earthquakes. The U.N. has predicted that around 200 million city dwellers in India will be exposed to storms and earthquakes by the year 2050. Most of the cities/towns in India do not have planned development and thus, it is the need of the hour to carryout the micro-zonation of area liable to liquefy during any future earthquake in order to get prepared well ahead of an actual disaster. This will help us in the adaptation of suitable mitigation techniques such as deep dynamic compaction and vibro-compaction/replacement.

The purpose of this paper is to review the I.S. code procedure documented in IS:1893–2016 for the evaluation of liquefaction potential using SPT blow counts (N-value). The I.S. code procedure is based on a "Simplified Method" developed by Seed and Idriss [8].

S. M. A. Jawaid (✉)
M. M. M. University of Technology, Gorakhpur, U.P., India
e-mail: smaj@rediffmail.com

T. G. Sitharam et al. (eds.), *Local Site Effects and Ground Failures*, Lecture Notes in Civil Engineering 117, https://doi.org/10.1007/978-981-15-9984-2_26

2 I.S. Code Procedure [I.S. 1983(Part I)-2016]

To evaluate the liquefaction potential of a site, the subsurface evaluation data should include SPT blow count (N-value) and the location of the water table. The cyclic stress ratio (CSR) induced by the earthquake may be evaluated using the following equation:

$$CSR = 0.65\left(\frac{a_{\max}}{g}\right)\left(\frac{\sigma_{v0}}{\sigma'_{v0}}\right)\cdot rd \tag{1}$$

where

a_{max} = Peak Ground Acceleration (PGA) which may be taken equal to seismic zone factor,

g = Acceleration due to gravity and

r_d = Stress reduction factor

The stress reduction factor r_d may be calculated by the following relationship proposed by Lio and Whiteman [5]:

$$r_d = \begin{cases} 1 - 0.00765\,Z & 0 < Z \leq 9.15\ m \\ 1.174 - 0.0267\,Z & 9.15 < Z \leq 23.0\ m \end{cases}$$

where Z is the depth below the ground surface in meters.

In order to obtain the cyclic resistance ratio (CRR), the SPT blow count must be converted to N_{60} using the following formula:

$$N_{60} = N\,C_{60} \tag{2}$$

where

$$C_{60} = C_{HT}.\,C_{HW}.C_{55}.\,C_{RL}.\,C_{BD} \tag{3}$$

If the SPT is conducted as per IS 2131 [2], it is suggested to take $C_{60} = 1.0$.

The computed N_{60} is normalized using the overburden correction factor C_N

$$(N_1)_{60} = C_N N_{60} \tag{4}$$

where

$$CN = \sqrt{\frac{P_a}{\sigma'_{v0}}} \leq 1.7 \tag{5}$$

where

σ'_{v0} = effective overburden pressure = 100 kPa

P_a = atmospheric pressure measured in the same unit.

$(N_1)_{60}$ must also be corrected for fines as per the following equation:

$$(N_1)_{60}\text{ corrected} = A + B\,(N_1)_{60} \tag{6}$$

where A and B are coefficients determined from the following relationships:

$$A = 0 \text{ for FC} \leq 5\% \tag{7}$$

$$A = \exp\left[1.76 - \left(190/\text{FC}^2\right)\right] \text{ for } 5\% < \text{FC} < 35\% \tag{8}$$

$$A = 5.0 \text{ for FC} \geq 35\% \tag{9}$$

$$B = 1.0 \text{ for FC} \leq 5\% \tag{10}$$

$$B = \left[0.99 - \left(\text{FC}^{1.5}/1000\right)\right] \text{ for } 5\% < \text{FC} < 35\% \tag{11}$$

$$B = 1.2 \text{ for FC} \geq 35\% \tag{12}$$

The cyclic resistance ratio (CRR) may be either estimated from Fig. 1 for an earthquake of magnitude 7.5 or using Rauch [6]'s relationship:

$$\text{CRR} = \frac{1}{34 - (N_1)_{60}} + \frac{(N_1)_{60}}{135} + \frac{50}{\left[10(N_1)_{60} + 45\right]^2} - \frac{1}{200} \tag{13}$$

The factor of safety against liquefaction may be defined as

$$\text{Factor of Safety against Liquefaction} = \text{CRR/CSA} \geq 1.00 \tag{14}$$

3 Discussion

If the Standard Penetration Test is conducted as per IS 2131 [2], the energy transfer to the drill rod is about 60% and then C_{60} may be taken as 1.0 [3]. The recommended standard SPT equipment/procedure as per IS 2131 [2] is given in Table 1.

As per the requirement given in Table 1, two wraps of rope around the pulley and blow count rates of 30–40 blows per minute are recommended which is not adopted

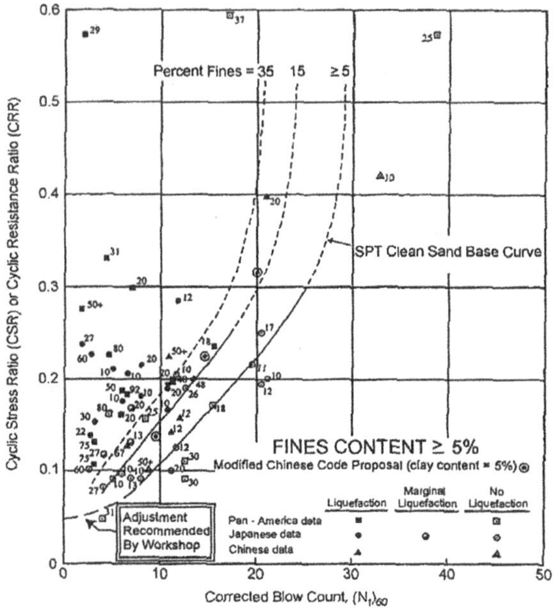

Fig. 1 Graph between CRR and $(N_1)_{60}$ for Sand for a magnitude 7.5 earthquake (courtesy [3, 9])

Table 1 Recommended standardized SPT equipment as per IS 2131 [2]

S. No.	Element	Detailed explanation
1	Sampler	"Standard split-spoon sampler with OD = 51 mm and ID = 35 mm (constant, that is, no room for liners in the barrel)"
2	Drill rods	"A or AW type for depth less than 15.2 m; N or NW type for greater depths. Standard (Safety Hammer) with (a) weight = 63.5 kg (b) drop Ht = 762 mm (delivered 60% of theoretical free-fall energy)"
3	Rope	"Two wraps of rope around the pulley"
4	Borehole	"100–130 mm diameter rotary borehole with bentonite mud for borehole stability"
5	Drill bit	Upward deflection of drilling mud
6	Blow count rate	"30 to 40 blows per minute"
7	Penetration resistant count	"Measured over a range of 150 mm–450 mm of penetration into the ground"

(*Courtesy* [3])

in most parts of India, thanks to the lack of automatic hammers in India. Normally, SPT is carried out manually using unskilled labors. They use more wraps of rope as well as the blow count rate is less than 15 blows per minute, thus taking C_{60} as unity is not correct.

In U.S. testing practice, Energy Ratio (ER) of 60% is accepted as a reference value for energy correction. ER depends on the type and weight of SPT hammer, anvil, lifting mechanism and method of hammer release. If N-value is measured in any other country using different weight, fall and lifting mechanism, hammer efficiency, etc., the correction should be made to SPT N-value for the difference in energy efficiency using the following equation [4]:

$$(N_1)_{60} = \left(\frac{ER_m}{ER_c}\right) N_m \tag{15}$$

where N_m and ER_m are the measured N-value and corresponding energy efficiency. $(N_1)_{60}$ and ER_c are the N-value at an energy ratio of 60% and energy efficiency equal to 60%. Since hammer efficiency and hammer used in India are different from the one used in the USA, it is necessary to normalize the measured N-value using energy ratio to get $(N_1)_{60}$.

Yoshimi et al. [11] had studied the effect of different methods of hammer release on SPT blow counts. They found that blow counts measured with the mechanical trip release system gave 13% lower blow counts than blow count measured with the rope and pulley method. The relationship between N-value measured with a mechanical trip system and manual system in China may be given by the following relationship [10]:

$$N_{mechanical} = 0.83\ N_{manual} \tag{16}$$

It shows that the rod energy ratio for manual type hammers is 20% less than that of a mechanical hammer and thus energy ratio for hammers operated with the rope and pulley system in China is about 50%. Seed et al. [9] had compiled the energy ratio (ER_m/ER_c) for different countries and the same is presented in Table 2.

Unfortunately, the standardized energy ratio (ER_m/ER_c) is not available for Indian SPT hammers. Gupta et al. [1] had proposed an energy ratio (the correction factor for 60% rod energy) of 0.75. However, some researchers proposed an energy ratio of 1.0 [3, 7]. Based on the above discussion, it is the need of the hour to correct the measured N-value based on actual energy transfer. This can be achieved by measuring the energy by the SPT analyzer at various sites and developing a correction factor for 60% energy transfer for India.

Some SPT tests were carried at different locations in and around Gorakhpur city using an automatic SPT hammer (make: AIMIL) by the author. The energy calibration rod is attached in between the top SPT A-rod and hammer guide rod and is then attached with a handheld unit of the SPT analyzer (make: Pile Dynamics, USA). Energy measurement was carried out at every 1.0 m of penetration. The SPT

Table 2 Energy correction factors for different countries

Country	Hammer type	Hammer Release	Estimated Rod Energy, %	Correction Factor for 60% Rod Energy
Japan	Donut	Free-fall	78	1.30
	Donut	Rope and Pulley with special throw release	67	1.12
United States	Safety	Rope and Pulley	60	1.00
	Donut	Rope and Pulley	45	0.75
Argentina	Donut	Rope and Pulley	45	0.75
China	Donut	Free-fall	60	1.00
	Donut	Rope and Pulley	50	0.83

(After [9])

analyzer had recorded energy transmitted per blow of the hammer. Only the blows for the last 30 cm increments were used in the analysis since these corresponding blows count as the N-value. The SPT data and energy measurement per site were recorded. It was found that the automatic hammer has energy efficiencies around 86.54%. The standard deviation was 0.869. Based on this study, it is suggested to apply the correction factor (C_{60}) of 0.86 for converting the measured SPT N-value in terms of N_{60} in the case of an automatic hammer in India. These values are within the range of correction factors reported in the literature for other countries. However, a broad study is needed in order to develop a universally accepted correlation for India.

4 Concluding Remarks

The Standard Penetration Test (SPT) is currently the most popular and economical means of obtaining subsurface information. In order to reduce the significant variability associated with the SPT N-value, it is recommended that N-values be standardized to N_{60}. This standardization is to be achieved by correcting the measured field N-values by the ratio of energy transfer to the standard 60% energy of an SPT hammer. This requires knowing the performance characteristics of the SPT system.

Various researchers had proposed an energy ratio (the correction factor for 60% rod energy) of 0.75–1.0 for India. It is the need of the hour to develop a standard correction factor based on the measurement of actual energy transfer. This can be achieved by measuring the energy by an SPT analyzer at various sites and developing a correction factor for 60% energy transfer for India. However, based on a pilot study, the author had found correction factors (C_N) of 0.86 in the case of automatic hammers available in India. These values are within the range of correction factors reported in the literature for other countries. However, a broad study is needed for India in order to develop a universally accepted correlation for India.

References

1. Gupta S, Sundaram R, Gupta S (2008) Liquefaction susceptibility study at commonwealth games village, Delhi. In: Proceedings of the Indian geotechnical conference 2008, vol 1. Bangalore, pp 436–440
2. IS 2131 (2002) Methods for standard penetration test for soils. Bureau of Indian Standards, New Delhi
3. IS 1893 (2016) Criteria for earthquake design of structures, vol 1. Bureau of Indian Standards, New Delhi
4. Japanese Geotechnical Society (1998) Remedial measures against soil liquefaction. Balkema, Rotterdam
5. Lio SSC, Whiteman RV (1986) Catalogue of liquefaction and non-liquefaction occurrences during earthquake. Research Report, Department of Civil Engineering, MIT, Cambridge, Mass
6. Rauch AF (1998) Evaluating cyclic liquefaction potential using cone penetration test. Can Geotech J 35(3):442–459
7. Saran S (2006) Soil dynamics and machine foundation. Galgotia Publications, New Delhi
8. Seed HB, Idriss IM (1971) Simplified procedure for evaluating soil liquefaction potential. J Geotech Eng Div, ASCE, 97(9):1249–1273
9. Seed HB, Tokimatsu K, Harder LF, Chung RM (1985) The influence of SPT procedures in soil liquefaction resistance evaluations. J Geotech Eng, ASCE, 111(2):1425–1445
10. ShiMing H (1982) Experience on standard penetration test. In: Proceedings of the second European symposium on penetration testing, Amsterdam
11. Yoshimi Y, Tokimatsu K, Oh-oka H (1983) Comparision of SPT N-values by cat-and-rope method and trip monkey method. In: Proceedings of the 18th annual meeting, JSSMFE

Assessment of Soil Liquefaction Safety Factors Under Different Earthquake Magnitudes

Shiva Shankar Choudhary, Avijit Burman, and Sanjay Kumar

Abstract In the present study, seismic soil liquefaction in terms of factor of safety against liquefaction (FS) is evaluated by the IS Code: (Criteria for earthquake resistant design of structures. Bureau of Indian Standards, New Delhi [5]) (Part-I, 2016) for a site of IIT Patna campus. The FS values against liquefaction are evaluated under three different earthquake magnitudes, namely, $M_w = 6.0$, 6.5 and 7.0. A design peak ground acceleration of 0.24 g used as the Patna city lies in the Zone IV of the seismic zoning map of India as per IS code. In this paper, an evaluation of the severity of liquefaction in the form of liquefaction potential index (LPI) is also determined. The LPI is determined at a single borehole location from the obtained factors of safety (FS) to predict the potential of liquefaction to cause damage at the site of interest. The evaluated FS and LPI values from IS Code are compared with Idriss and Boulanger (Soil Dyn Earthq Eng 26:115–130 [4]) procedures to investigate the liquefaction behavior for cohesionless soils. The FS and LPI results of both SPT-based semi-empirical procedures indicate that the soil liquefaction probability increases during earthquakes with magnitude $\geq$7.0 at this site. It is also observed that the values of FS from IS code match with Idriss and Boulanger (Soil Dyn Earthq Eng 26:115–130 [4]) results under high earthquake magnitude.

Keywords Earthquake magnitudes · IS code · Liquefaction potential index · Soil liquefaction

S. S. Choudhary (✉) · A. Burman · S. Kumar
National Institute of Technology Patna, Patna 800005, Bihar, India
e-mail: shiva@nitp.ac.in

A. Burman
e-mail: avijit@nitp.ac.in

S. Kumar
e-mail: sanjay@nitp.ac.in

T. G. Sitharam et al. (eds.), *Local Site Effects and Ground Failures*, Lecture Notes in Civil Engineering 117, https://doi.org/10.1007/978-981-15-9984-2_27

1 Introduction

Soil liquefaction can be defined as a temporary loss in shear strength of saturated sandy and silty soil due to the significant swelling in the pore-water pressure loudening under the influence of cyclic/dynamic loading generated due to incoming stress waves reaching the site in the form of seismic loading. Liquefaction is generally observed in the case of saturated cohesionless soil; however, the clayey soil may also undergo huge deformation/spreading. Liquefaction could cause serious damage to the buildings, earth structures, and sloping grounds due to the loss of strength of the soil, which eventually leads to the failure of the structure. The failure of the soil may be in the form of vertical ground settlements and slope instability, sand boil, and lateral spreading. To quantify the severity of soil liquefaction an index known as liquefaction potential index (LPI) could be computed for the determination of the degree of severity of liquefaction.

The liquefaction potential index (LPI) and factor of safety against liquefaction (FS) provide a technique to assess the level of risk of liquefaction-induced damage to the foundation of a structure. It is the basic and initial investigation from a geotechnical point of view and it is more important when the structural project site is under a higher seismic zone or near river/sea. The potential for soil liquefaction to occur at a certain depth at a site is quantified in terms of the factors of safety against liquefaction (FS). Seed and Idriss [9] suggested a simplified approach to evaluate the liquefaction resistance of soils in terms of factor of safety against liquefaction by taking the ratio of the capacity of a soil element to resist soil liquefaction to the seismic demand imposed on it. The liquefaction potential index (LPI) as proposed by Iwasaki et al. [6] was also used to illustrate the liquefaction vulnerability, which takes into account the effect of thickness and depth along with the FS of the liquefiable layer. Capacity to resist soil liquefaction is computed as the cyclic resistance ratio (CRR), and seismic demand is computed as the cyclic stress ratio (CSR). Factor of safety of a soil layer can be calculated with the help of several field tests such as standard penetration test (SPT), conic penetration test (CPT), Becker penetration test (BPT), and shear wave velocity (V_s) test. Factor of safety along the depth of soil profile is generally calculated using the surface-level peak ground acceleration (PGA), earthquake magnitude (Mw), and SPT data, namely, SPT blow counts (N), overburden pressure, fines content (FC), liquid limits and grain size distribution which is discussed by many researchers [7, 10, 12, 11]. Bao et al. [1] and Bwambale and Andrus [2] analyzed and discussed the characteristics of the newly developed methods including the mitigating mechanism and effectiveness, and possible executive problems to understand the progress of liquefaction resistance. Olson et al. [8] performed a broad parametric study to assess the viability of using nonlinear site response analysis with validated constitutive and pore-water pressure generation models to evaluate level-ground liquefaction. In this study, the nonlinear site response analysis with pore-water pressure generation correctly predicted liquefaction for dynamic centrifuge tests and field cases, demonstrating that this approach can assess level ground liquefaction while avoiding highly uncertain correction factors required in the cyclic stress method. An approach for soil liquefaction evaluation using a probabilistic method based on the SPT databases has been presented by Goharzay et al. [3]. In this paper, the uncertainties of the parameters for liquefaction probability have been taken into account. A calibrated mapping function is developed using Baye's theorem in order to capture the failure probabilities in the absence of the knowledge of parameter uncertainty.

To study the liquefaction vulnerability with different earthquake magnitudes for possible building construction and design at the Indian Institute of Technology Patna (IITP) site, initial liquefaction check is done in terms of FS and LPI. Patna (Bihta) city falls under seismic zone IV and established near the Ganga River. So, it is absolutely necessary to make foundation soil improve before construction to eliminate the liquefaction risk. There are many empirical relations are developed based on in situ and laboratory test results. In the current study, two different SPT-based semi-empirical procedures [4, 5] are used for the liquefaction vulnerability check under earthquake magnitude (M_w) of 6.0, 6.5, and 7.0.

2 Site Location and Soil Condition

Soil sample of Indian Institute of Technology Patna campus site (located at Bihta and 35 km from Patna) is investigated. The borehole site lies in the latitude of 25° 32′ 03.1″ North and longitude of 84° 50′ 54.5″ East. The borehole location and soil property with varying depth is presented in Fig. 1 and Table 1, respectively.

At this site, many borehole tests are performed. However, in this study, one borehole test data is used for the investigation of soil liquefaction potential. The standard penetration test (SPT) is performed up to the depth of 15 m in order to determine the SPT-N values in the field. The laboratory tests are also performed on collected soil

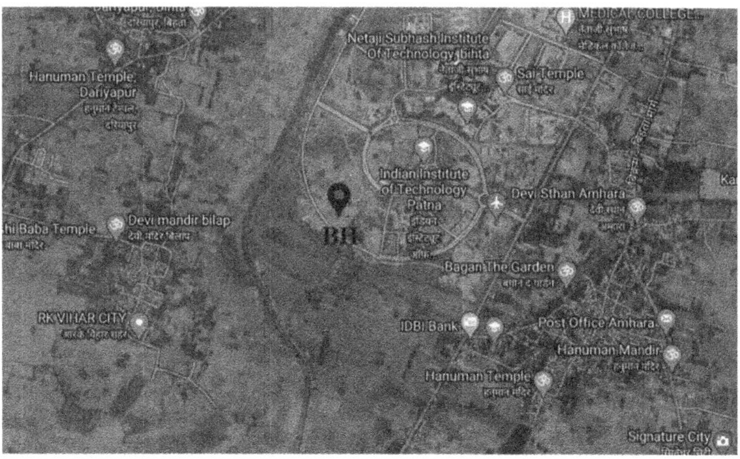

Fig. 1 Borehole location at IIT Patna campus site

Table 1 Soil properties at IIT Patna campus

Depth (m)	Fine Content, *FC* (75 η)	γ_d (g/cc)	*W* (%)	SPT-N value
0–1.5	6.50	1.570	12.00	8
1.5–3.0	3.75	1.581	13.50	12
3.0–4.5	5.75	1.602	14.10	17
4.5–6.0	0.75	1.612	8.20	24
6.0–7.5	6.50	1.588	9.30	26
7.5–9.0	4.50	1.595	10.30	29
9.0–10.5	1.75	1.615	7.50	30
10.5–12.0	1.00	1.620	8.50	32
12.0–13.5	88.60	1.534	23.25	32
13.5–15.0	87.30	1.512	24.53	30

samples at the National Institute of Technology Patna (NIT Patna) to determine the soil properties at different depths. From the SPT and laboratory tests, it is found that the subsurface soil layers mainly consist of sandy soil.

3 Evaluation of Soil Liquefaction Potential

The empirical relations based on SPT and CPT are widely used in practice for the determination of factors of safety against liquefaction (FS) and liquefaction potential index (LPI). In this paper, empirical relations based on the SPT test [4, 5] are used. The values of FS are determined by comparing the seismic demand expressed in terms of cyclic stress ratio (CSR) to the capacity of liquefaction resistance of the soil expressed in terms of cyclic resistance ratio (CRR). CSR distinguishes the seismic demand produced by a given earthquake and it is evaluated from peak ground surface acceleration that is based upon the ground motions. CRR requires fines content (FC) of the soil to correct updated SPT blow count $(N_1)_{60}$ to an equivalent clean sand standard penetration resistance value $(N_1)_{60cs}$.

From IS Code: [5] (Part-I), 2016,

$$CSR = 0.65\left(\frac{a_{\max}}{g}\right)\left(\frac{\sigma_v}{\sigma_v{}'}\right)r_d \tag{1}$$

$$CRR = \frac{1}{34-(N_1)_{60cs}} + \frac{(N_1)_{60cs}}{135} + \frac{50}{\left[10\times(N_1)_{60cs}+45\right]^2} - \frac{1}{200} \tag{2}$$

From Idriss and Boulanger [4],

$$CSR = 0.65\left(\frac{a_{\max}}{g}\right)\left(\frac{\sigma_v}{\sigma_v'}\right)r_d\frac{1}{MSF}\frac{1}{K_\sigma} \tag{3}$$

$$CRR = exp\left\{\frac{(N_1)_{60cs}}{14.1} + \left(\frac{(N_1)_{60cs}}{126}\right)^2 + \left(\frac{(N_1)_{60cs}}{23.6}\right)^3 + \left(\frac{(N_1)_{60cs}}{25.4}\right)^4 - 2.8\right\} \tag{4}$$

where $a_{\max}$ = peak horizontal ground acceleration;   = acceleration due to gravity; r_d = depth-dependent stress reduction factor; σ_v and $\sigma_{v'}$= total vertical overburden stress and effective vertical overburden stress, respectively, at a given depth below the ground surface; MSF = magnitude scaling factor and K_σ = overburden correction factor. For the soil profiles with the depth less than 20 m, LPI is calculated using the following expression:

$$LPI = \sum_{i=1}^{n} \omega_i F_i H_i \ with\left\{\frac{F_i = 1 - FS_i \ for \ FS_i < 1.0}{F_i = 0 \ for \ FS_i \geq 1.0}\right\} \tag{5}$$

where H_i = thickness of the discretized soil layers; n = number of layers; F_i = liquefaction severity for i-th layer; FS_i = factor of safety for i-th layer; w_i = weighting factor (=10–0.5 z_i); and z_i = depth of i-th layer.

4 Results and Discussion

In this study, FS and LPI for earthquake magnitudes (Mw) of 6.0, 6.5, and 7.0 are determined with the use of SPT-based IS code: [5], 2016 and Idriss and Boulanger [4] methods. For calculation of soil liquefaction potential, the earthquake magnitudes with a peak ground acceleration level of 0.24 g are considered. Two basic definitions of factors of safety against liquefaction (FS) for soil liquefaction potential are used in the present study: liquefiable (FS ≤ 1.0) and non-liquefiable (FS > 1.0). However, it cannot be established that FS just below 1.0 will certainly liquefy or FS above 1.0 will not, as it may vary based on the method adopted and the level of safety desired. From in situ and laboratory tests, the typical subsoil profiles of the existing site indicate sandy soil and the groundwater table is found at a depth of 3.0 m.

The type of soil most susceptible to liquefaction is one in which the resistance to deformation is mobilized by friction between particles. The LPI and FS with three different earthquake magnitudes are summarized in Tables 2 and 3 for IS code: [5] and Idriss and Boulanger [4], respectively. It is found from Tables 2 and 3 that the values of factor of safety against soil liquefaction are decreases with the increase of seismic magnitude. This type of increasing pattern indicates that the soil liquefaction vulnerability increases with an increase in the intensity of seismic events.

The FS values with M_w = 6.0 and 6.5 indicate no zone of liquefaction and M_w = 7.0 indicate two zones of potential liquefaction in the poorly graded sand layer at depths of 3.0 m from ground surface as per IS code. However, in case of

Table 2 Factors of safety (FS) against soil liquefaction potential [5]

Depth (*m*)	CSR	CRR_6	$CRR_{6.5}$	CRR_7	FS			LPI		
					$M_w =$ 6	$M_w =$ 6.5	$M_w =$ 7	$M_w =$ 6	$M_w =$ 6.5	$M_w =$ 7
1.5	0.15	0.18	0.15	0.15	1.19	0.97	0.80	–	–	2.71
3.0	0.15	0.22	0.18	0.15	1.45	1.18	0.98	–	–	0.27
4.5	0.18	0.29	0.24	0.19	1.57	1.28	1.06	–	–	–
6.0	0.21	0.45	0.37	0.30	2.17	1.76	1.46	–	–	–
7.5	0.22	0.46	0.38	0.31	2.08	1.69	1.40	–	–	–
9.0	0.23	0.48	0.39	0.32	2.08	1.70	1.40	–	–	–
10.5	0.23	0.51	0.42	0.35	2.20	1.79	1.48	–	–	–
12.0	0.23	0.52	0.42	0.35	2.26	1.84	1.52	–	–	–
13.5	0.21	0.62	0.51	0.42	2.93	2.38	1.97	–	–	–
15.0	0.21	0.51	0.42	0.34	2.46	2.01	1.66	–	–	–
$LPI = \Sigma_{wi} F_i H_i =$								0.00	0.00	2.98

Table 3 Factors of safety (FS) against soil liquefaction potential [4]

Depth (*m*)	CRR	CSR_6	$CSR_{6.5}$	CSR_7	FS			LPI		
					M_w = 6	M_w = 6.5	M_w = 7	M_w = 6	M_w = 6.5	M_w = 7
1.5	0.10	0.13	0.10	0.08	1.34	1.03	0.79	–	–	2.91
3.0	0.11	0.18	0.14	0.10	1.65	1.25	0.93	–	–	0.89
4.5	0.16	0.29	0.22	0.15	1.80	1.40	0.93	–	–	0.81
6.0	0.16	0.38	0.29	0.21	2.40	1.84	1.33	–	–	–
7.5	0.17	0.40	0.30	0.23	2.38	1.78	1.33	–	–	–
9.0	0.17	0.44	0.32	0.24	2.60	1.89	1.41	–	–	–
10.5	0.15	0.41	0.31	0.23	2.70	2.06	1.52	–	–	–
12.0	0.19	0.52	0.38	0.27	2.73	2.00	1.41	–	–	–
13.5	0.20	0.73	0.53	0.38	3.67	2.65	1.88	–	–	–
15.0	0.20	0.64	0.48	0.32	3.20	2.40	1.59	–	–	–
$LPI = \Sigma_{wi} F_i H_i =$								0.00	0.00	4.61

Mw = 7.0, top three layers (up to 4.5 m depth) are found liquefiable as per Idriss and Boulanger [4] calculation. The results of soil liquefaction potential indicated that soil liquefaction probability increases when the $M_w \geq 7.0$, whereas the higher the seismic magnitude, the greater the liquefaction potential and the more damage it causes. Comparisons of FS values evaluated from IS code and Idriss and Boulanger [4] method is presented in Fig. 2 for different earthquake magnitude (M_w). It is found from the comparison curves that the evaluated values of FS from IS code are similar to Idriss and Boulanger [4] results for $M_w = 7.0$. Liquefaction susceptibility for

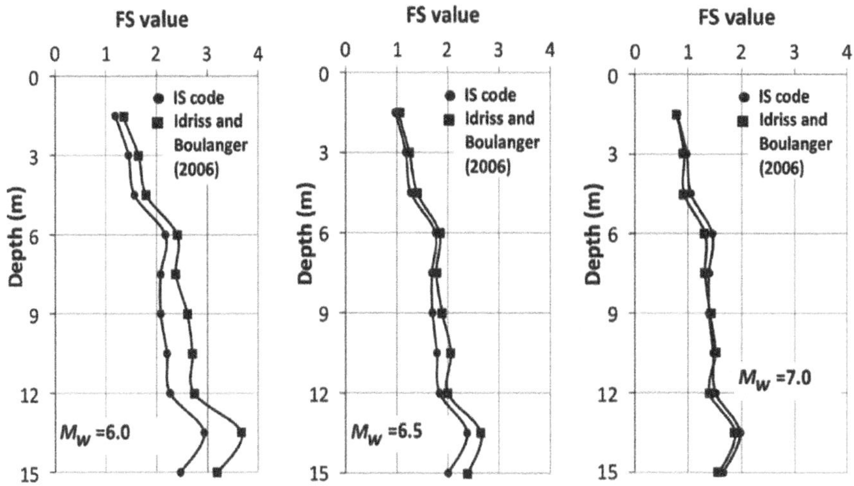

Fig. 2 Comparison of FS values between IS code and Idriss and Boulanger [4]

sites with LPI > 15 is very high, and the liquefaction is very unlikely at sites with LPI < 5. From this study, it is found that the site is in a liquefaction-susceptible area, indicating the possibility of damage during high earthquake.

5 Conclusions

In the present study, SPT-based semi-empirical procedures [4, 5] are used to evaluate the liquefaction potential of cohesionless soils with three different earthquakes magnitude ($M_w = 6.0$, 6.5, and 7.0). The SPT-based soil liquefaction in terms of FS and LPI, revealed that the soil liquefaction probability increases during earthquakes with a magnitude greater and equal to 7.0 at the IIT Patna campus site. So, mitigation techniques such as densification methods, modification of site geometry, or drainage to lower the groundwater table are needed for this site. It is also found that the values of FS from IS code matches with Idriss and Boulanger [4] results under higher earthquake magnitude.

References

1. Bao X, Jin Z, Cui H, Chen X, Xie X (2019) Soil liquefaction mitigation in geotechnical engineering: an overview of recently developed methods. Soil Dyn Earthq Eng 120:273–291
2. Bwambale B, Andrus RD (2019) State of the art in the assessment of aging effects on soil liquefaction. Soil Dyn Earthq Eng 125:1–13

3. Goharzay M, Noorzad A, Ardakani AM, Jalal M (2020) Computer-aided SPT-based reliability model for probability of liquefaction using hybrid PSO and GA. J Comput Des Eng 7:1–21
4. Idriss IM, Boulanger RW (2006) Semi-empirical procedures for evaluating liquefaction potential during earthquakes. Soil Dyn Earthq Eng 26:115–130
5. IS: 1893 (Part 1): Criteria for earthquake resistant design of structures. Bureau of Indian Standards, New Delhi
6. Iwasaki T, Tokida K, Tatsuoka F, Watanabe S, Yasuda S, Sato H (1982) Microzonation for soil liquefaction potential using simplified methods. In: Proceedings of 2nd international conference on microzonation, pp 1319–1330. Seattle
7. Karakan E, Eskişar T, Altun S (2018) The liquefaction behavior of poorly graded sands reinforced with fibers. Adv Civ Eng 1–14, Hindawi. Article ID 4738628 (2018)
8. Olson SM, Mei X, Hashash YMA (2020) Nonlinear site response analysis with pore-water pressure generation for liquefaction triggering evaluation. J Geotech Geoenviron Eng 146(2):1–17. ASCE (2020)
9. Seed HB, Idriss IM (1971) Simplified procedure for evaluating soil liquefaction potential. J Soil Mech Found 97(9):1249–1273. ASCE (1971)
10. Seed HB, Tokimatsu K, Harder LF, Chung RM (1985) The influence of SPT procedures in soil liquefaction resistance evaluations. J Geotech Eng. ASCE 111(12):1425–1445
11. Shelley EO, Mussio V, Rodríguez M, Chang JGA (2015) Evaluation of soil liquefaction from surface analysis. Geofísica Int 54(1):95–109
12. Youd TL, Idriss IM (2001) Liquefaction resistance of soils. Summary report from the (1996) NCEER and 1998 NCEER/NCF workshops on evaluation on liquefaction resistance of soil. J Geotech Geoenviron Eng 297–313

Reliability-Based Assessment of Liquefaction Potential Using SPT Approach

G. S. Bhatia, K. Bhargava, and A. Mondal

Abstract Earthquake-induced liquefaction is one of the most complex and interesting phenomenon in Geotechnical Earthquake Engineering. Liquefaction is a very significant phenomenon in alluvial soil deposits consisting of silty sand or sandy silt type of soils. Many different methods are available at present to assess the liquefaction potential of soil using in situ field test data. Generally, field tests like SPT are carried out and liquefaction assessment is done for each borehole. Different boreholes provide different depths up to which soil has potential for liquefaction for particular peak ground acceleration (PGA) and magnitude of Earthquake. Maximum liquefaction depth obtained from different boreholes is used to convey the liquefaction depth of the area in consideration. The greatest drawback of this approach is that results of only one borehole are used to conservatively predict the liquefaction potential of the entire area and results of remaining boreholes are neglected as they have predicted lesser depth of liquefaction. In case some more boreholes are drilled and assessment is done then there is quite a good probability that liquefaction depth of that area will change depending upon the SPT results from new boreholes. In this paper liquefaction potential assessment of alluvial soil site is carried out using SPT-based approach as proposed by NCEER [5]. SPT data of 25 boreholes are analyzed and converted into the equivalent single borehole using probabilistic approach to assess the liquefaction potential of alluvial soil site. Probabilistic approach is used to assess 95 percentile values of all variables required to assess Liquefaction potential like SPT blow counts, percentage fines, and soil density. As there is uncertainty

G. S. Bhatia (✉)
Nuclear Power Corporation of India Limited, Hisar, Haryana, India
e-mail: bhatiasgurpreet09@gmail.com

K. Bhargava
Nuclear Recycle Board, Bhabha Atomic Research Centre, Mumbai, India

Homi Bhabha National Institute, Mumbai, India

K. Bhargava
e-mail: kapilesh_66@yahoo.co.uk

A. Mondal
Nuclear Power Corporation of India Limited, Mumbai, India
e-mail: apurba@npcil.co.in

T. G. Sitharam et al. (eds.), *Local Site Effects and Ground Failures*, Lecture Notes in Civil Engineering 117, https://doi.org/10.1007/978-981-15-9984-2_28

present in the evaluation of these parameters at the site and in the laboratory, it is appropriate to evaluate these parameters based on a probabilistic approach using the best fit probabilistic distribution curve. Parameters required for assessing CSR (Cyclic Stress Ratio) like overburden pressure and effective overburden pressure are also analyzed using probabilistic distribution curves considering data from 25 boreholes and converting them into an equivalent single borehole. Factor of Safety (FOS) obtained using 50, 95, and 98 percentile values of different parameters is compared for converted equivalent single borehole.

Keywords Liquefaction · Cyclic stress ratio · Cyclic resistance ratio · 95 percentile values

1 Introduction

Liquefaction is one of the most spectacular phenomenon associated with an earthquake that has caused trouble to engineers and mankind. Some of the most devastating damages to structures in history have occurred when soil deposits have lost their strength during seismic loading. In liquefaction, the soil strength is reduced drastically to the point where the soil is unable to support structures over it. It is essential to clearly predict the liquefaction potential of soil deposits before the construction of any particular structure in order to mitigate the effects of liquefaction by the improvement of strata. There are various in situ methods available to assess the liquefaction potential of soil like SPT, CPT, BPT, and Shear Wave Velocity. These methods have been developed and refined over the years by different researchers to predict the liquefaction potential with quite a good probability. SPT-based liquefaction assessment is quiet popular worldwide as it is one of the most convenient and effective ways of determining liquefaction potential.

Seed and Idriss [6] developed a simplified procedure for evaluating soil liquefaction potential. This procedure was further refined by various researchers including some landmark studies by Seed and Idriss [7], Cetin et al. [2], Idriss and Boulanger [4]. Youd and Idriss [9] published a paper that summarizes the recommendation from NCEER [9] workshop that has incorporated research findings on liquefaction assessment since many decades. The proceedings of this workshop on liquefaction potential procedures [5] have shaped the state of practice to evaluate liquefaction potential.

There are no clear guidelines regarding the extent of geotechnical investigation required to assess the liquefaction potential of area with a high degree of certainty. For the construction of important structures like Nuclear Power Plants, Dams, Chemical storage facilities, etc., the need for stringent geotechnical investigation studies is essential to the safety of the complete structure. Currently, liquefaction assessment of a particular area based on SPT field data is carried out using the most conservative approach, wherein liquefaction potential depth of soil deposit is calculated using the data of worst borehole. Generally, liquefaction assessment of individual boreholes is carried out and the results of the worst borehole are used to define the liquefaction potential depth of soil deposit. In case some new boreholes are drilled in the same area at later stage, there are possibilities that the liquefaction depth of the area might change depending upon data from new boreholes. This approach gives all due attention to the results of single boreholes and neglects the data from other boreholes to conservatively predict the liquefaction depth. Here the word 'conservative approach' is misleading, as there are chances that some other borehole may give much more worst results and the liquefaction depth may change considerably thus eliminating the importance of earlier boreholes.

In order to eliminate this problem of pseudo-conservative results and to give due respect to the data of every borehole, reliability-based assessment has been performed on SPT data. Reliability-based assessment is helpful in dealing with uncertainties associated with the field as well as laboratory data. The data were collected by a single operator using the same equipment in order to reduce the data uncertainty at the site. Data from 25 boreholes are analyzed at each depth and different probabilistic distribution curves (like normal, lognormal, and gamma) are used to find the best suitable curve for each parameter at each depth under consideration. This approach is adopted for all variables associated with the determination of cyclic resistance ratio (CRR) like SPT blow counts, soil density and percentage fines. 95 percentile values of these parameters are determined at each depth thereby converting the data from 25 boreholes to an equivalent single borehole for all analysis purposes. Factor of Safety has been determined using this approach and also compared with the FOS values determined using 50 percentile and 98 percentile values of these parameters.

2 Probabilistic Determination of Field Parameters Required for Liquefaction Assessment

Liquefaction is a state, in which effective shear strength of saturated cohesionless soil is reduced to negligible value for the all engineering purpose due to an increase in pore water pressure. In this condition, the soil tends to behave like a fluid mass. In order to assess liquefaction resistance of soils, estimation of two variables is required: (1) the seismic demand on a soil layer, expressed in terms of CSR (Cyclic Stress Ratio); and (2) the capacity of the soil to resist liquefaction, expressed in terms of CRR (Cyclic Resistance Ratio). CSR is evaluated using the formulation given by Seed and Idriss [6], which involves parameters like PGA and density of soil layers determined from laboratory tests. On the other hand, CRR is evaluated using the formulations given by NCEER [5] modified from Seed et al. [8]. The parameters involved in the evaluation of CRR are determined from in situ field tests like corrected SPT blow counts and from laboratory tests like percentage fines and soil density.

Data obtained from the field tests have randomness due to various reasons like local conditions, faulty equipment, manual errors, various environmental factors, etc. Randomness means that more than one outcome is possible, or it can be said that usually a range of measured or observed values is possible. Also, it is possible that within this range some values may occur more frequently than others. Representation of a random variable or mathematical modeling is thus a major step in any probabilistic formulation, which needs to be conducted systematically.

The first step in converting all boreholes to an equivalent single borehole is to group the data of all boreholes at each depth. Data from each borehole like corrected SPT blow counts; percentage fines and soil densities are grouped together for each depth and analyzed separately. Figure 1 shows the distributed data of SPT values at the same depth from 25 different boreholes in a particular area after making the required corrections as suggested by NCEER [5]. While designing any structure values of different soil parameters are required at a depth where the structure will rest and also below it. To quantify such data and extract the meaningful value from this is not an easy task because the mean value of data will not give the overall estimate of the data obtained. As can be seen from Fig. 1, the mean of corrected N-values at depth 16.5 m is 32.48. However, at this depth, the corrected SPT blow counts have maximum and minimum values as 79 and 12, respectively. This shows a considerable difference between the mean values with extreme values of SPT at a particular depth. It is therefore very much essential to estimate a value that can be used conservatively for the estimation of required parameters for further evaluations. Thus, probabilistic determination is used to determine the non-exceedance probability value at a particular depth from available borehole data.

Corrected N-values as shown in Fig. 1 are arranged in Table 1 for better understanding and evaluation of essential parameters. In Table 1, the maximum value for SPT is 79 and the minimum value is 12. The information on minimum and maximum value is very useful, but if one of these values is used for design purposes or for the

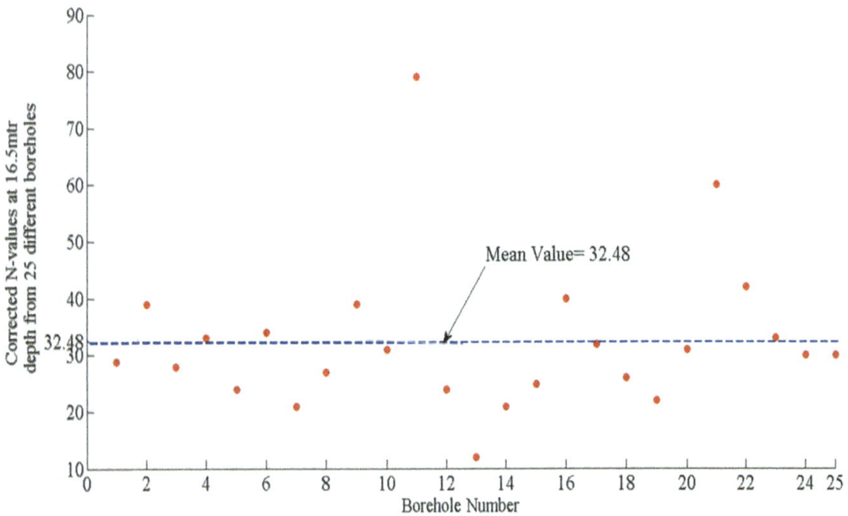

Fig. 1 Plot of corrected N-values obtained from 25 different bore holes at a depth of 16.5 m

Table 1 Corrected SPT blow counts at 16.5 m depth

Bore hole No.	$(N_1)_{60}$	Bore hole No.	$(N_1)_{60}$	Bore hole No.	$(N_1)_{60}$	Bore hole No.	$(N_1)_{60}$	Bore hole No.	$(N_1)_{60}$
BH-1	29	BH-6	34	BH-11	79	BH-16	40	BH-21	60
BH-2	39	BH-7	21	BH-12	24	BH-17	32	BH-22	42
BH-3	28	BH-8	27	BH-13	12	BH-18	26	BH-23	33
BH-4	33	BH-9	39	BH-14	21	BH-19	22	BH-24	30
BH-5	24	BH-10	31	BH-15	25	BH-20	31	BH-25	30

determination of other ground parameters, it may not predict the actual behavior of soil, since the actual N-value will probably be something different. The behavior of soil, in that case, will be either underestimated or overestimated. Furthermore, it is possible that these values may not be the absolute minimum or maximum values. If more samples are collected, the minimum and maximum values can change. Thus, working with minimum and maximum values may not be desirable and are usually not accepted as a standard design practice.

To overcome the deficiency of minimum–maximum value approach, one common practice is to go for average or mean or expected mean values of SPT data obtained for design purpose. For 25 samples of SPT values in Table 1, the mean N-value is 32.48. However, the mean value alone cannot provide the entire information. Obviously, information on the dispersion of the values with respect to the mean is needed. The measure of dispersion can be expressed in terms of variance, standard deviation, or coefficient of variation. It is also helpful to know if the dispersion is symmetrical or unsymmetrical. The degree of symmetry can be measured using the concept of skewness.

Therefore, for corrected SPT blow counts at 16.5 m depth as given in Table 1, the following information can be calculated. Mean is 32.48; Variance is170.81; Standard Deviation is 13.07; COV is 0.402; Skewness is 0.433×10^4; Skewness Coefficient is 1.941. Thus for the given corrected N-value data, the uncertainty in SPT values is relatively high, the randomness is unsymmetrical, and the dispersion is more above the mean than below the mean.

It is always better to depict data in pictorial form. Plotting of information graphically in form of histograms describes it completely. Figure 2 and Fig. 3 shows the

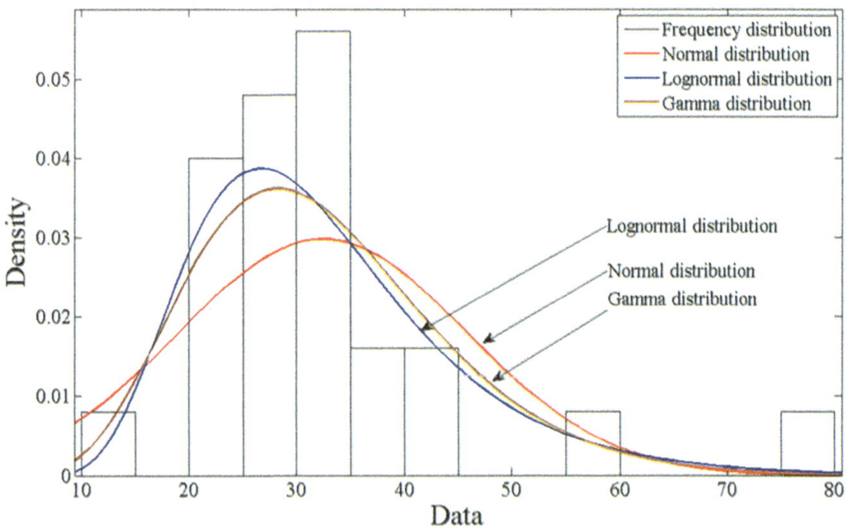

Fig. 2 Histogram and frequency distribution curve of SPT values at 16.5 m depth from 25 boreholes

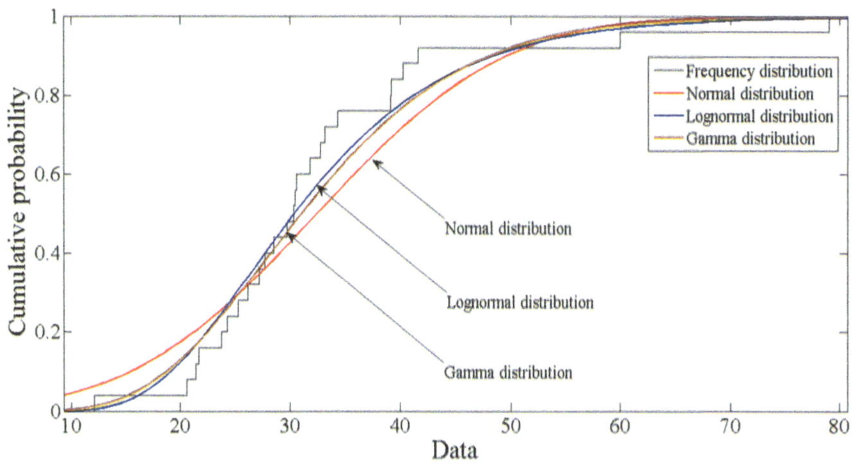

Fig. 3 CDF plot of SPT values at 16.5 m depth from 25 boreholes showing different distributions

PDF and CDF plots respectively for the data given in Table 1 using different distributions. It can be seen from Fig. 2, that it is difficult to deduce the best distribution (among normal, lognormal, and gamma) for the corrected SPT data at 16.5 m depth. In order to find out the 95 percentile value of data from these distributions, it is first required to plot the PDF and CDF of the random data using all the probability distribution functions. In case random data satisfies all the probability functions, it is then required to choose the best function amongst the available functions that best suits the randomness of data. Comparing probability functions with each other can be done by performing statistical tests like the Chi-square test, Kolmogorov–Smirnov test, or Anderson-Darling test.

The K-S and A-D tests are based on the error between the observed and assumed CDF of the distribution, while Chi-square test is based on the error between the observed and assumed PDF of the distribution. Chi-square goodness of fit test requires a large number of data points for the test statistic to converge to its underlying Chi-square distribution, something that K-S and A-D tests do not require. Therefore, K-S test is used to estimate the best distribution available for parameters like corrected N-values, density, and % fines as data from only 25 boreholes are used in this study. K-S test has the advantage over Chi-square test that it is not necessary to divide the data into intervals; thus the error or judgment associated with the number or size of the interval is avoided.

Different probability distribution functions are used to determine the best fit curve for each parameter at each depth. 95 percentile values of corrected SPT blow counts, soil density, and percentage fines at each depth are determined from the data of 25 bore holes at each depth. This 95 percentile value of each parameter at each depth becomes an input for an equivalent single borehole. The equivalent single borehole concept is used to assess the liquefaction potential of area with fairly good probability. In case some more boreholes are drilled at later stage, results are not affected drastically.

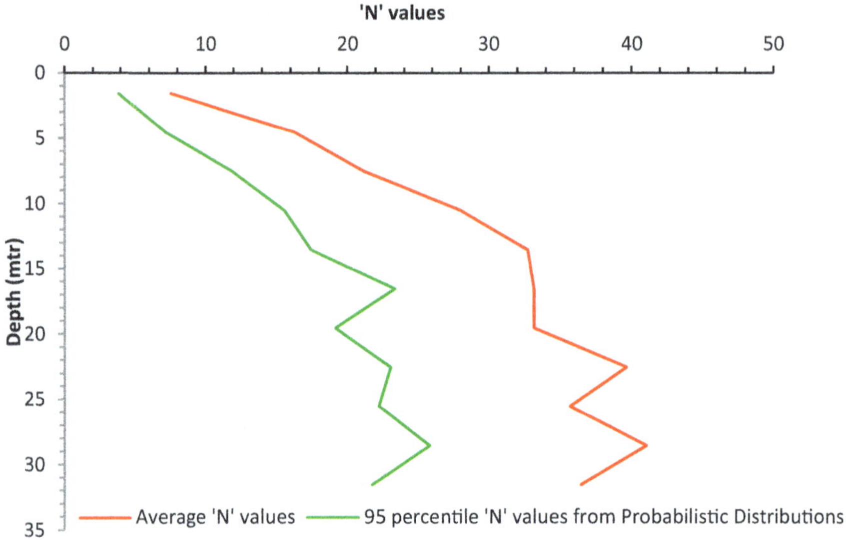

Fig. 4 Comparative plot between corrected SPT blow counts obtained using average values and 95 percentile values using the best probabilistic distribution curve at each depth

Comparison has been made between the 95 percentile values obtained from the best probabilistic function as determined using K-S test and mean values of these parameters at every depth. It can be observed from Fig. 4 that values obtained from probabilistic distribution are more conservative in comparison to mean values of these parameters. While assessing liquefaction potential for critical facilities like nuclear power plants, hydropower plants, military bases, etc., it is always recommended to go for reliability-based approach to assess critical parameters so as to deal with the uncertainties associated with the estimation of such parameters. Therefore, it is beneficial to use reliability method to assess liquefaction potential rather than going for deterministic or mean value approach.

Figure 4 shows the graphical comparison between corrected average N-values and corrected 95 percentile N-values at each depth. It can be observed from Fig. 4, that N-values obtained from probabilistic distributions are almost half of the N-values obtained from mean of N-values at each depth. This is due to the large variability present in N-values collected from different boreholes at each depth. Minimum and maximum N-values obtained at each depth from different boreholes have large variations which affect the mean value as already explained above. On the other hand, values obtained from probabilistic distribution are 95 percentile values which give more reliable values as compared to average values. A similar comparison is shown for calculated values of density and percentage fines content in Fig. 5 and Fig. 6, respectively.

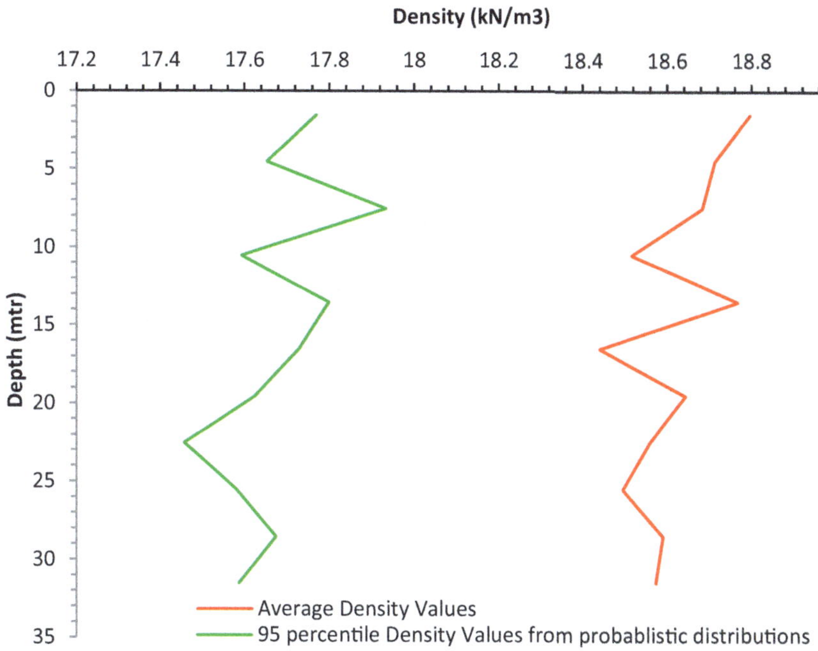

Fig. 5 Comparative plot between soil densities obtained from average values and 95 percentile values using the best probabilistic distribution curve at each depth

3 Assessment of CRR and CSR Using Inputs from Probabilistic Distributions

SPT was conducted in 25 boreholes up to the depth of 110 m from ground level. However, Liquefaction assessment is carried out up to 31.5 m depth from ground level. The probability of liquefaction decreases drastically below 20 m depth due to the overburden pressure of soil layers. The soil at shallower depths is more prone to liquefaction as compared to the soil layers at deeper depths. Laboratory tests were conducted on the soil samples obtained from the locations of SPT to determine percentage fines, dry density, and natural moisture content of the soil samples. Water table in the area varies from 2 to 3 m depth below ground level. However, for analysis purpose, water table is considered at ground level.

SPT values (N) as obtained from the in situ testing are corrected to obtain the $(N_1)_{60}$ by using correlation given in NCEER [5]. These corrected N-values from 25 boreholes are then used to determine the 95 percentile values at each depth as explained earlier. A similar assessment has been made for soil densities and percentage fines. Corrected SPT values $(N_1)_{60}$ are further corrected to account for fines content present in the soil using the approach given in NCEER [5].

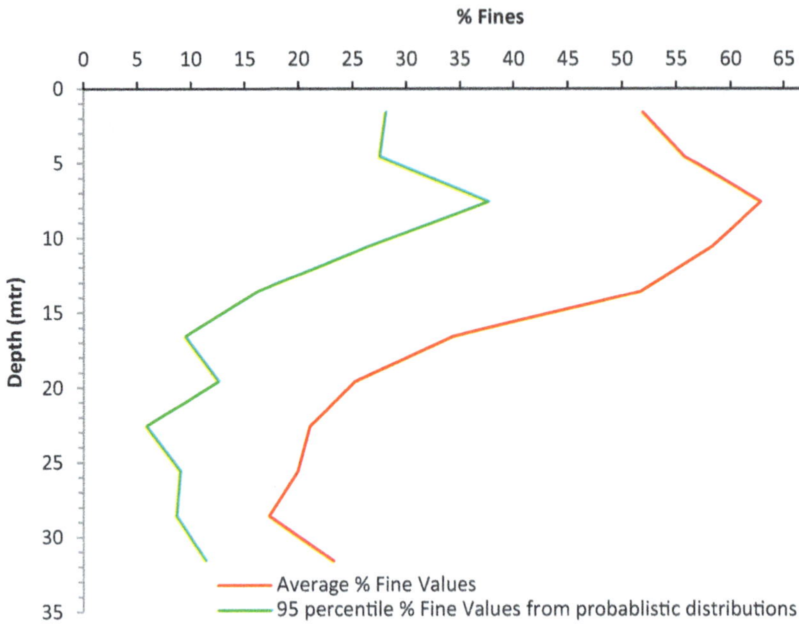

Fig. 6 Comparative plot between % Fines obtained using average values and 95 percentile values using the best probabilistic distribution curve at each depth

3.1 Evaluation of CRR

SPTs are generally preferred to evaluate liquefaction potential of soil due to the vast database available and past experiences. The added advantage of SPT is that soil samples can be retrieved during the process and can be tested at the laboratory to assess important soil properties. CRR is evaluated using the curve recommended by NCEER [5]. This curve gives CRR for a magnitude 7.5 earthquake. The clean-sand base curve as suggested by NCEER [5] can be approximated with Eq. (1). Scaling factors are used to adjust the CRR for different earthquake magnitudes. Different formulations for scaling factors are given by various researchers.

$$\mathrm{CRR}_{7.5} = \frac{1}{34 - (\mathrm{N}_1)_{60}} + \frac{(\mathrm{N}_1)_{60}}{135} + \frac{50}{[10.(\mathrm{N}_1)_{60} + 45]^2} - \frac{1}{200} \tag{1}$$

The above equation is valid for $(N_1)_{60} < 30$. NCEER [5] suggest that for $(N_1)_{60} \geq 30$, clean sands are too dense to liquefy and hence termed as non-liquefiable. Table 2 shows the data of an equivalent single borehole that will be used to assess CRR at different depths. This data is derived from 25 boreholes using the probabilistic distribution curves as explained above. This is the 95 percentile data for corrected SPT blow counts, soil density, and percentage fines. In order to assess the liquefaction potential of the same area more conservatively even 98 percentile values of these

Table 2 Data for equivalent borehole to assess liquefaction potential

Depth (m)	95 percentile values of $(N_1)_{60}$ for equivalent borehole	95 percentile values of % fines for equivalent borehole	95 percentile values of field density for equivalent borehole (kN/m^3)
1.5	3.8	28.07	17.8
4.5	7.2	27.49	17.7
7.5	11.9	37.65	17.9
10.5	15.5	26.34	17.6
13.5	17.4	16.15	17.8
16.5	23.3	9.39	17.7
19.8	19.1	12.56	17.6
22.5	23.0	5.84	17.5
25.5	22.2	8.98	17.6
28.5	25.8	8.62	17.7
31.5	21.7	11.40	17.6

parameters can be adopted. 95 percentile values of any parameter suggest that the possibility of exceeding this value is only 5%. Therefore, evaluation of Factor Safety using this approach suggests that there is only a 5% probability that liquefaction in this area can occur beyond the calculated depth. Thus all uncertainties associated with the parameters required for estimation of CRR can be dealt with using this approach with fairly good probability.

3.2 *Evaluation of CSR*

Seismic demand on the soil layer, expressed in terms of CSR is calculated using the formulation given by Seed and Idriss [6] as given in Eq. (2).

$$\mathrm{CSR} = 0.65\left(\frac{a_{max}}{g}\right)\left(\frac{\sigma_{vo}}{\sigma'_{vo}}\right) r_d \tag{2}$$

where a_{max} = peak horizontal acceleration at the ground surface generated by the earthquake; g = acceleration of gravity; σ_{vo} and σ'_{vo} are total and effective vertical overburden stresses, respectively; and r_d = stress reduction coefficient. This coefficient accounts for the flexibility of the soil profile.

Overburden and effective overburden stresses are calculated using soil density of equivalent single borehole as determined using the probabilistic approach explained earlier. CSR has been calculated for peak ground acceleration values of 0.2g.

4 Evaluation of Factor of Safety (FOS)

Factor of safety is calculated as the ratio of CRR and CSR. Soil liquefaction and subsequently depth of treatment will be assessed on the basis of this factor of safety. As per NCEER [5] soil is non-liquefiable for FOS > 1. The CRR curves (or clean-sand base curves) for SPT apply only to magnitude 7.5 earthquakes. To adjust the clean-sand curves to magnitudes smaller or larger than 7.5, Seed and Idriss [7] introduced correction factors termed "Magnitude Scaling Factors (MSFs)". These factors are used to scale the CRR base curves upward or downward on CRR versus $(N_1)_{60}$ plots as given in Eq. (3).

$$CRR' = (CRR_{7.5}).MSF \tag{3}$$

where CRR′ is CRR for earthquake magnitudes other than 7.5.

To illustrate the influence of magnitude scaling factors on calculated hazard, the equation for factor of safety (FOS) against liquefaction is written in terms of CRR, CSR, and MSF as follows:

$$FOS = \left(\frac{CRR_{7.5}}{CSR}\right).MSF \tag{4}$$

where CSR is calculated cyclic stress ratio generated by the earthquake shaking; $CRR_{7.5}$ is cyclic resistance ratio for magnitude 7.5 earthquakes as calculated earlier.

NCEER [5] proposed MSF to adjust liquefaction resistance for magnitudes (M) other than 7.5 as given in Eq. (5).

$$MSF = {10^{2.24}}/{M_w^{2.56}} \tag{5}$$

where M_w is the Magnitude of Earthquake used in the analysis.

Figure 7 shows the Factor of Safety for equivalent borehole using 50 percentile, 95 percentile and 98 percentile values at a_{max} equal to 0.2g and 7.5 Magnitude of Earthquake.

5 Conclusions

Standard penetration test (SPT) was conducted in 25 boreholes up to the depth of 110 m from the existing ground level. However, analysis for liquefaction potential was done up to 31.5 m depth as the probability of liquefaction potential decreases beyond 20 m depth due to overburden pressure of soil layers. Undisturbed and disturbed soil samples were collected from the same boreholes to determine the soil density and percentage fines of soil. There are many uncertainties involved in the assessment

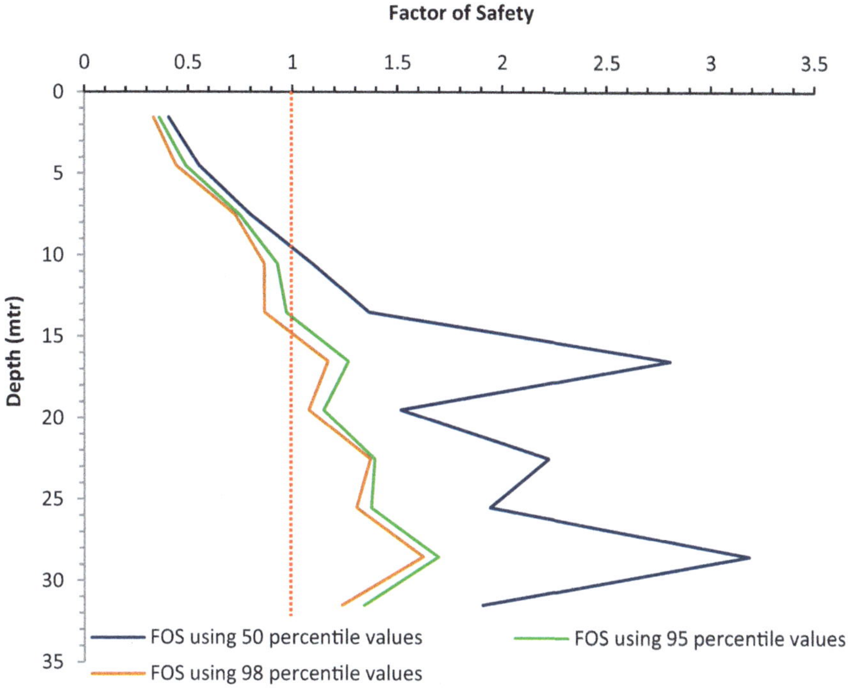

Fig. 7 Comparison between FOS using parameters obtained from different percentile values

of input parameters involved in the determination of liquefaction potential (like N-values, density of soil, %fines of soil). Therefore, to assess these parameters the probabilistic approach has been followed in this study. 95 percentile values of all parameters are calculated based on the best distribution followed by the input parameters obtained from the field as well as laboratory results. These values were then used to convert data from 25 boreholes to an equivalent single borehole, which was used to determine liquefaction potential of the entire area. The analysis was done at a_{max} equal to 0.2g and at 7.5 magnitude of earthquake.

Cyclic resistance ratio (CRR) for this equivalent borehole was determined using the relationship given in NCEER [5] using the 50, 95, and 98 percentile values of different parameters. A comparison was made between the FOS obtained using the different percentile values of these parameters. 95 percentile value means that there are only 5% chances that the value of a particular parameter will go beyond the calculated value. Therefore, higher percentile values predict more conservative values as compared to lesser percentile values. This can also be seen from Fig. 7 that a more conservative assessment of Factor of Safety was done using higher percentile values. The major advantage of this approach is that results from all boreholes have been used to assess the values of parameters with a certain probability. In case some more bore holes will be drilled at a later stage, values from these boreholes will not affect the overall liquefaction potential depth of that area significantly.

References

1. BIS (1981) IS 2131:1981, Method for standard penetration test for soils. Bureau of Indian Standards, New Delhi
2. Cetin KO, Seed RB, Kiureghian AD, Tokimatsu K, Harder LF Jr, Kayen RE, Moss RE (2004) Standard penetration test-based probabilistic and deterministic assessment of seismic soil liquefaction potential. J Geotech Geoenviron Eng 130(12):1314–1340
3. Haldar A, Mahadevan S (2000) Probability, reliability and statistical methods in engineering design. Wiley, New York
4. Idriss IM, Boulanger RW (2008) Soil liquefaction during earthquakes. Earthquake Engineering Research Institute, USA
5. NCEER (1997) Proceedings of the NCEER workshop on evaluation of liquefaction resistance of soils. National Centre for Earthquake Engineering Research, Report No. NCEER-97-0022, State University of New York at Buffalo, Buffalo, NY
6. Seed HB, Idriss IM (1971) Simplified procedure for evaluating soil liquefaction potential. Am Soc Civ Eng J Soil Mech Found Div 97(SM9):1249–1273
7. Seed HB, Idriss IM (1982) Ground motions and soil liquefaction during earthquakes. Monograph series. Earthquake Engineering Research Institute, University of California, Berkeley, CA
8. Seed HB, Tokimatsu K, Harder LF, Chung R (1985) Influence of SPT procedures in soil liquefaction resistance evaluations. Am Soc Civ Eng J Geotech Eng Div 111(12):861–878
9. Youd TL, Idriss IM (2001) Summary report from the 1996 NCEER and 1998 NCEER/NSF workshops on evaluation of liquefaction resistance of soils. J Geotech Geoenviron Eng 127(4):297–313

Building Performance and Geotechnical Failures in 7.5M Palu Earthquake and Tsunami 28 September 2018

Sumedha Koul and Harish Mulchandani

Abstract On 28 September 2018, a 7.5 magnitude earthquake and subsequent tsunami hit Palu and Donggala in Central Sulawesi, Indonesia killing at least 2,245 people. Thousands of people were missing and over 10,000 were injured, of whom over 4,000 were with severe injuries. Nearly 75,000 were displaced in the three most affected areas: Donggala, Palu City, and Sigi. This paper discusses the performance of reinforced concrete buildings, bridges, port facilities, and lifelines in the Palu Earthquake. It was observed that a number of residential and commercial buildings were partially or fully collapsed and many sustained significant structural damages. The iconic twin steel arch cable-suspended Palu Bridge IV over the Palu River was collapsed. Lateral spreading induced by liquefaction resulted in huge housing damages and farm fields. It was seen from field investigations that possible reasons for building failure include the lack of confinement bars, improper confinement in beam and column joints, Ssrong beams weak columns and presence of soft story in multi-story buildings.

Keywords Palu earthquake · Building performance · Geotechnical failure

1 Introduction

The Palu earthquake with a magnitude of 7.5 Mw occurred on 28 September, 2018, at 18:02 WITA time. Palu-Koro Fault experienced a strike with a mechanism of "strike-slip". The earthquake epicenter is located 72 km north of the city of Palu with an earthquake depth of 10 km. This earthquake triggered a tsunami, a shift in the fault (rupture) along The Palu-Koro Fault and its secondary fault, and the occurrence of extensive and massive landslides in the Petobo, Jono Oge, Sibalaya, Balaroa and Lolu villages [1–7]. There is evidence of earthquake rupture occurring due to supershear velocities [8]. The earthquake triggered an instantaneous tsunami that reached up to the height of 5 m [9, 10].

S. Koul · H. Mulchandani (✉)
Department of Civil Engineering, Birla Institute of Technology & Science Pilani, Pilani, India
e-mail: mulchandani1010@gamil.com

T. G. Sitharam et al. (eds.), *Local Site Effects and Ground Failures*, Lecture Notes in Civil Engineering 117, https://doi.org/10.1007/978-981-15-9984-2_29

Major building failures occurred in the region were due to soft storey shear failure, beam-column joint failure, buildings with strong beam and weak column. It leads to complete collapse, discontinuity in lateral load resisting systems, and due to landslides resulting from wet rice fields [11].

2 Building Performance

Many of the buildings suffered huge losses because of the earthquake. A few notable multi-storey buildings included the Tatura shopping mall, Roa Roa Hotel and several mosques. The major reasons for structural failure have been listed below.

2.1 Lack of Confinement in the Joints

The majority of buildings suffered beam-column joint failures. It occurred due to a lack of confinement reinforcement bars at the joints. The design check for joints is not very crucial for buildings designed to sustain normal gravity loads. However, this becomes an essential criterion for buildings situated in earthquake-prone areas. Most of the reinforced concrete structures in Palu also collapsed due to the same reason.

Shown below are the images of various failed joints of the Tatura shopping complex. The shopping complex was a four-storeyed building that suffered a partial collapse of the roof and third floor due to the failure of multiple columns and beam-column joints.

The absence of stirrups in beam column joints led to buckling of the vertical column reinforcement due to which unconfined concrete in these joints failed. The lack of shear stirrups also caused many columns to fail due to shear. Some interior columns also collapsed due to shear or flexure failure in the hinging region. However, there was no damage in the frame or beams due to these columns as can be seen in Fig. 1.

The Dunia Baru restaurant, a three-storeyed building, close to the shopping complex also suffered complete collapse since the bars used for reinforcement were smooth. Despite the complete collapse of the structure, the beams were not damages hinting toward a strong beam weak column condition like in the Tatura shopping complex. Similarly, the Mecure hotel also faced structural damages due to the lack of shear stirrups in the beam-column joints and the strong beam weak columns condition.

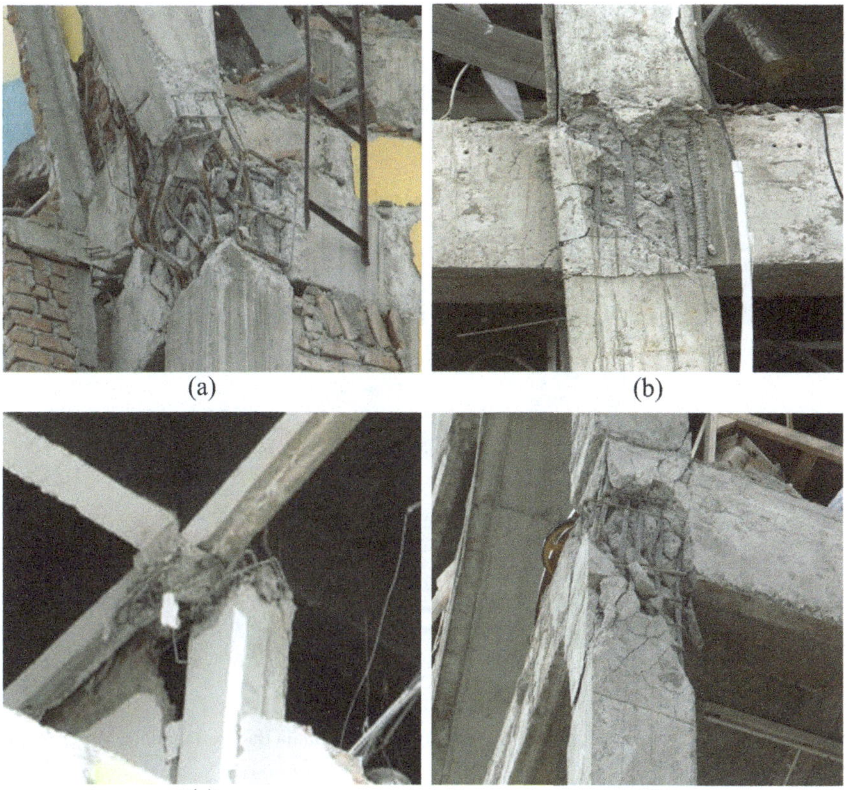

(a) (b) (c) (d)

Fig. 1 Beam-column joint failure due to lack of confinement bars

2.2 *Soft Story Failure*

A soft storey refers to a multi-storey building in which at least one floor has openings in places like windows, wide doors, etc., where shear walls would have been necessary for stability in earthquake engineering design. Technically, a soft storey can be defined as if a floor has a stiffness less than 70% stiffness of subsequent floor or less than 80% stiffness in comparison to the average stiffness of the upper three floors. Such buildings are more vulnerable to damage due to a moderate or severe earthquake. Due to the inadequate bracing, the floors are less resistant to the lateral forces due to the earthquake.

Many buildings have collapsed in Palu due to this phenomenon. The Tatura shopping complex suffered from many columns and joint failures since these were not appropriately braced. The failure of the columns led to the partial collapse of the third floor and the roof. The figure below shows the damaged building.

Fig. 2 Soft storey failure at **a** Ramayana Shopping centre, **b** Dunia Baru restaurant, **c** Mecure Hotel at Palu Indonesia [1]

The Dunia Baru restaurant, a three-storeyed building was also completely collapsed due to the failure of the columns which led to the failure of the floors as can be seen in the image. The Mecure Hotel, a five-storeyed building, also had weak columns that failed leading to the complete collapse of the ground floor and partial collapse of all the other floors as shown in Fig. 2.

3 Damage Due to Lateral Spreading

The earthquake ground shaking induced liquefaction over three large areas in and around Palu City, namely the Balaroa neighbourhood, the Petobo sub-district and Jono Oge Village. Despite the slope being as low as 1%, these areas experienced lateral spreading that extended up to 3.5 km.

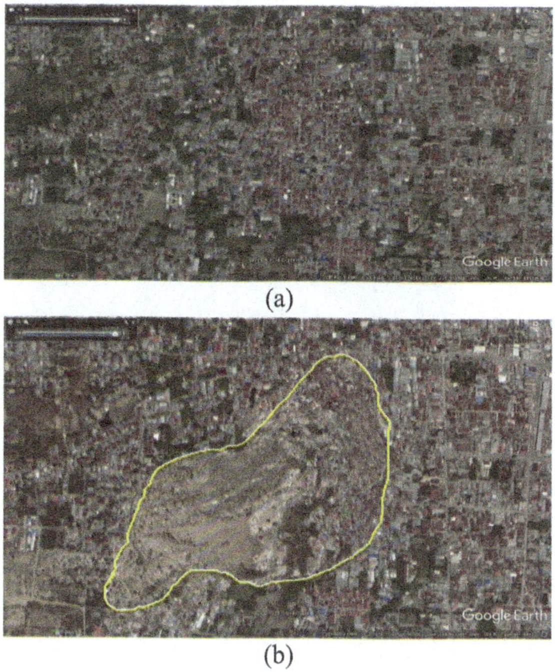

Fig. 3 Google Earth images of Balaroa region of the Palu **a** before the lateral spreading **b** after the lateral spreading [1]

3.1 Balaroa

Balaroa is a district in Palu which experienced a lateral spread of up to 1000 m on an average slope of 3.4%. Figure 3 shows an aerial view of the site before and after the lateral spreading. The upper layer of the soil along with the structures moved downslope due to the liquefaction of the soil caused due to the earthquake.

3.2 Petobo Sub-district

The Petobo sub-district lies towards the south of the Palu airport and 7.2 km east of Palu-Koro fault line. This region is a residential area that experienced destruction due to liquefaction which had a lateral spreading of up to 2200 m on an average slope of only 2%. Lateral spreading lead to massive damage to the building in the region, which were either buried or moved along with the liquefied soil. Figure 4 shows the aerial view of the site before and after the earthquake transpired.

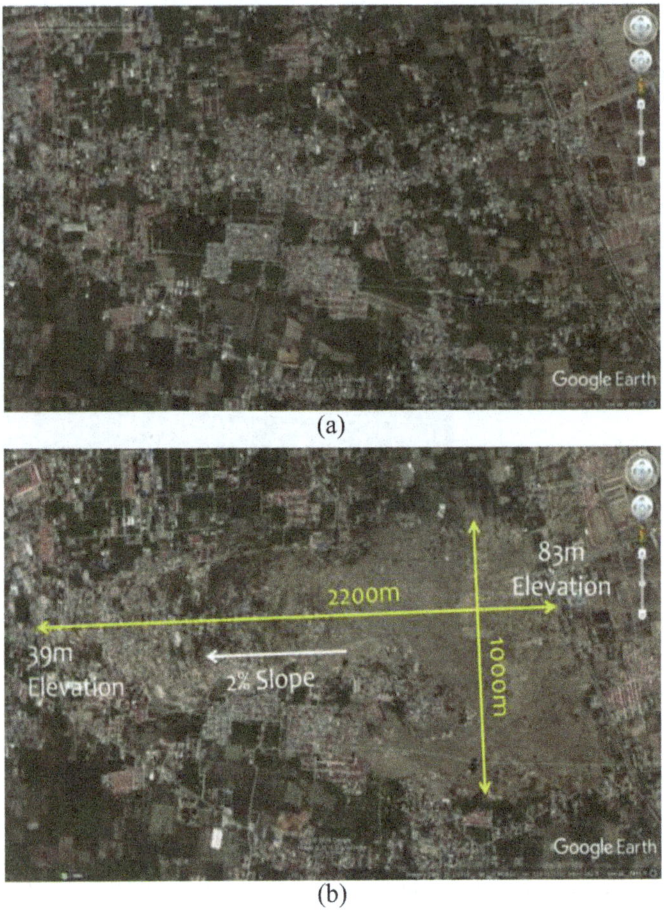

Fig. 4 Google Earth images of Petobo sub-district region of the Palu **a** before the lateral spreading **b** after the lateral spreading [1]

3.3 Jono Oge Village

Due to the liquefaction, the lateral spread in this area spread up to 3500 m on a terrain that had an average slope of only 1.2%. Despite it being a rural setup, many structures were destroyed because of the severity of the slide. Figure 5 shows the village before and after the slide occurred.

(a)

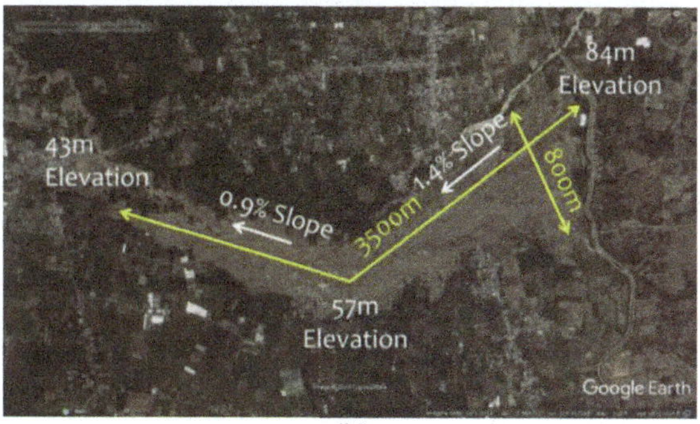

(b)

Fig. 5 Images of Jono Oge village of the Palu **a** before the lateral spreading **b** after the lateral spreading [1]

4 Damage to Palu Bridge IV

Palu bridge IV was a prominent steel double arch suspension bridge that connected the east and west sides of the bay of Palu. This bridge, as per eye witness, got damaged before the tsunami arrived. Some of the possible reasons for its failure may be due to the lateral collapse because of the impact caused by heavy concrete deck suspended from the arches by steel cables on the base of the arches or due to the lateral deck motion which could have been sufficient to fail the structure. Figure 6 shows the images of Palu bridge IV before and after damage.

5 Conclusion

This paper discusses the damages caused to infrastructure due to the Palu earthquake which hit Indonesia on September 28th, 2018. In this study, the major focus has been laid on the Tatura shopping complex, the Dunia Baru restaurant and, the Mecure hotel amongst other RCC buildings that were damaged. Particular emphasis was put on these buildings since all three of them failed more or less due to the same reasons

Fig. 6 Palu bridge IV **a** before the earthquake **b** after the earthquake [1]

i.e. weak-column-strong-beams and the lack of shear reinforcements in the beam-column joints. Along with these, the paper also lays emphasis on damages caused due to the liquefaction of the soil because of the earthquake. Balaroa, Petobo and Jono Oge village were completely washed out by great distances despite the gentle slope highlighting the magnitude of the earthquake.

By studying the causes of damage of structures, one can understand the need for earthquake-resistant designs. We can reduce the damage caused to life to infrastructure by an earthquake, by carrying out proper geotechnical analysis of the area and studying its liquefaction tendency and designing the foundations accordingly.

References

1. Robertson I et al (2019) StEER—Palu earthquake and tsunami, Sulawesi, Indonesia: Field assessment team 1 (FAT-1) early access reconnaissance report (EARR). DesignSafe-CI. https://doi.org/10.17603/DS2JD7T
2. Harnantyari AS, Takabatake T, Esteban M, Valenzuela P, Nishida Y, Shibayama T, Kyaw TO (2020) Tsunami awareness and evacuation behaviour during the 2018 Sulawesi Earthquake tsunami. Int J Disaster Risk Reduct. https://doi.org/10.1016/j.ijdrr.2019.101389
3. Putra PS, Aswan A, Maryunani KA, Yulianto E, Kongko W (2019) Field survey of the 2018 Sulawesi Tsunami deposits. Pure Appl Geophys. https://doi.org/10.1007/s00024-019-02181-9
4. Paulik R, Gusman A, Williams JH, Pratama GM, Lin S, Prawirabhakti A, Suwarni NWI (2019) Tsunami hazard and built environment damage observations from Palu City after the September 28th 2018 Sulawesi Earthquake and Tsunami. Pure Appl Geophys. https://doi.org/10.1007/s00024-019-02254-9
5. Widiyanto W, Santoso PB, Hsiao S-C, Imananta RT (2019) Post-event field survey of September 28th 2018 Sulawesi Earthquake and Tsunami. Nat Hazards Earth Syst Sci Discuss. https://doi.org/10.5194/nhess-2019-91
6. Syamsidik B, Umar M, Margaglio G, Fitrayansyah A (2019) Post-tsunami survey of the September 28th 2018 tsunami near Palu Bay in Central Sulawesi, Indonesia: impacts and challenges to coastal communities. Int J Disaster Risk Reduct. https://doi.org/10.1016/j.ijdrr.2019.101229
7. Rajindra R, Wekke IS, Sabara Z, Pushpalal D, Samad MA, Yani A, Umam R (2019) Diversity, resilience, and tragedy: three disasters in Palu of Indonesia. Int J Innov Creat Change

8. Socquet A, Hollingsworth J, Pathier E, Bouchon M (2019) Evidence of supershear during the 2018 magnitude 7.5 Palu earthquake from space geodesy. Nat Geosci. https://doi.org/10.1038/s41561-018-0296-0
9. Sassa S, Takagawa T (2019) Liquefied gravity flow-induced tsunami: first evidence and comparison from the 2018 Indonesia Sulawesi earthquake and tsunami disasters. Landslides. https://doi.org/10.1007/s10346-018-1114-x
10. Carvajal M, Araya-Cornejo C, Sepúlveda I, Melnick D, Haase JS (2019) Nearly instantaneous tsunamis following the Mw 7.5 2018 Palu Earthquake. Geophys Res Lett. https://doi.org/10.1029/2019GL082578
11. Bradley K, Mallick R, Andikagumi H, Hubbard J, Meilianda E, Switzer A, Hill EM (2019) Earthquake-triggered 2018 Palu Valley landslides enabled by wet rice cultivation. Nat Geosci. https://doi.org/10.1038/s41561-019-0444-1

Numerical Modeling of Three-Tiered Reinforced Soil Wall with Different Offset Distances Subjected to Dynamic Excitation

Sudipta Sikha Saikia and Arup Bhattacharjee

Abstract Reinforced retaining wall using geogrids is an effective method to deal with high and steep soil slope under complicated geological engineering conditions. There have been comparatively lesser studies on multi-tiered walls due to its limited application. This paper aims at making a comparative study of the behavior of three-tiered reinforced soil wall with small and large offset distances subjected to seismic excitations. A 2.8 m high numerical model of reinforced soil wall is developed using finite element software PLAXIS 2D. The numerical model is subjected to dynamic excitations of 0.4 g Kobe earthquake and results of the response of the numerical model are validated with shake table tests results of Ling et al. [7]. A 9 m high three-tiered wall with small offset distances (i.e., 0.5 m and 0.75 m) and large offset distances (1.5 m and 2.5 m) is simulated with validated model parameters. The tiered walls are subjected to seismic excitations of the Kobe earthquake at a peak acceleration of 0.4 g and the variation of lateral pressure, maximum reinforcement loads, and acceleration amplification factors of three-tiered walls with various offset distances are compared. The results showed that the lateral stress, maximum reinforcement load, and acceleration amplification factor decreases with the increasing tier offset.

Keywords Reinforced retaining wall · Finite element model · Dynamic excitation

1 Introduction

Soil reinforcement is a modern and efficient technique for improving the mechanical properties of soil, which takes advantage of the frictional interaction between the soil and reinforcement. The most common application of reinforced earth is in the construction of geosynthetic reinforced soil (GRS) retaining walls. Retaining walls are the structures that are built to retain vertical or nearly vertical earth banks or any

S. S. Saikia (✉) · A. Bhattacharjee
Jorhat Engineering College, Jorhat 785007, India
e-mail: sudiptasikhasaikia@gmail.com

A. Bhattacharjee
e-mail: bhatta_arup@yahoo.com

T. G. Sitharam et al. (eds.), *Local Site Effects and Ground Failures*, Lecture Notes in Civil Engineering 117, https://doi.org/10.1007/978-981-15-9984-2_30

other materials (Murthy [1]). Geo-synthetically reinforced soil wall is cost-effective, durable, and perform better during earthquakes as compared to conventional gravity retaining walls. They are also suitable for uneven settlements and can adapt to new conditions. The tensile stress in the reinforcement increases rapidly with height while designing single-tiered walls. To take advantage of both the aesthetics and the economics of GRS walls while considering high heights, multi-tiered walls are often used.

The study of multi-tiered GRS wall has not achieved much progress as compared to the single-tiered GRS wall. Though few guidelines and numerical studies are available [1–8] etc., for the design of multi-tiered walls but still these are not adequate for the execution of multi-tiered reinforced soil wall in different conditions (like fill height, quality of retaining soil, etc.).

The objective of the paper is to understand the behavior of three-tiered reinforced soil walls with small as well as large offset distances under dynamic loading conditions and making a comparative study on the variation of lateral stress of backfill, maximum reinforcement load, acceleration amplification factor with tier height through numerical simulations.

2 Development of Numerical Model of Reinforced Soil Wall in PLAXIS 2D

2.1 Full-Scale Model: Background

Ling et al. (2005) [9] conducted large-scale shaking table tests on modular block geosynthetic reinforced soil retaining wall subjected to Kobe earthquake motions. The wall was 2.8 m high, 4 m long, and 2 m wide constructed on a 20 cm thick soil foundation. The facing blocks were 20 cm high, 30 cm deep, and 45 cm wide. The wall was backfilled with medium dense Tokachi port sand ($D_r = 55\%$) and reinforced with PET geogrid. Geogrids length of $h = 205$ cm were placed at vertical intervals of 60 cm and its ultimate strength was 35 KN/m.The foundation soil has the same properties as the backfill soil but at a relative density of 90%. To prevent waves reflecting from the steel walls during shaking, 10 cm thick expanded polystyrene (EPS) boards had been placed at the front and back ends of the steel container. The geometry of the wall is shown in Fig. 1. The wall was subjected to horizontal shakings of Kobe earthquake (1995) scaled to a peak acceleration of 0.4 g.

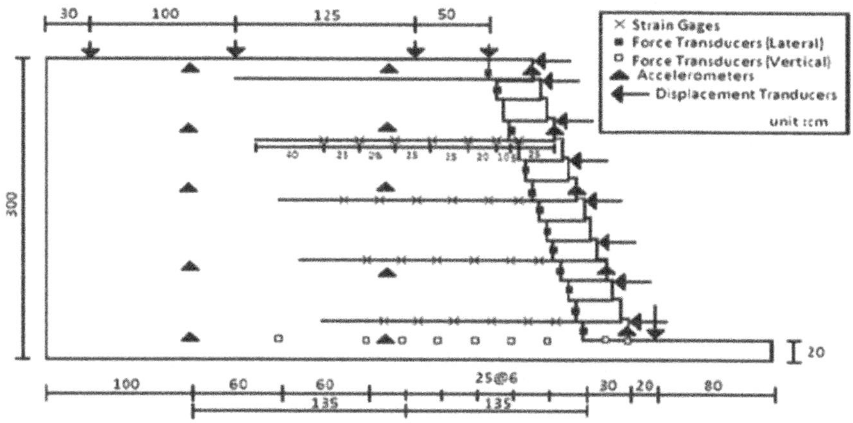

Fig. 1 Geometry of the shake table test used for validation (after Ling et al. [9])

2.2 *Numerical Modeling in PLAXIS 2D*

The components of shake table test as reported by Ling et al. (2005) [10] are simulated using finite element program PLAXIS 2D. The analysis is conducted under two-dimensional plane strain conditions and 15-noded triangular elements are selected [11]. The input model parameters used in PLAXIS 2D are given in Table 1. The model was fixed at the base and assumed to have roller boundaries at the sides. In the finite element model, the bottom boundary is fixed in both horizontal and vertical directions while the side boundaries are fixed in only horizontal directions. To allow for absorption of stress waves, the absorbent boundary condition in PLAXIS 2D is created by placing the boundaries far apart. For dynamic excitation prescribed displacement option is used to define the seismic load. The wall is excited with a maximum horizontal acceleration of 0.4 g. The acceleration time history for 0.4 g is shown in Fig. 2.

The geometry of the model in PLAXIS 2D is shown in Fig. 3.

Table 1 Input model parameters used in finite element program PLAXIS 2D

Properties	Backfill	Foundation	Facing wall	Geogrid
Material model	Mohr–Coulomb	Mohr–Coulomb	Linear elastic	Linear elastic
Elastic modulus (kPa)	156E3	156E3	2E6	
Cohesion (kPa)	1	1		
Angle of friction (∘)	38	40		
Dilatancy angle (ψ)	8	8		
Mass density (kN/m^3)	14.30	14.30		
Poisson's ratio	0.33	0.33	0.2	

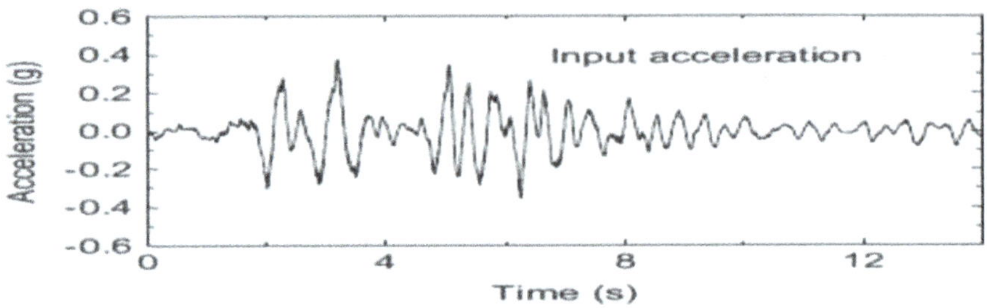

Fig. 2 Input dynamic excitation at peak acceleration of 0.4 g

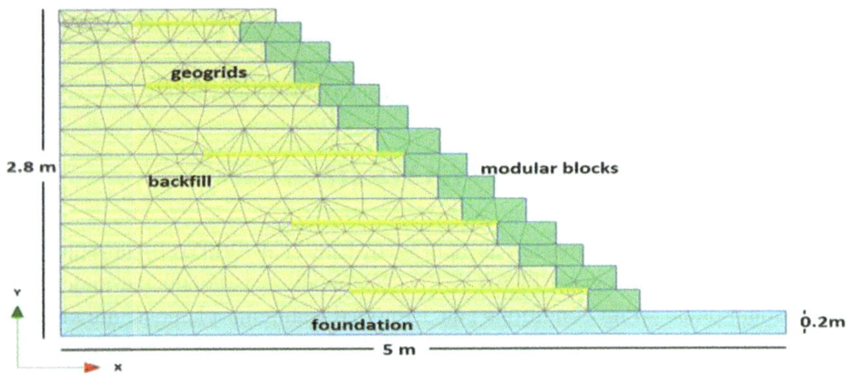

Fig. 3 Geometry of numerical model used for validation

2.3 *Validation of the Numerical Model with the Full Scale Model*

To verify the accuracy of the finite element model developed in the study, the results of the shake table test as reported by Ling et al. [9] are compared with the results obtained from the finite element model.

Horizontal displacement of facing: The comparison between measured and predicted results of horizontal displacement of the facing wall is shown by in Fig. 4.

Fig. 4 Horizontal displacement measured by Ling et al. [9] and FE analysis

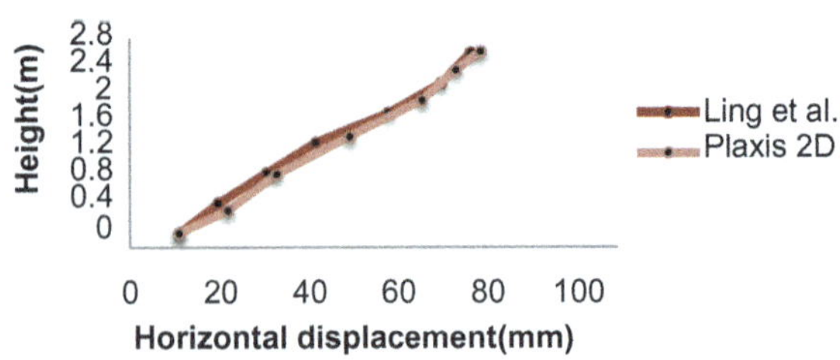

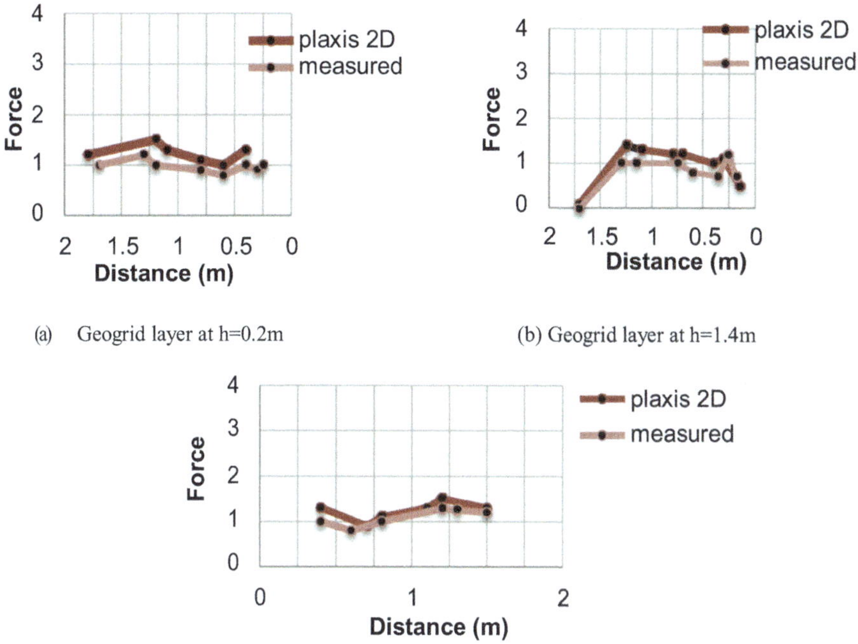

(a) Geogrid layer at h=0.2m (b) Geogrid layer at h=1.4m

(c) Geogrid layer at h=2.6m

Fig. 5 Strain in various primary geogrid layers

The maximum horizontal displacement at the top of the wall is found to be in reasonably good agreement with the measured value as discussed by Ling et al. [9].

Strain in primary geogrid layers: The strain distributions in the geogrid layers at wall height $h = 0.2$ m, 1.4 m, 2.6 m are shown in Fig. 5. The tensile strain measured in the various geogrid layers were converted to tensile force. Although the increase in tensile stress under dynamic load is not proportional to the increase in wall height; but the distribution within the wall height shows similar trends with the measured results.

3 Numerical Modeling of Three-Tiered Reinforced Soil Walls

A 9 m high wall is selected for numerical analysis of three-tiered geosynthetic reinforced soil retaining walls. Three-tiered walls with small offset (0.5 m, 0.75 m) as well as large offset distances (1.5 m, 2.5 m) were considered for the analysis. The

Table 2 Reinforcement lengths for different tiers of a three-tiered wall as per FHWA [3]

No. of tiers	Position of tier	Reinforcement length
Three tier ($H = 9$ m, $H_1 = H2 = 3$ m)	Upper tier	0.7 H2 = 2.1 m
	Middle tier	0.7 H1 = 2.1 m
	Lower tier	0.6 H = 5.4 m

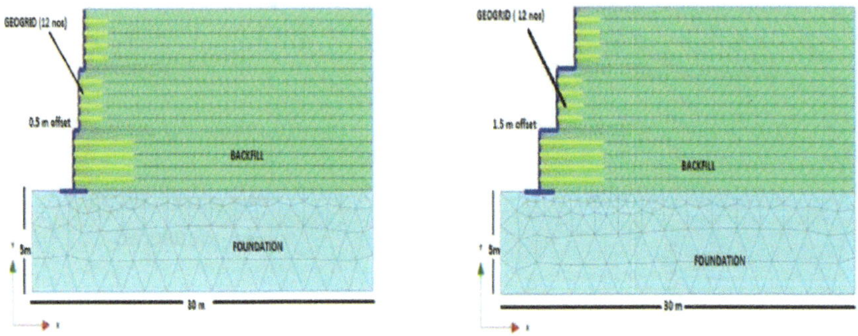

(a) (b)

Fig. 6 Geometry of three-tiered walls with **a** 0.5 m offset, **b** 1.5 m offset distances

reinforcement lengths of the three-tiered walls are calculated as per FHWA [3] and are shown in Table 2.

The different models of three-tiered soil walls are reinforced using 12 number of geogrid layers at a spacing of 0.6 m. The model parameters and boundary conditions used for modeling the three-tiered walls are the same as that used in the validation analyses. The numerical models are subjected to a maximum horizontal acceleration of 0.4 g at the base of the foundation (Fig. 6).

3.1 Comparison of Results

Lateral stress of backfill: Fig. 7 shows the distribution of lateral soil pressure at the face of the wall for small as well as large offsets of three-tiered walls. From the plot, it can be inferred that the lateral stress decreases almost linearly with an increase in wall height, however at the junction of the tiers the magnitude of stresses increases. The maximum lateral stress on the facing wall is not found exactly at the base of the wall, but somewhere near the position of the first geogrid layer. The reason behind such behavior may be due to the rigid interaction between the foundation and the lowermost facing panel as considered in the analysis.

Maximum reinforcement load: Fig. 8 shows the comparison of maximum reinforcement load of three-tiered walls for different offsets. With an increase in the tier offset, the maximum reinforcement load decreases but due to the small reinforcement

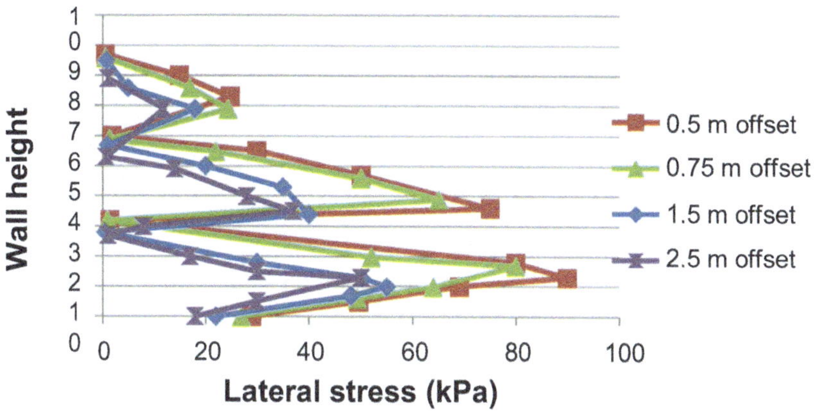

Fig. 7 Variation of lateral stress versus wall height for different offsets of three-tiered walls subjected to Kobe excitations

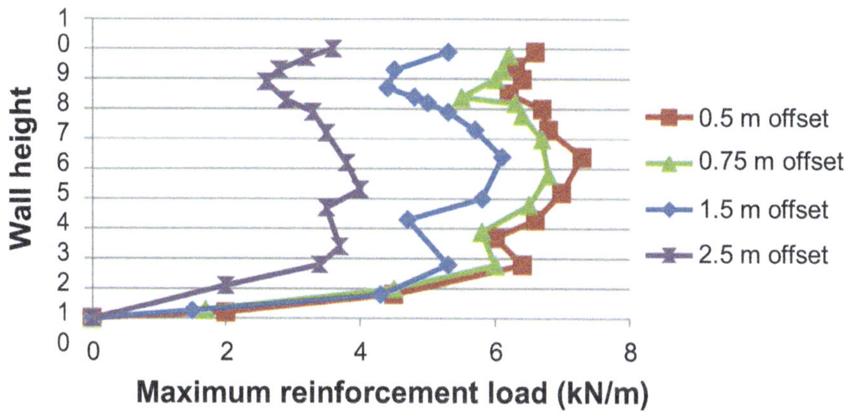

Fig. 8 Maximum reinforcement load versus tier height for different offsets of three-tiered walls subjected to Kobe excitations

length in upper tiers, there is not much decrease in reinforcement load for different offsets in three-tiered walls. The maximum reinforcement load at the mid-height is found to be 7.2 kN/m, 6.9 kN/m, 6 kN/m, and 3.9 kN/m for offset distances of 0.5 m, 0.75 m, 1.5 m, and 2.5 m, respectively.

Acceleration amplification: The amplification factor is given for the ratio of maximum acceleration in the backfill, typically at the top of backfill, to the acceleration at the foundation level. The excitation was given at the base of the wall. After the excitation, the acceleration amplification in the backfill was determined behind the facing blocks. The acceleration amplifications at different elevations of the wall are quantified as root mean square acceleration (RMSA) amplification factor. The

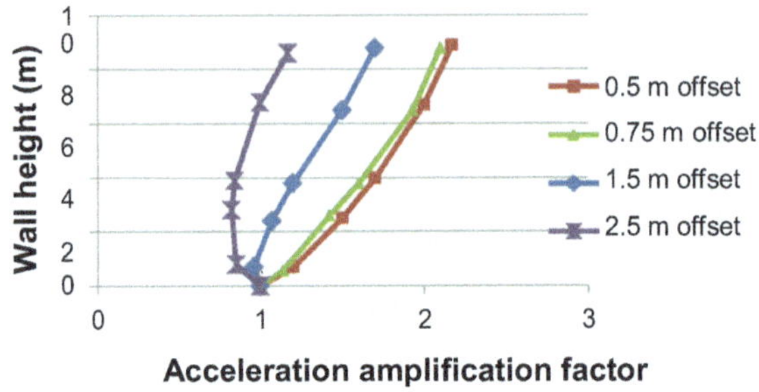

Fig. 9 Variation of acceleration amplification factor with wall height for three-tiered walls subjected to Kobe excitations

RMS acceleration can be calculated as

$$\mathrm{RMS} = \left[1/\mathrm{t_d} \int_0^{\mathrm{t_d}} \mathrm{a(t)^2dt}\right]^{1/2}$$

where $a(t)$ is acceleration time history, t_d is the duration of the acceleration record. dt is the time interval of the acceleration record.

Figure 9 shows the acceleration amplification of three-tiered walls at 10 m away from the toe. Acceleration amplification factors decreased with increasing tier offset in the model walls analyzed in the study. As can be seen in the figure, the amplification factor is high for smaller offset and for higher offsets amplification factor is decreasing.

4 Important Outcomes

The following important outcomes are drawn from the study:

- A numerical model of shake table test was developed in finite element program PLAXIS 2D and validated by comparing the numerical results with physical measurements taken from shake table test on reinforced soil wall as reported by Ling et al. [9]. The closed agreement of numerical results with the physical data indicated that finite element model is capable of simulating the construction sequence of geosynthetic reinforced soil structures.
- For three-tiered walls, the values of lateral stresses are smaller for larger offsets as compared to that of small offset distances. The lateral stress decreases almost linearly with height except at the junction of the tiers where it shows a contradictory behavior.
- With the increase in tier offset, the maximum reinforcement load in the three-tiered walls with different tier-offset decreases. For all the walls considered in the analysis, the maximum reinforcement load is found almost at the midheight of the walls.

- The acceleration amplification factor was found to increase along the height of the wall. For smaller offsets, the values are found slightly higher for three-tiered walls as compared to larger offset distances.

References

1. AASHTO (1998) Standard specifications for highway bridges, American Association of State Highway and Transportation Officials, Washington, DC.
2. FHWA Publication (1995) Geosynthetic design and construction guideline. FHWA Publication No: FHWA-HI-95–038
3. FHWA Publication (2010) Mechanically stabilized earth walls and reinforced soil slopes design and construction guidelines, vol I & II. Publication No. FHWA-NHI-10–024, US Department of Federal Highway Administration (FHWA)
4. Leshchinsky D, Han J (2004) Geosynthetic reinforced multitiered walls. J Geotech Geoenviron Eng 130(12)
5. Liu H, Yang G, Ling HI (2014) Seismic response of multi-tiered reinforced soil retain- ing walls. Soil Dyn Earthquake Eng 61–62, 1–12
6. Mohamed SBA, Yang K-H, Hung W-Y (2014) Finite element analyses of two tier geosynthetic-reinforced soil walls: Comparison involving centrifuge tests and limit equi- librium results. Comput Geotech 61:67–84
7. NCMA (1997) Design manual for segmental retaining walls. In: Collin J (ed) National concrete masonry association, 2nd edn. Herndon, Virginia, USA
8. Yoo C, Kim S (2008) Performance of two-tier geosynthetic reinforced segmental retaining wall under a surcharge load: full scale load test and 3D finite element analysis. Geo-Textiles Geomemb 26:460–472
9. Ling et al (2005) Large scale shaking table tests on modular block reinforced soil retaining walls. J Geotech Geoenviron Eng ASCE
10. Kramer SL (1996) Geotechnical earthquake engineering. Prentice Hall, Upper Saddle River, NJ
11. Kumari S, Bhattacharjee A (2018) Numerical stimulation of tiered reinforced soil retaining wall subjected to dynamic excitation. In: Proceedings of the Indian geotechnical conference 2018, Bengaluru, India

Study on Earth Pressure Distribution and Displacements of Narrow Reinforced Earth (RE) Wall Under Static and Cyclic Loading

Shivani R. Patel, P. J. Mehta, and M. V. Shah

Abstract To study the reduction in lateral earth pressure due to narrow reinforced earth (RE) wall, which is used for various earth retaining structures such as bridge abutments, retaining walls, and also where the available space for the reinforced earth (RE) walls is less than required. In a narrow RE wall, interface connection will be provided to prevent extensive pressure and cracks which are developed in-between existing wall and narrow RE wall. The main objective of this paper is to evaluate the earth pressure distribution for a narrow RE wall under static and cyclic loading considering the elastic (Flexible) and non-elastic (Rigid) behavior of the wall keeping the same relative density. Two major conditions, i.e., rigid boundary condition and flexible condition, which is used to perform a series of load–displacement and load-settlement test on the RE wall model using uniaxial geogrid reinforcement. The load–displacement-settlement is measured by using a conventional high capacity compressive mechanical jack and dial gauges. To validate the experimental results, the earth pressure distribution using Arching theory, Rankine theory, Coulomb theory. From results and analysis, there was a considerable variation is determined in load–displacement characteristics for both flexible and rigid boundary conditions. The percentage reduction in earth pressure was also observed in the case of a flexible RE wall as compared to a rigid narrow RE wall.

Keywords Narrow RE wall · Displacement · Static and cyclic loading

S. R. Patel · P. J. Mehta (✉) · M. V. Shah (✉)
Applied Mechanics Department, L.D. College of Engineering, Ahmedabad 380015, Gujarat, India
e-mail: mehta.priti67@gmail.com

M. V. Shah
e-mail: drmvs2212@gmail.com

S. R. Patel
e-mail: shivani.patel2996@gmail.com

T. G. Sitharam et al. (eds.), *Local Site Effects and Ground Failures*, Lecture Notes in Civil Engineering 117, https://doi.org/10.1007/978-981-15-9984-2_31

1 Introduction

The concept of associating two materials of different strength characteristics to form a composite material of greater strength is quite familiar with civil engineering practices. Reinforced earth technique is commonly used in earth retaining structures such as bridge abutments, retaining walls, steep terrain, etc. It comprises of incorporating reinforcement such as steel strips, fibers, geogrids, etc., into soil mass in such a way that it enhances overall structures stability. Mechanically stabilized earth (MSE) walls, are commonly used structures for retaining the earth, especially in urban areas [1]. These walls are constructed from the bottom-up by placing alternating layers of soil and reinforcement. This creates friction between soil and adjacent reinforcement and thus increasing the strength of the composite material as a whole [2].

As the population has increased, transportation demand has increased so that the development of urban areas has become a priority, which has led to the widening of existing highways to improve traffic flow. Widening has been placed in front of existing stable walls.

The existing state of practice suggests a minimum wall width and MSE reinforcement length equal to 70 percent of the wall height [2]. Due to the high cost of additional rights-of-way and limited space available at job sites, construction of those walls is often done under a constrained space. This leads to MSE walls narrower than those in current design guidelines. Narrow MSE walls are referred to as MSE walls having an aspect ratio L/H (ratio of wall width L to wall height H) of less than 0.7, the minimum value suggested in Federal Highway Administration (FHWA) MSE-wall design guidelines [3]. This design guideline for shored mechanically stabilized earth (SMSE) wall systems [4] provides design methods specifically for wall aspect ratios ranging from 0.3 to 0.7. A minimum value of L/H (L being the length of reinforcement and H is the height of the wall) equal to 0.3 was recommended for constructing narrow walls.

The behavior of narrow walls is different from those of conventional walls because of constrained space and interaction with stable walls. Such differences include the magnitude of earth pressures and the location of failure planes [5].

Woodruff [6] performed a series of centrifugal tests on RE wall with confined space [7]. A total of 24 tests were undertaken of wall aspect ratio ranging from 0.17 to 0.9. He observed that when L/H > 0.6, the wall fails in an internal mode with weaker reinforcement or no failure with stronger reinforcement. If a wall fails, the critical failure plane is linear and passes through the entire reinforced area. When L/H is between 0.25 and 0.6, the wall fails internally in a compound mode with bilinear failure surfaces that pass through the reinforced soil and the edge between the reinforced backfill and stable wall. The inclination angle of the failure plane in a compound mode is less than that predicted by the Rankine theory. The compound failure has a failure surface formed partially through the reinforced soil and partially along with the interface between the MSE and stabilized wall faces. When L/H < 0.25, failure mode changes from internal to external resulting in overturning failure

Table 1 Summary of wall failure modes and corresponding design guidelines [7]

L/H	L/H < 0.25	0.25 < L/H < 0.3	0.3 < L/H < 0.6	0.6 < L/H < 0.7	L/H > 0.7
Failure mode	External	Compound failure		Internal failure	
Design method	Cement stabilized wall	FHWA SMSE-wall design guidelines [4]		FHWA MSE-wall design guidelines [3]	

Table 2 Geotechnical properties of sand

Soil properties	Test	Determination
Index properties	Specific gravity	2.65
	Grain size analysis	Cc = 1.07 Cu = 2.77 SP
	Relative density	ϒmax = 19.10 kN/m^3 ϒmin = 16.95 kN/m^3
Engineering properties	Direct box shear	c = 0.05 kg/cm^2 $\phi = 36.9°$

of RE wall. The dominant failure modes and current design methods with various aspect ratios are summarized in Table 1.

2 Laboratory Investigation

Laboratory investigation includes material properties and experimental setup.

Material properties

See Table 2.

Geogrid

Uniaxial geogrid having ultimate tensile strength 40 kN/m in machine direction and 20 kN/m in the cross-machine direction. Mechanical properties of geogrid are shown in Table 3.

3 Experimental Setup

The model tank was fabricated of size 900 mm * 400 mm * 930 mm using angle sections and metal sheets. A metal sheet of size 395 mm * 925 mm was fixed at the base. Two sides of the model tank were fixed with acrylic sheets so that the behavior

Table 3 Mechanical properties of geogrid

Mechanical properties	Value
Ultimate tensile strength-MD (kN/m)	40
Ultimate tensile strength-CMD (kN/m)	20
Creep reduction factor	1.49
Partial factor-installation damage	1.11
Partial factor-environmental effects	1.15
Pull-out interaction coefficient	0.8

of the narrow RE wall can be observed during the test. The front side of the wall is flexible and the backside of the wall is rigid. The front side of the wall is kept open for measuring the displacement of the narrow RE wall. The plywood sheet was used to construct a hollow shoring wall model. Then the hollow wall model is filled with some filler material to achieve the desired weight. The front side of the shoring wall has facing better of 1H: 9 V. The backside of the shoring wall is kept vertical. For proper positioning of the shoring wall inside the tank, capsule welding was done on both sides of the tank and two angle sections were bolted with a model tank having the same slope as the shoring wall. Aspect ratio (L/H) for reinforced earth fill is decided based on the SMSE wall design guideline which states that aspect ratio for narrow RE wall should range from 0.25 to 0.6. Aspect ratio of the minimum geogrid reinforcement length (measured at the bottom of the wall) was 400 mm, corresponding to an aspect ratio (L/H) of approximately 0.44 [8]. The reinforcement length increased from bottom to top with the largest reinforcement provided at the top of length 500 mm, corresponding to an aspect ratio (L/H) of 0.55 (Figs. 1, 2, 3 and 4).

Fig. 1 Schematic diagram of narrow RE wall model

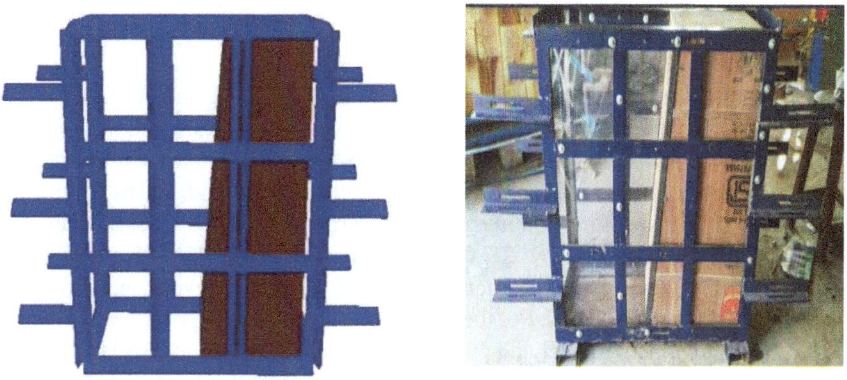

Fig. 2 Side view of model tank

Fig. 3 Dial gauge arrangement

Fig. 4 Fitting of geogrid with facing panel

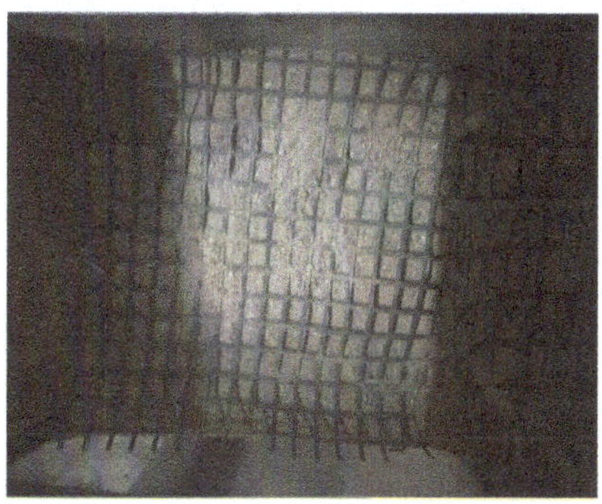

4 Test Procedure

Load–displacement and load-settlement tests will be carried out in a narrow RE wall model to know displacement and settlement characteristics. First, the tank will be filled in layer by layer by compacting the backfill material to achieve the desired density. In each layer, the volume of the sand was calculated ahead, and by controlling the weight of corresponding soil mass to achieve relative density. Geogrid is used as the reinforcement in the narrow RE wall and placing at the 0.4H (225 mm) vertical distance of the facing panel. The tank will be filled in four layers and geogrid is placed in between sand layers. For the unconnected system, geogrid is mechanically connected with a facing panel and curtail down near the shoring wall. This process was continued for all three layers of geogrid. Footing plate of size 360 mm × 150 mm and 25 mm thick was placed at the top of the material having firm contact with the backfill material to have the uniform transfer of load. Displacements and settlement measurements can be made using dial gauges of suitable capacity. Totally 6 dial gauges will be set up in a narrow RE wall model out of which 2 dial gauges will be set up to measure vertical settlement, 4 dial gauges will be set up at the front side of the model tank to measure lateral displacement. In flexible condition 1, dial gauge will be set up at the backside of the model tank. To carry out tests on the narrow RE wall model, the vertical downward load was applied using a proving ring of 5 tones capacity. A seating load of 7 kN/m^2 was applied on the plate using a mechanical jack. Proving ring of capacity 50 kN was used to maintain the loading at the constant interval of the 1.66 kN. When there is no perceptible increase in settlement or when the rate of settlement reduces to 0.02 mm per minute as given in IS 1888-1982, the dial gauge reading was noted (Table 4).

Table 4 Dial gauge location

Dial gauge	H1 (front side)	H2 (front side)	H3 (front side)	H4 (front side)	H5 (backside)
Horizontal distance from left to right (mm)	245	155	245	155	200
Vertical distance from top to bottom (mm)	105	305	505	705	300

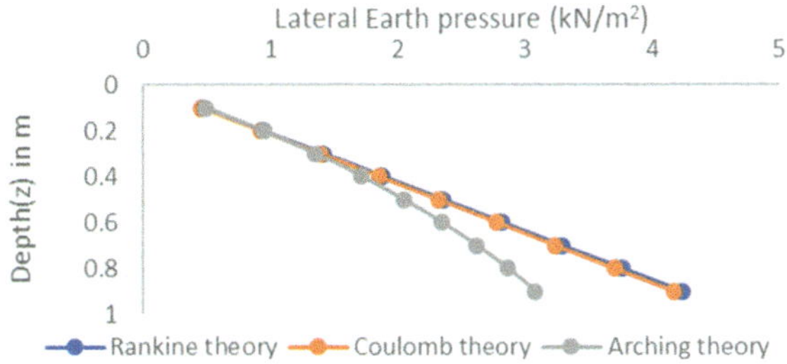

Fig. 5 Variation of lateral earth pressure with depth

Table 5 Value of lateral earth pressure in kN/m^2

Z (in m)	σh (Rankine's theory)	σh (Coulomb's theory)	σh (Arching theory)
0.1	0.47	0.463	0.493
0.2	0.941	0.926	0.941
0.3	1.411	1.389	1.345
0.4	1.882	1.852	1.712
0.5	2.353	2.315	2.042
0.6	2.823	2.778	2.342
0.7	3.294	3.241	2.612
0.8	3.764	3.7	2.864
0.9	4.235	4.167	3.081

5 Result and Discussion

5.1 Lateral Earth Pressure (σh) Distribution

Figure 5 illustrated the distribution of lateral earth pressure by Rankine's theory, Coulomb's theory, and Arching theory for $\phi = 36.9°$. From the figure, it is observed that as depth increases, the value of lateral earth pressure increases linearly in Rankine's and coulomb's theory thus shows a straight-line diagram. While in arching theory, the lateral earth pressure distribution is nonlinear, thus the shape of lateral earth pressure distribution is curve type (Table 5).

5.2 Load Versus Displacement Curve for Static Loading

The test was performed on a narrow RE wall model with sand as a backfill with reinforcement. The result obtained from displacement and settlement at a particular location of dial gauge is shown in various graphs (Figs. 6, 7, 8 and 9).

At location of dial gauge H5 (Fig. 10).

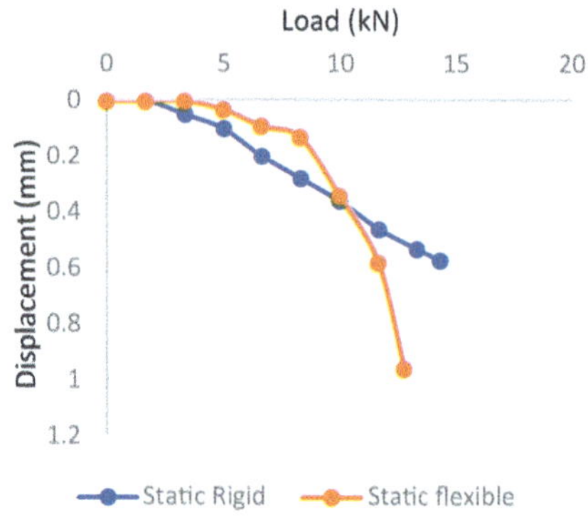

Fig. 6 Load versus displacement characteristics for static loading in rigid condition and flexible condition with reinforcement at dial gauge H1

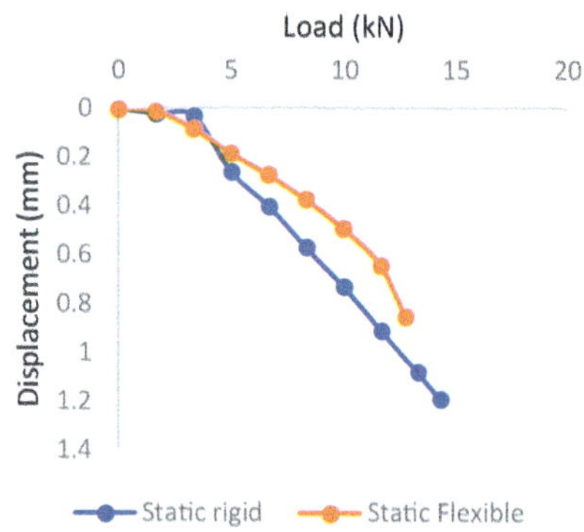

Fig. 7 Load versus displacement characteristics for static loading in rigid condition and flexible condition with reinforcement at dial gauge H2

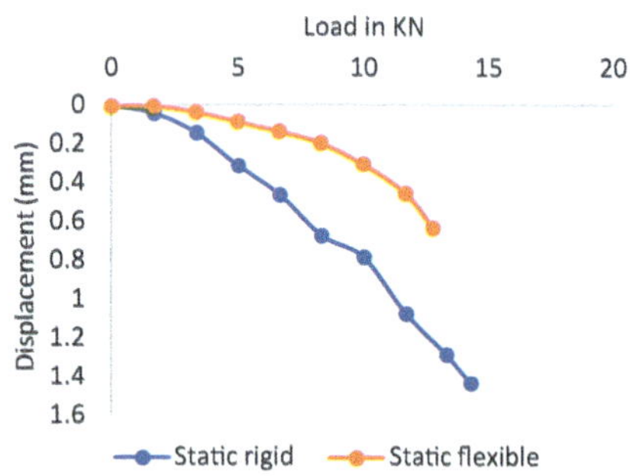

Fig. 8 Load versus displacement characteristics for static loading in rigid condition and flexible condition with reinforcement at dial gauge H3

The displacement behavior narrow of RE wall is shown from the above graph at the location of dial gauge H1, H2, H3, and H4 for backfill material sand with geogrid as reinforcement. The total displacement for an applied load of 12.78 kN is observed as for rigid condition and flexible condition is 0.55 mm, 1.15 mm, 1.36 mm, 0.68 mm and 0.85 mm, 0.96 mm, 0.63 mm, 0.23 mm, respectively. In flexible conditions the total displacement at the backside of the wall for an applied load of 12.78 KN is observed as 3.78 mm. In flexible condition, dial gauge shows maximum displacement

Fig. 9 Load versus displacement characteristics for static loading in rigid condition and flexible condition with reinforcement at dial gauge H4

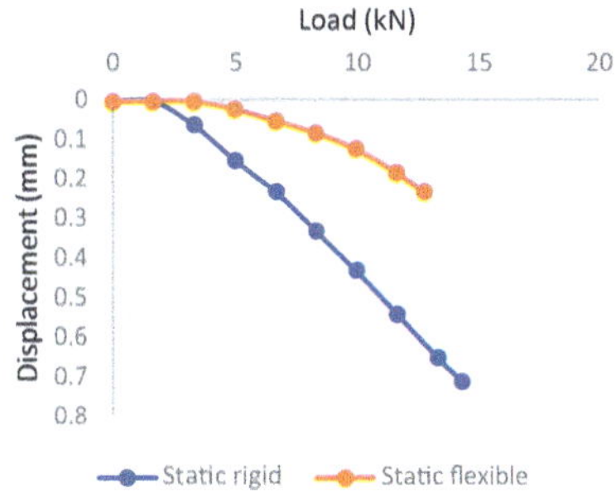

Fig. 10 Load versus displacement characteristics for static loading in flexible condition with reinforcement at dial gauge H5

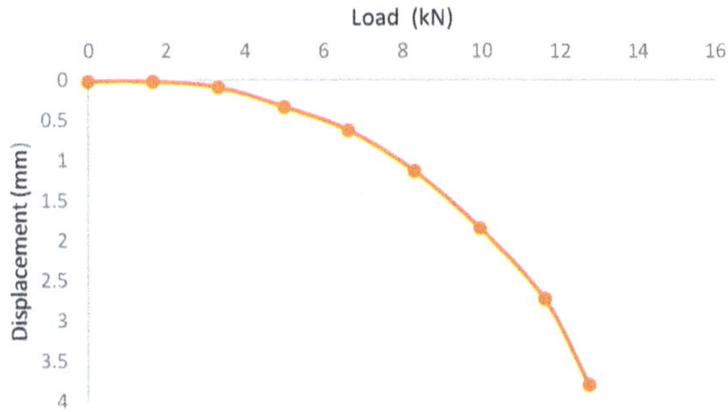

at the backside of the wall and the front side of wall dial gauge shows minimum displacement compared to a rigid condition.

5.3 *Load Versus Displacement Curve for Cyclic Loading*

See Figs. 11, 12, 13 and 14.

At the location of dial gauge H5 (Fig. 15).

The displacement behavior of the narrow RE wall is shown from the above graph at the location of dial gauge H1, H2, H3, and H4 for backfill material sand with geogrid as reinforcement. The total displacement for an applied load of 11.629 kN is observing as, for rigid condition and flexible condition is 0.34 mm, 1.09 mm, 1.09 mm, 0.52 mm and 0.5 mm, 0.89 mm, 0.91 mm, 0.37 mm, respectively. In flexible condition, the total displacement at the backside of the wall for an applied load of 12.78 KN is observed as 3.57 mm. In flexible condition, dial gauge shows maximum displacement at the backside of the wall and the front side of the wall dial gauge shows minimum displacement compared to the rigid condition.

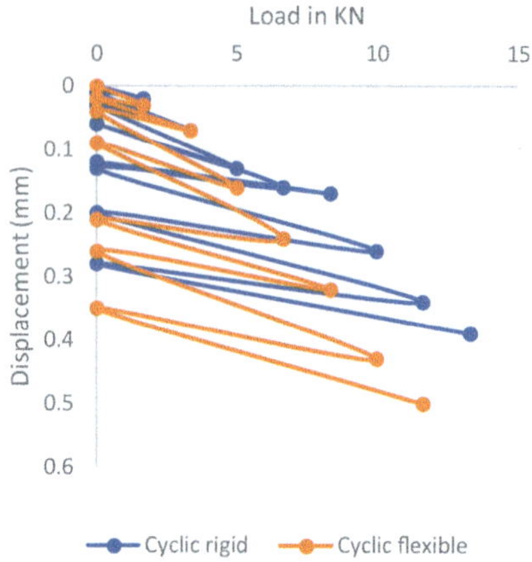

Fig. 11 Load versus displacement characteristics for cyclic loading in rigid condition and flexible condition with reinforcement at dial gauge H1

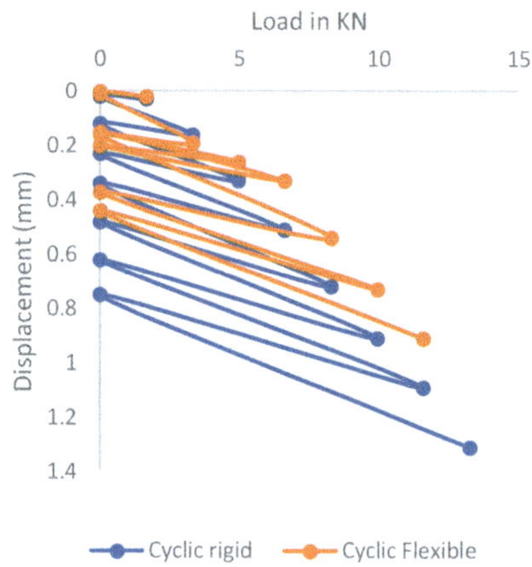

Fig. 12 Load versus displacement characteristics for cyclic loading in rigid condition and flexible condition with reinforcement at dial gauge H2

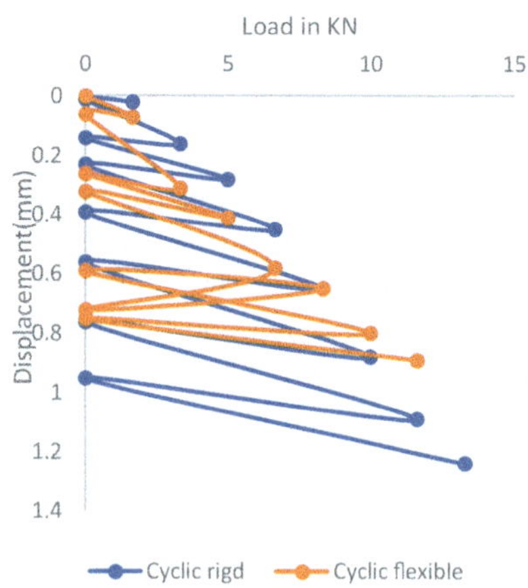

Fig. 13 Load versus displacement characteristics for cyclic loading in rigid condition and flexible condition with reinforcement at dial gauge H3

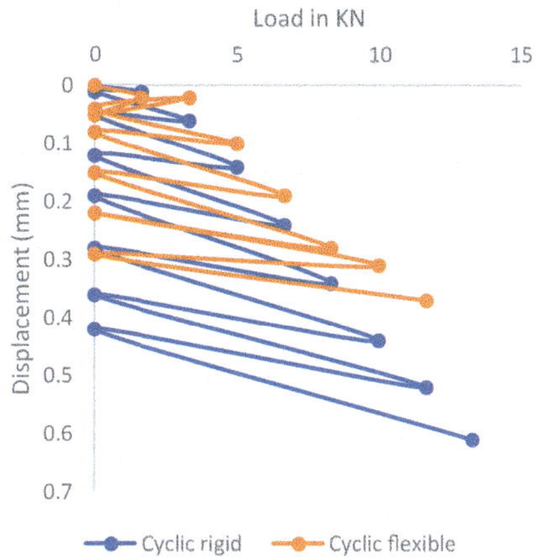

Fig. 14 Load versus displacement characteristics for cyclic loading in rigid condition and flexible condition with reinforcement at dial gauge H4

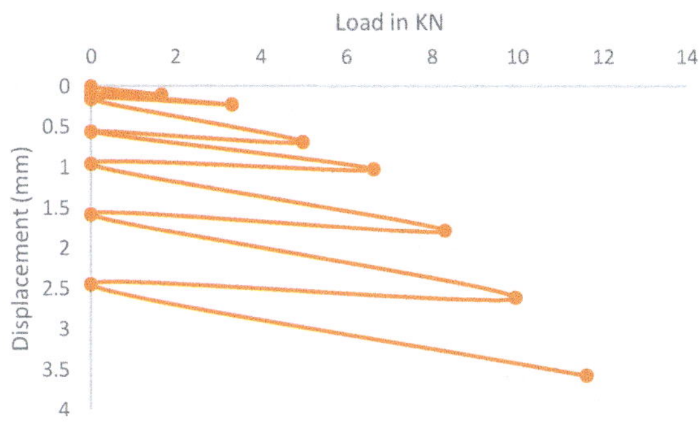

Fig. 15. Load versus displacement characteristics for cyclic loading in flexible condition with reinforcement at dial gauge H5

5.4 Load Versus Settlement Curve for Static Loading and Cyclic Loading

The settlement behavior of the narrow RE wall is shown from the above graph. The total displacement in static loading for an applied load of 12.78 KN is observing as, for rigid condition and flexible condition is 6.05 and 13.71 mm. In flexible condition, dial gauge shows maximum settlement as compared to rigid condition (Figs. 16 and 17).

The total displacement in cyclic loading for an applied load of 11.629 KN is observing as, for rigid condition and flexible condition is 5.66 and 11.24 mm. In flexible condition, dial gauge shows maximum settlement as compared to rigid condition.

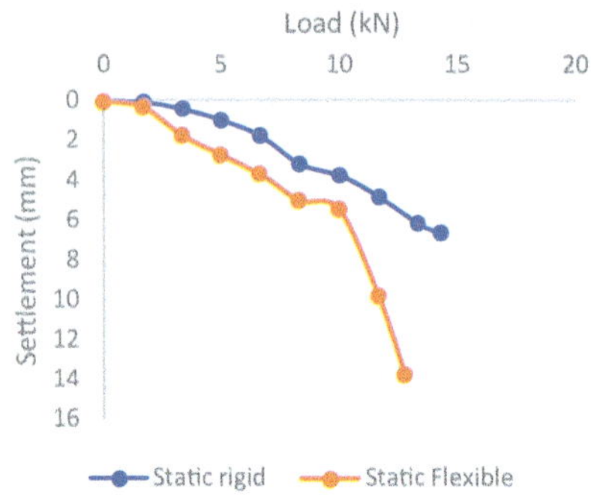

Fig. 16 Load versus settlement characteristics for static loading in rigid condition and flexible condition with reinforcement at dial gauge V1

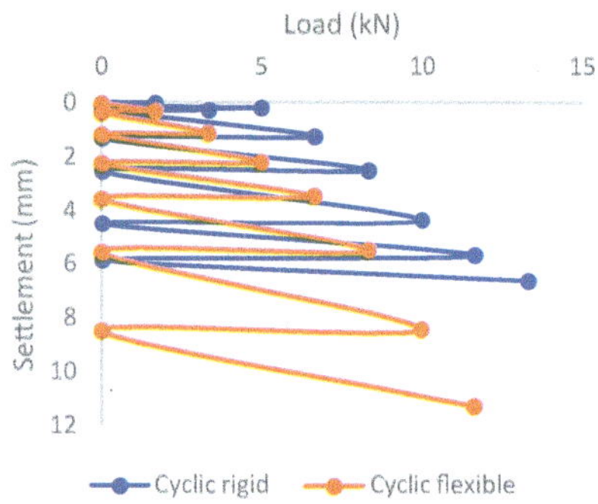

Fig. 17 Load versus settlement characteristics for static loading in rigid condition and flexible condition with reinforcement at dial gauge V1

6 Conclusion

In this paper, a comparison between cyclic loading and static loading is carried out on a narrow RE wall model with two conditions, i.e., rigid condition, flexible condition. In flexible condition settlement and displacement of soil is more compared to rigid condition [9]. Also, displacement of backside of the tank plays a major role in flexible condition. The dial gauge at the backside of the tank shows the highest displacement among all dial gauges.

In this study, Janssen's arching theory was used to provide a reasonable estimate of the lateral earth pressures on narrow RE wall and compare the results of lateral earth pressure with conventional earth pressure theories such as Rankine's theory and Coulomb's theory. It is observed that as depth increases, the value of lateral earth pressure increases linearly in Rankine's and coulomb's theory thus shows a straight-line diagram. While in arching theory, the lateral earth pressure distribution is nonlinear, thus shape of lateral earth pressure distribution is curve type.

Acknowledgements The authors are very thankful to Prof. (Dr.) R. K. Gajjar, Principal, L. D. College of Engineering, Ahmedabad, for providing all the necessary research facilities.

References

1. Chao X, Luo Y-S (2016) Effects of interface connections on narrow mechanically stabilized earth walls. Environ Earth Sci 75:1411
2. Tavakolian R, Grien M (2011) Narrow shored reinforced earth wall with friction-based reinforcing strip connection, as an innovative solution to expand urban highways. Geo-Frontiers
3. Elias V, Christopher BR, Berg RR (2001) Mechanically stabilized earth walls and reinforced soil slopes design and construction guidelines. Report No. FHWA-NHI-00-043, National Highway Institute, Federal Highway Administration, Washington, D.C. March
4. Morrison KF, Harrison FE, Collin JG, Dodds A, Arndt B (2006) Shored Mechanically Stabilized Earth (SMSE) wall systems design guidelines. Report No. FHWA-CFL/TD-06-001, Federal Highway Administration, Central Federal Lands Highway Division.
5. Yang KH, Zornberg JG, Hung WY, Lawson CR (2011) Location of failure plane and design considerations for narrow geosynthetic reinforced soil wall systems. J Geoeng
6. Woodruff R (2003) Centrifuge modeling of MSE-shoring composite walls. Master Thesis, University of Colorado, Boulder, Colorado
7. Yang KH, Kniss KT, Zornberg JG, Wright SG (2007) Finite element analyses for centrifuge modelling of narrow MSE walls
8. Xiao C, Han J, Zhang Z (2015) Experimental study on the performance of geosynthetic reinforced soil model walls on rigid foundations subjected to static footing loading. Geotext Geomem 1–14
9. Paik KH, Salgadoy R(2003), Estimation of active earth pressure against rigid retaining walls considering arching effects. Geotechnique 53(7):643–653

Seismic Mitigation Liquefaction—An Extensive Study on New Concepts

Jiji Krishnan and Shruti Shukla

Abstract Liquefaction of soil significantly affects the life of buildings as well as the life of humans. Liquefaction develops when the shear strength of the soil is lesser to resist the shear stresses induced when subjected to dynamic loadings during vibration or an earthquake. In order to improve the load-bearing capacity so as to mitigate the liquefaction characteristics, the addition of chemicals in soil can also be used. Addition of chemicals can be done in two ways. The first method in which void spaces can be filled by grouting/stabilising material, whereas the second method is the mechanical stabilisation of external materials. The modifications of soil to upgrade its properties through grouting technologies are extensively popular these days. The present study is an attempt to study the non-conventional seismic liquefaction mitigation methods. Also, this study reviews the most significant laboratory tests with respect to liquefaction mitigation and compares colloidal silica with many other recent liquefaction mitigation techniques such as bentonite suspension grouting, bio-cementation, colloidal silica grout and sand–rubber tire shred mixtures. The current study revealed that the two main grouts, which can be used as a prospective liquefaction reduction materials in the upcoming era are colloidal silica and bentonite suspensions.

Keywords Liquefaction mitigation · Grouting · Ground improvement · Colloidal silica

1 Introduction

Soil liquefaction is a foremost threat for structures built on saturated sandy soils. More than 300 lives and loss of 200 billion dollars were estimated in the 1995 Kobe earthquake due to liquefaction [1]. The significantly functional technologies for liquefaction mitigation are vibro compaction, vibro flotation and dynamic compaction, mixing with wet soil, vertical earthquake drains, and compacted stone

J. Krishnan (✉) · S. Shukla
Sardar Vallabhbhai National Institute of Technology, Surat, India
e-mail: jijiktu@gmail.com

T. G. Sitharam et al. (eds.), *Local Site Effects and Ground Failures*, Lecture Notes in Civil Engineering 117, https://doi.org/10.1007/978-981-15-9984-2_32

columns [2–6]. In order to progress the load-bearing capacity so as to mitigate the liquefaction characteristics, the addition of chemicals in soil also can be used. Addition of chemicals can be done in two ways. The first method in which void spaces can be filled by grouting/stabilising material, whereas the second method is the mechanical stabilisation of external materials. The modifications of soil to upgrade its properties through grouting technologies are extensively popular these days. Improving the seismic responses of the ground by various new concepts such as grouting with colloidal silica and bentonite suspension, bio-cementation, using tire chips have been popular for the past two decades with the advances of science and technology. Site remediation without disturbing the existing structures should be the main concern while selecting a suitable stabiliser for chemical grouting as well as bio grouting. The research works on liquefaction mitigation by new concepts in ground improvement is still in its flourishing phase. Regardless of such problems, many researchers have made an effort to discover the prospective research zones associated with the new concepts in seismic liquefaction mitigation. In order to improve the research in this field, every work nevertheless of the amount of work is significant. The current study makes an effort to summarise all the new phases of soil grouting and its economic viability. Also, this study reviews the most significant laboratory tests with respect to liquefaction mitigation and compares colloidal silica with many other recent liquefaction mitigation techniques such as bentonite suspension grouting, bio-cementation, colloidal silica grout and sand–rubber tire shred mixtures.

2 Motivation and Objective(s)

Traditional liquefaction mitigation methods such as Vibro flotation, dynamic compaction, Vibro compaction cause disturbance to the existing structures with limited areas of applications. Recently developed new methods in mitigating the hazards of seismic liquefaction such as grouting with colloidal silica grouting and bentonite, bio-cementation, microfine cement grout, extended set control admixtures grout, sodium silicate grout, acrylate grout, iron precipitation grout, epoxy grout and polysiloxane grout are non-disruptive and environmental friendly. The new methods mentioned above suggest new perceptions and methodologies for improving the research and explaining present-day engineering problems for mitigating liquefaction.

The primary purpose of the current study is to review the new soil enhancement approaches for mitigating liquefaction. This study also aims to review the economic viability of various grouting methods in mitigating liquefaction for passive site stabilisation. Passive site stabilisation is a technique in ground stabilisation of liquefiable soil under already developed structures. Moreover, the significance of these novel methods in resolving the reduction of seismic hazard in ground improvement and future research orientations are also discussed.

3 Practical Records of New Soil Improvement Methods

3.1 Dynamic Properties of Sand Mixed with Shredded Rubber

Rubber fibres in soil foundations can absorb low amplitude mechanical waves generated by vibration caused by machines, locomotives and micro-tremors. Shredded rubber tires are used in soil mixtures due to its high damping behaviour [7, 8].

Madhusudhan et al. [9] investigated the shear modulus and damping ratio of the sand mixed with shredded tires at around seven number of loading cycles. They conducted the strain-controlled two-way cyclic triaxial tests in the sand mixed with shredded tires under undrained conditions. In order to avoid the segregation on rubber-sand mixtures and to ensure homogeneity and uniformity, researchers selected the sizes smaller than 2 mm for both sand as well as rubber particles. The mixtures of the tires used for the experiment were 0, 10, 30, 50 and 100%. They observed a reduction in shear modulus until the 5th cycle for all the mixtures, after which it remains constant (Fig. 1). Also, Fig. 1 shows that during critical earthquakes, sand mixed with rubber tires becomes softer significantly at large strains which in turn results in period lengthening. They reported that the 10 wt.% rubber tire mixture in sand mixture improved static and dynamic properties. The enhancement in properties of treated sand could be made use for seismic base isolation in low-rise buildings.

Bahadori and Manafi [10] conducted experiments to calculate the properties of rubber tire chips on the dynamic properties of saturated sands. In their study, four

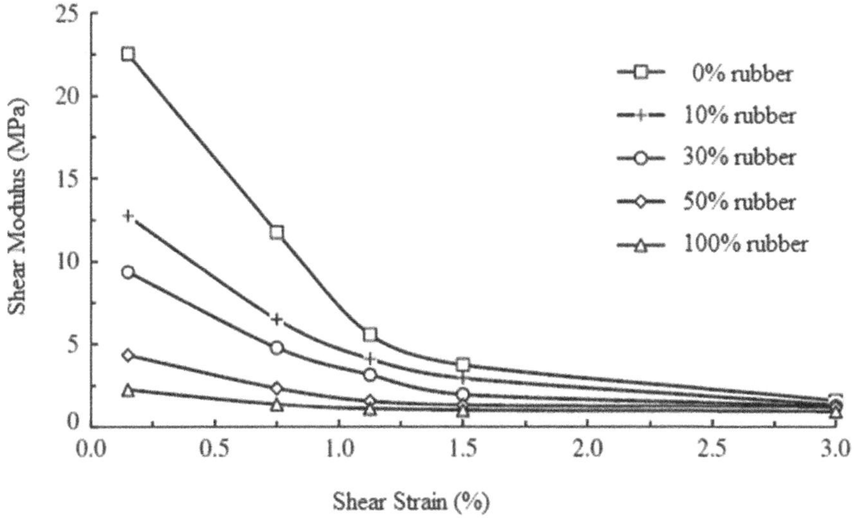

Fig. 1 Shear strain versus shear modulus of rubber-sand mixtures at different rubber volume fractions [9]

different mixture ratios of 10, 20, 30 and 40 were used. In order to study the improvement in mitigation as well of sand-tire mixtures, 1 g shaking table prototypical tests were conducted. They noticed a decrease in settlement caused by liquefaction with the addition of tire chips in sands. Also, they noticed an increment of mean damping ratio and decrement of shear modulus with the increment of tire particles in the reinforced soil.

Senetakis et al. [11] conducted experiments to measure the dynamic properties of rubber mixed in dry sand as well as gravel in an extensive range of shearing strain amplitudes. They proposed the generic normalised shear modulus and damping ratio versus shearing strain amplitude curves for dry mixtures of sand-rubber (0–35%) and gravel-rubber (0–40%) appropriate for the engineering practice. They also mentioned the different factors which affect the strain shear modulus and damping ratio of soil-rubber and gravel-rubber mix. The factors are rubber content, grain size, the comparative size of soil and rubber and also the dynamic properties of the intact soil.

Ali et al. [12] reported the dynamic properties of dense sand-rubber mixtures with small particle size ratio. In order to determine the damping ratio and shear modulus, bender elements were inserted into a triaxial cell. The rubber chips used for the study were having sizes changing from 0.06 to 0.60 mm and also had the shape of angular to sub-angular. They investigated the results which had the confining pressures ranging from 50 to 500 kPa with the addition of less than 50% rubber volume fractions in the sand. They notice a decrement in shear modulus and increment in damping ratio with the addition of rubber chips in the sand for all the confining pressures selected. They also proposed that a 10% rubber volume fraction in sand increases the damping ratio without substantial loss of rigidity and negligible sensitivity to confining pressure.

3.2 *Dynamic Properties of Colloidal Silica Grouted Sand*

Colloidal silica is considered to be a potential stabiliser in ground engineering due to its comparable costs with other grouting methods, fewer disturbances to structures while stabilising, controllable gel time as well as low viscosity [13–15].

Gallagher and Mitchell [16] investigated the effect of colloidal silica suspension in loose sand on the improvement of liquefaction potential and cyclic undrained behaviour. The experimental tests were conducted on sand samples treated with colloidal silica at four different concentrations (5, 10, 15 and 20%). They investigated an enhancement in deformation behaviour by adding colloidal silica grout in sand. In order to study the effects of the curing period, they have conducted tests on 10% colloidal silica-treated sand samples for different curing periods of 11, 21, 32 and 56 days. They noticed that strain level reached during cyclic loading decreases with an increase in the curing period (Fig. 2). They concluded that cyclic resistance of colloidal silica-treated sand will progress with time. They also recommended a concentration of up to 10 wt.% colloidal silica solution in sand to offer satisfactory stabilisation with tolerable strain levels.

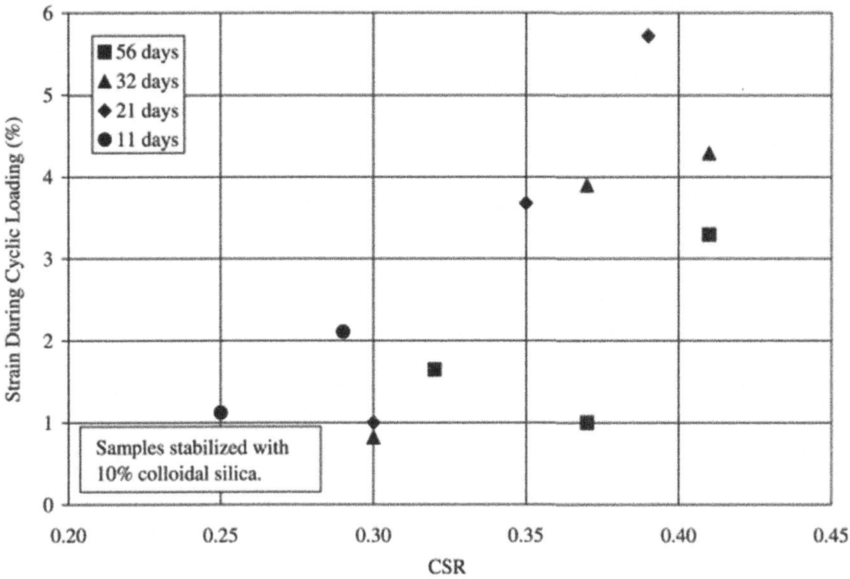

Fig. 2 Cyclic stress ratio versus strain during cyclic loading at different curing times [16]

Kodaka et al. [17] conducted laboratory tests such as monotonic and cyclic torsional shear test in colloidal silica-treated sand. They noticed that the treated colloidal silica sand exhibits greater liquefaction strength than the untreated sample. They have also proposed an elastoplastic model to define the performance of treated sand. This proposed model as then employed in a finite element analysis program, which is based on fully coupled effective stress. From the FEM analysis, they proposed that treating sand with colloidal silica grout is an active method in increasing the resistance against liquefaction.

Díaz-Rodríguez et al. [18] studied the cyclic strength characteristics of sand stabilised with colloidal silica grout. They have conducted a series of experimental tests on cyclic simple shear apparatus on liquefiable silty sand with and without sample enhancement with colloidal silica grout. The tests were conducted on poorly graded sand having non-plastic silt. They have noticed that the sand sample treated with colloidal silica-treated grout substantially improved the liquefaction resistance compared with untreated soil specimens. Also, upon cyclic loading, a considerable reduction in shear strain and pore pressure generation was also observed.

Conlee et al. [19] conducted centrifuge tests to study the liquefaction mitigation using colloidal silica stabiliser in sands. The study was conducted on model geometry in order to determine the properties of lateral spreading in a 4.8 m thick liquefiable layer, which was overlain by silty clay sloping towards a central channel. They have noticed an increment in cyclic resistance ratios and decrement in cyclic shear strains with the increase in colloidal silica concentrations.

Spencer et al. [20] performed resonant column experiments in order to notice the dynamic properties such as shear modulus and damping ratios in colloidal silica-treated sand samples. They have selected a colloidal silica grout percentage of 5, 7, 9 wt.% in sand to study the dynamic properties. They noticed only a lesser improvement in the shear modulus with the cumulative concentration of colloidal silica in sand. And, they observed a negligible variation in damping ratio with the addition of colloidal silica in sand over the strain range examined.

3.3 Dynamic Properties of Bentonite Sand Suspensions

Mohtar et al. [21] investigated the cyclic resistance of sand bentonite specimens. The sand specimens were mixed with different proportions of the dry mass of bentonite (0, 3 and 5%) and were tested to find out the cyclic shear strength of bentonite-treated sand specimens. Also, undrained cyclic triaxial tests were conducted on sand specimens as well as sand treated with 10% bentonite grout with the addition of 0.5% sodium pyrophosphate. They noticed an increment in resistance of cyclic shear strength of sand with the incorporation of bentonite suspensions in sand. The tests of 10% bentonite suspension treated sand show a tenfold increment in the number of cycles to liquefaction with reference to untreated sand. Also, it has been noted that after a delay of 12 h, bentonite permeated sand was capable to reconstruct the resistance beyond the first loading.

Mohtar et al. [22] investigated the effects of cyclic response on bentonite-treated sand. They have conducted cyclic triaxial as well as resonant column experiments on sand treated with bentonite grout. Cyclic tests were conducted at the rate of 1 cycle/s with the reversal in stress. Also, in order to simulate the dynamic earthquake loading in the field, undrained conditions were chosen. They noticed an increase in the order of magnitude in the number of cycles that produce liquefaction at any given cyclic stress ratio. Figure 3 shows the cyclic stress ratio in bentonite treated as well as untreated sand vs the number of cycles to liquefaction obtained from experimental results as a function of the cyclic stress ratio applied. Also, it compared the test results of Salgado et al. [23] in which tests were conducted in the same sand. Resonant column tests indicated that there is a need for strain increment to initiate excess pore pressure generation.

Rugg et al. [24] investigate the effect of static liquefaction mitigation with the help of undrained shear tests in sand permeated with bentonite suspensions. A reduction in excess pore water generation was observed by conducting consolidated undrained tests. Also, they noticed hardly any undrained instability state, which indicates that static loading didn't cause any strain softening. The results from their findings concluded that permeation grouting by adding 2% sodium pyrophosphate mixed in bentonite suspensions could be an operational technique for static liquefaction mitigation.

Yoon and Mohtar [25] conducted tests on sodium pyrophosphate modified bentonite suspensions to study the dynamic rheological properties for liquefaction

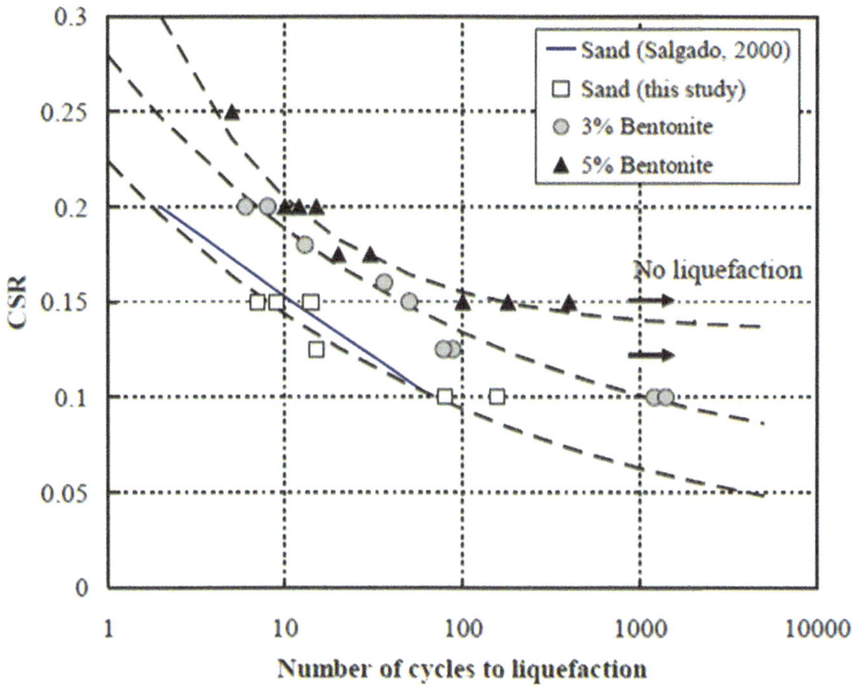

Fig. 3 Cyclic stress ratio in bentonite treated as well as untreated sand versus number of cycles to liquefaction [22]

mitigation. Sequences of oscillatory shear experiments were performed at a constant oscillation rate of 1 Hz. Investigations were conducted on varying weight percentages of bentonite suspensions (5, 7.5, 10 and 12 wt.%) modified with sodium pyrophosphate. The results showed a decrement in critical storage modulus with the increase in sodium pyrophosphate. Also, they have noticed an increase in critical storage modulus with the increase in resting time. The normalised modulus of the treated sand specimens was degraded more gradually than in the untreated suspensions, which is an appropriate property of the suspensions for mitigation of liquefaction.

3.4 Dynamic Response of Liquefiable Sand Induced by Bio-cementation

Montoya et al. [26] investigated the improvement in dynamic properties of liquefiable sand with the addition of microbial-induced calcite precipitation by conducting dynamic centrifuge model experiments. Cyclic strength, as well as stiffness of the sand, showed significant improvement after bio-cementation treatment in sands. Also, a reduction in excess pore water generation and settlements were observed in treated sand during dynamic loading. Treatment in sand increases the maximum accelerations at the ground level as well. So great care to be taken while designing the treatment level to upgrade liquefaction resistance and reducing unwanted amplified soil surface motions.

Burbank et al. [27] performed tests on sands treated by calcite precipitation to study the effects of seismic-induced liquefaction. Cyclic triaxial shear testing was conducted to determine the increment in soil resistance to liquefaction by bio-induced calcite precipitation. A substantial increment in CSR was noticed in cyclic triaxial shear tests in treated sands. Calcite precipitation ranges from 2.1 to 2.6% in sand increases the cyclic resistance ratio to twofold. Whereas calcite precipitation ranges from 3.8 to 7.4% in sand increases the cyclic resistance ratio to four- to fivefold.

Xiao et al. [28] investigated the effect of bio-cementation on the relative density and cyclic response of sand with the help of cyclic triaxial tests. Also, 0.5 mol/L urea mixed with 0.5 mol/L $CaCl_2$ was used for the bio-cementation purpose. Cyclic triaxial experiments were executed on treated sand with varying relative densities (10, 50 and 80%). The experiments were performed at a frequency of 1 Hz cyclic stress ratios varied from 0.125 to 0.500. They noticed that bio-cementation reduces the generation of excess pore pressure in sand specimens with the advancement of axial strain. Also, during liquefaction in treatment of calcareous sand, the number of cycles in sand increases. The increase in relative density, as well as the increase in the magnitude of bio-cementation, increases the number of cycles to liquefaction. Figure 4 shows the variation in the cyclic resistance ratio of bio-cemented sand at different relative densities by varying unit weight. However, they mentioned that the

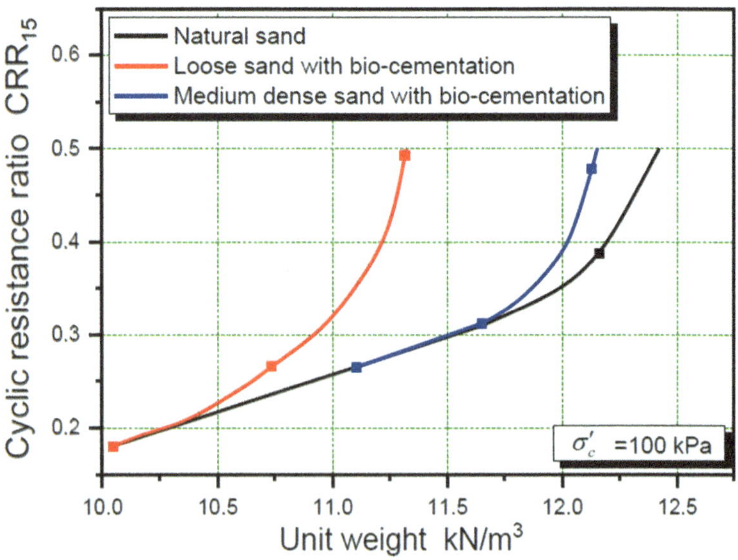

Fig. 4 Variation in cyclic resistance ratio of bio-cemented sand at different relative densities by varying unit weight [28]

Table 1 Colloidal silica grouting method versus non-conventional method [29]

	Bio-cementation	Bentonite suspension grouting	Rubber tires	Colloidal silica grout
Passive site stabilisation	No	Yes	No	Yes
Non-toxic	Yes	Yes	Yes	Yes
Process for liquefaction mitigation	Calcification by the microorganisms	Bentonite gel formation	Reduction in soil saturation by mixing rubber tires	Colloidal silica gel formation
Cost-effective	Yes	Yes	Yes	Yes

degree of increase in cyclic resistance with the technique of bio-cementation could be a more successful soil improvement method than densifying the sand strata for a given variation in relative density (not including cost considerations).

4 Feasibility Study Analysis of Colloidal Silica Grouting Method with Other Non-conventional Methods of Liquefaction Mitigation

The two main grouts which can be used as prospective liquefaction mitigation materials in the near future are colloidal silica and bentonite suspensions. Huang and wen [29] performed large-scale tests, and the cost-effective comparison was made as per the tests in comparison with conventional techniques of soil improvement and liquefaction mitigation (Table 1).

5 Conclusions

The current research attempts to sweep the entire new concepts and methods on liquefaction mitigation.

1. New liquefaction mitigation methods are environmental friendly and non-disruptive, which in turn rectifies the limitations in traditional methods.
2. Economic viability study of the new methods in liquefaction mitigation recommends the use of bio-cementation-bentonite suspensions, shredded rubber tires and colloidal silica as preferable grouting material.
3. Colloidal silica and bentonite suspensions could be scrutinised as a prospective grouting material in liquefaction mitigation for passive site stabilisation techniques.

4. Colloidal silica can be used as a prospective liquefaction mitigation materials in the near future since it is environmental friendly, comparable cost, less disturbance to existing structures during injection, controllable gel time, low viscosity as well as a lifetime in excess of 25 yrs.

References

1. Werner SD, Dickenson SE, Taylor CE (1997) Seismic risk reduction at ports: case studies and acceptable risk evaluation. J Waterway Port Coastal Ocean Eng 123(6):246–337
2. Mitchell JK, Baxter CDP, Munson TC (1995) Performance of improved ground during Earthquakes. In: Soil improvement for liquefaction hazard mitigation, Geotech special Pub No 49, ASCE, pp 1–36
3. Marcuson WF, Hadala PF, Ledbetter RH (1996) Seismic rehabilitation of earth dams. J Geotech Eng ASCE 122(1):7–20
4. Elias V, Welsh J, Wareen J, Lukas R, Collin J, Berg R (2006) Ground improvement methods: reference manual, vol I. NHI Course No 13204, Federal Highway Administration, Washington, DC
5. Mitchell JK (2008) Mitigation of liquefaction potential of silty sands. In: From research to practice in geotechnical engineering, ASCE, pp 433–451
6. Chu J, Varaksin S, Klotz U, Menge P (2009) Construction Process. In: Proceedings of 17th international conference on soil mechanics and geotechnical engineering, IOS Press, Fairfax, VA, pp 3006–3135
7. Hazarika H, Yasuhara K, Kikuchi Y, Karmokar AK, Mitarai Y (2010) Multifaceted potentials of tire-derived three dimensional geosynthetics in geotechnical applications and their evaluation. Geotext Geomembranes. https://doi.org/10.1016/j.geotexmem.2009.10.011
8. Tafreshi SNM, Norouzi AH (2013) Bearing capacity of a square model footing on sand reinforced with shredded tire—an experimental investigation. Constr Build Mater (2012). https://doi.org/10.1016/j.conbuildmat.2012.04.092
9. Madhusudhan BR, Boominathan A, Banerjee S (2017) Static and large-strain dynamic properties of sand-rubber tire shred mixtures. J Mater Civ Eng. https://doi.org/10.1061/(ASCE)MT.1943-5533.0002016
10. Bahadori H, Manafi S (2015) Effect of tyre chips on dynamic properties of saturated sands. Int J Phys Model Geotech. https://doi.org/10.1680/ijpmg.13.00014
11. Senetakis K, Anastasiadis A, Pitilakis K (2012) Dynamic properties of dry sand/rubber (SRM) and gravel/rubber (GRM) mixtures in a wide range of shearing strain amplitudes. Soil Dyn Earthq Eng. https://doi.org/10.1016/j.soildyn.2011.10.003
12. Brara A, Brara A, Daouadji A, Bali A, Daya EM (2017) Dynamic properties of dense sand-rubber mixtures with small particles size ratio. Eur J Environ Civ Eng. https://doi.org/10.1080/19648189.2016.1139509
13. Krishnan J, Shukla S (2019) The behaviour of soil stabilised with nanoparticles: an extensive review of the present status and its applications. Arab J Geosci 12:436. https://doi.org/10.1007/s12517-019-4595-6
14. Gallagher PM, Pamuk A, Abdoun T (2007) Stabilisation of liquefiable soils using colloidal silica grout. J Mater Civ Eng 19:33–40. https://doi.org/10.1061/(ASCE)0899-1561(2007)19:1(33)
15. Consoli (2004) Effect of material properties on the behavior of sand-cement–fiber composites. Proc Inst Civ Eng Gr Improv 8:77–90. https://doi.org/10.1680/Grim.2004.8.2.77
16. Gallagher PM, Mitchell JK (2002) Influence of colloidal silica grout on liquefaction potential and cyclic undrained behavior of loose sand. Soil Dyn Earthq Eng 22:1017–1026. https://doi.org/10.1016/S0267-7261(02)00126-4

17. Kodaka T, Oka F, Ohno Y, Takyu T, Yamasaki N (2005) Modeling of cyclic deformation and strength characteristics of silica treated sand. Geotech Spec Pub. https://doi.org/10.1061/40797(172)11
18. Díaz-Rodríguez JA, Antonio-Izarraras VM, Bandini P, López-Molina JA (2008) Cyclic strength of a natural liquefiable sand stabilized with colloidal silica grout. Can Geotech J 45:1345–1355. https://doi.org/10.1139/T08-072
19. Conlee CT, Gallagher PM, Boulanger RW, Kamai R (2012) Centrifuge modeling for liquefaction mitigation using colloidal silica stabilizer. J Geotech Geoenviron Eng 138:1334–1345. https://doi.org/10.1061/(ASCE)GT.1943-5606.0000703
20. Spencer L, Rix GJ, Gallagher P (2008) Colloidal silica gel and sand mixture dynamic properties. Geotech Earthq Eng Soil Dyn IV:1–10. https://doi.org/10.1061/40975(318)101
21. Mohtar CSE, Bobet A, Santagata MC, Drnevich VP, Johnston CT (2013) Liquefaction mitigation using bentonite suspensions. J Geotech Geoenviron Eng. https://doi.org/10.1061/(ASCE)GT.1943-5606.0000865
22. El Mohtar CS, Clarke J, Bobet A, Santagata M, Drnevich V, Johnston C (2008) Cyclic response of a sand with thixotropic pore fluid. Geotech Spec Pub. https://doi.org/10.1061/40975(318)63
23. Salgado R, Bandini P, Karim A (2000) Shear strength and stiffness of silty sand. J Geotech Geoenviron Eng. https://doi.org/10.1061/(ASCE)1090-0241(2000)126:5(451)
24. Rugg DA, Yoon J, Hwang H, El Mohtar CS (2011) Undrained shearing properties of sand permeated with a bentonite suspension for static liquefaction mitigation. Geotech Spec Pub. https://doi.org/10.1061/41165(397)70
25. Yoon J, Mohtar CE (2013) Dynamic rheological properties of sodium pyrophosphate-modified bentonite suspensions for liquefaction mitigation. Clays Clay Miner. https://doi.org/10.1346/CCMN.2013.0610411
26. Montoya BM, DeJong JT, Boulanger RW (2013) Dynamic response of liquefiable sand improved by microbial-induced calcite precipitation. Geotechnique. https://doi.org/10.1680/geot.SIP13.P.019
27. Burbank M, Weaver T, Lewis R, Williams T, Williams B, Crawford R (2013) Geotechnical tests of sands following bioinduced calcite precipitation catalysed by indigenous bacteria. J Geotech Geoenviron Eng. https://doi.org/10.1061/(ASCE)GT.1943-5606.0000781
28. Xiao P, Liu H, Stuedlein AW, Evans TM, Xiao Y (2019) Effect of relative density and biocementation on cyclic response of calcareous sand. Can Geotech J. https://doi.org/10.1139/cgj-2018-0573
29. Huang Y, Wen Z (2015) Recent developments of soil improvement methods for seismic liquefaction mitigation. https://doi.org/10.1007/s11069-014-1558-9

The manufacturer's authorised representative in the EU is Springer Nature Customer Service Centre GmbH, Europaplatz 3, 69115 Heidelberg, Germany. If you have any concerns regarding our products, please contact ProductSafety@springernature.com

Printed and bound by CPI Group (UK) Ltd, Croydon, CR0 4YY
30/09/2026
02225287-0014